**New Strategies Combating
Bacterial Infection**

*Edited by
Iqbal Ahmad and Farrukh Aqil*

Related Titles

Ahmad, I., Aqil, F., Owais, M. (eds.)

Modern Phytomedicine

Turning Medicinal Plants into Drugs

2006
ISBN: 978-3-527-31530-7

Brogden, K. A., Minion, K. F. C., Cornick, N., Stanton, T. B., Zhang, Q., Nolan, L., Wannemuehler, M. J. (eds.)

Virulence Mechanisms of Bacterial Pathogens

2007
ISBN: 978-1-55581-469-4

Gillespie, S. H., Bamford, K. B.

Medical Microbiology and Infection at a Glance

2007
ISBN: 978-1-4051-5255-6

Kaufmann, S. H. E., van Helden, P., Rubin, E., Britton, W. J. (eds.)

Handbook of Tuberculosis

2008
ISBN: 978-3-527-31683-0

Frosch, M., Maiden, M. C. J. (eds.)

Handbook of Meningococcal Disease

Infection Biology, Vaccination, Clinical Management

2006
ISBN: 978-3-527-31260-3

New Strategies Combating Bacterial Infection

Edited by
Iqbal Ahmad and Farrukh Aqil

WILEY-
BLACKWELL

WILEY-VCH Verlag GmbH & Co. KGaA

The Editor

Dr. Iqbal Ahmad
Aligarh Muslim University
Faculty of Agricultural Sciences
Department of Agricultural Microbiology
Aligarh 202002
India

Dr. Farrukh Aqil
Aligarh Muslim University
Faculty of Agricultural Sciences
Department of Agricultural Microbiology
Aligarh 202002
India

All books published by Wiley-VCH are carefully produced. Nevertheless, authors, editors, and publisher do not warrant the information contained in these books, including this book, to be free of errors. Readers are advised to keep in mind that statements, data, illustrations, procedural details or other items may inadvertently be inaccurate.

Library of Congress Card No.: applied for

British Library Cataloguing-in-Publication Data
A catalogue record for this book is available from the British Library.

Bibliographic information published by the Deutsche Nationalbibliothek
The Deutsche Nationalbibliothek lists this publication in the Deutsche Nationalbibliografie; detailed bibliographic data are available on the Internet at http://dnb.d-nb.de.

© 2009 WILEY-VCH Verlag GmbH & Co. KGaA, Weinheim

Cover Design   Adam Design, Weinheim
Typesetting   Thomson Digital, Noida, India
Printing   Strauss Gmbh, Mörlenbach
Binding   Litges & Dopf GmbH, Heppenheim

Printed in the Federal Republic of Germany
Printed on acid-free paper

ISBN: 978-3-527-32206-0

Contents

New Strategies Combating Bacterial Infection. Edited by Iqbal Ahmad and Farrukh Aqil
Copyright © 2009 WILEY-VCH Verlag GmbH & Co. KGaA, Weinheim
ISBN: 978-3-527-32206-0

x | *Contents*

Preface

The remarkable success of antimicrobial drugs had generated a feeling in the 1960s and early 1970s that infectious diseases were under control. However, even today they remain the third leading cause of death in the United States, and the second leading cause of mortality and morbidity worldwide. This is primarily due to the emergence of new infectious agents and the development of drug resistance in microbial pathogens. The development and spread of multidrug-resistant bacterial pathogens such as methicillin-resistant *Staphylococcus aureus*, extended-spectrum β-lactamase-producing enteric bacteria, *Mycobacterium tuberculosis* and several others are on the rise. Recent trends in the development of plasmid-encoded resistance against fast-acting flouroquinolones in *Escherichia coli* is a dangerous trend, and is an example of great genetic adaptation and acquisition of resistance genes in bacteria. This has necessitated reviewing the scope and future prospects of old antibiotics in combating bacterial infections in the present scenario of antibiotic therapy and declining trend of new antibiotic discovery with novel modes of action. Several books have been published highlighting mainly the molecular mechanism of antibiotic resistance and drug discovery. In recent years, a considerable amount of data and literature has appeared on the efforts of academic institutions and some pharmaceutical companies in their continued efforts on antibacterial drug discovery. On the other hand, there is a declining trend in the registration of new chemical entities against infectious diseases. Therefore, the development of strategies for developing new anti-infective drugs with novel modes of action with intelligent design of test systems from microbial, plant and animal sources is to be encouraged. Similarly, targeting drug resistance mechanisms, virulence and pathogenicity, enhancing host immunity, and use of probiotics, plant-derived natural products and phage therapy are viable strategies for combating bacterial infections. However, to minimize the excessive use of antibiotics, both in medical and nonmedical applications, there is a great need to develop awareness in the community and in hospitals. Further development and strict implementation of antibiotic use policy at local and national levels is needed.

Here, we have made an attempt to bring together recent work and current trends in the field of antibacterial drug discovery. Classical and modern approaches to

New Strategies Combating Bacterial Infection. Edited by Iqbal Ahmad and Farrukh Aqil
Copyright © 2009 WILEY-VCH Verlag GmbH & Co. KGaA, Weinheim
ISBN: 978-3-527-32206-0

combat the drug resistance problem, and alternative therapy and the use of natural products are discussed. Chapter 1 highlights the molecular mechanism of dug resistance in bacteria and indicates the need for new drug discovery. Chapter 2 elaborates the novel approaches attempted by various groups of researchers in the search for new antibiotics or anti-infective drugs. Chapter 3 gives special attention to drug-resistant *M. tuberculosis* and current treatment strategies. Other chapters (Chapters 4–11) discuss the potential role of nonantibiotics, plant-derived compounds/extracts, essential oils, and natural products like honey and probiotics in combating bacterial infections. Experts from different parts of the world have contributed chapters on the above aspects.

It is intended that this book will be useful for students, teachers and researchers in universities, research and development institutions, pharmaceutical industries, and health care organizations.

With great pleasure and respect, we extend our sincere thanks to all the contributors for their timely responses, excellent and updated contributions, and cooperation. We express our deep sense of gratitude to all our respected teachers, scientific collaborators, colleagues and friends for their guidance, cooperation and healthy criticisms. The cooperation received from research scholars, especially Dr Farah Ahmad, Mr Mohd Imran, Mr Mohd Sajjad A. Khan and Miss Maryam Zahin, is thankfully acknowledged.

The financial assistance rendered by the University Grants Commission (New Delhi) in the form of a major research project is greatly appreciated.

The technical assistance and support received from the excellent book publishing team at Wiley-VCH (Weinheim, Germany) and from Mrs Vallika Devi Katragadda (Wiley, New Delhi) is appreciated and thankfully acknowledged.

Finally, we acknowledged the Almighty God who provided all the thoughts and channels to the successful completion of this task.

Aligarh, August 2008

Iqbal Ahmad
Farrukh Aqil

List of Contributors

Iqbal Ahmad
Aligarh Muslim University
Department of Agricultural
Microbiology
Anoopshahar Road
Aligarh 202 002
India

Shamim Ahmad
Aligarh Muslim University
J.N. Medical College
Institute of Ophthalmology
Ramghat Road
Aligarh 202 002
India

Bolanle A. Adeniyi
University of Illinois at Chicago
Department of Pharmacy Practice
College of Pharmacy
Pan American Health Organization/
World Health Organization
Collaborating Center for Traditional
Medicine
833 South Wood Street
Chicago, IL 60612
USA

Farrukh Aqil
Aligarh Muslim University
Department of Agricultural
Microbiology
Anoopshahar Road
Aligarh 202 002
India

Sabulal Baby
Tropical Botanic Garden and Research
Institute
Phytochemistry and Phytopharmacology
Division
Pacha-Palode
Thiruvananthapuram 695 562
India

Sujit K. Bhattacharya
ICMR Virus Unit
Infectious Diseases and
Beliaghata General Hospital
57 Dr Suresh C. Banerjee Road
Kolkata 700 010
India

Boyan B. Bonev
University of Nottingham
School of Biomedical Sciences and the
Institute of Infection and Immunity
Queen's Medical Centre
Nottingham NG7 2UH
UK

New Strategies Combating Bacterial Infection. Edited by Iqbal Ahmad and Farrukh Aqil
Copyright © 2009 WILEY-VCH Verlag GmbH & Co. KGaA, Weinheim
ISBN: 978-3-527-32206-0

Debprasad Chattopadhyay
ICMR Virus Unit
Infectious Diseases and
Beliaghata General Hospital
57 Dr Suresh C. Banerjee Road
Kolkata 700010
India

María Carmen Collado
Instituto de Agroquímica y Tecnología
de Alimentos (CSIC)
Apartado de Correos 73
46100 Burjassot/Valencia
Spain

Larry H. Danziger
University of Illinois at Chicago
Department of Pharmacy Practice
College of Pharmacy
Pan American Health Organization/
World Health Organization
Collaborating Center for Traditional
Medicine
833 South Wood Street
Chicago, IL 60612
USA

Soumen Kumar Das
ICMR Virus Unit
Infectious Diseases and
Beliaghata General Hospital
57 Dr Suresh C. Banerjee Road
Kolkata 700010
India

Jacobus N. Eloff
University of Pretoria
Faculty of Veterinary Science
Department of Paraclinical Sciences
Phytomedicine Programme
Private Bag X04
Onderstepoort 0110
South Africa

Varughese George
Amity Institute of Herbal and
Biotech Products Development
Mannamoola, Peroorkada
Thiruvananthapuram 695 005
India

Kamanzi Atindehou Kagoyire
University of Cocody-Abidjan (UFR
Biosciences)
Centre Suisse de Recherches
Scientifiques (CSRS)
BP 1303 Abidjan 01
Ivory Coast

M. Sajjad A. Khan
Aligarh Muslim University
Department of Agricultural
Microbiology
Anoopshahar Road
Aligarh 202 002
India

Gail B. Mahady
University of Illinois at Chicago
Department of Pharmacy Practice
College of Pharmacy
Pan American Health Organization/
World Health Organization
Collaborating Center for Traditional
Medicine
833 South Wood Street
Chicago, IL 60612
USA

Lindy J. McGaw
University of Pretoria
Faculty of Veterinary Science
Department of Paraclinical Sciences
Phytomedicine Programme
Private Bag X04
Onderstepoort 0110
South Africa

Peter Molan
University of Waikato
Department of Biological Sciences and
the Honey Research Institute
Private Bag 3105
Hamilton 3240
New Zealand

Arup Ranjan Patra
ICMR Virus Unit
Infectious Diseases and
Beliaghata General Hospital
57 Dr Suresh C. Banerjee Road
Kolkata 700010
India

Yolanda Sanz
Instituto de Agroquímica y Tecnología
de Alimentos (CSIC)
Apartado de Correos 73
46100 Burjassot/Valencia
Spain

Mohammed Shahid
Aligarh Muslim University
Jawaharlal Nehru Medical College
and Hospital
Department of Medical Microbiology
Medical Road
Aligarh 202 002
India

Marcus Vinícius Nora de Souza
Instituto de Tecnologia em
Fármacos-Far-Manguinhos
Rua Sizenando Nabuco
100 Manguinhos
21041-250 Rio de Janeiro RJ
Brazil

John Dale-Skinner
University of Nottingham
School of Biomedical Sciences and the
Institute of Infection and Immunity
Queen's Medical Centre
Nottingham NG7 2UH
UK

Christine M. Slover
University of Illinois at Chicago
Department of Pharmacy Practice
College of Pharmacy
Pan American Health Organization/
World Health Organization
Collaborating Center for Traditional
Medicine
833 South Wood Street
Chicago, IL 60612
USA

Koné Mamidou Witabouna
University of Abobo-Adjamé /UFR SN
Centre Suisse de Recherches
Scientifiques (CSRS)
BP 1303 Abidjan 01
Ivory Coast

Maryam Zahin
Aligarh Muslim University
Department of Agricultural
Microbiology
Anoopshahar Road
Aligarh 202 002
India

1

Molecular Mechanisms of Antibiotic Resistance: The Need for Novel Antimicrobial Therapies

John W. Dale-Skinner and Boyan B. Bonev

Abstract

Despite the enormous success of antibiotics as chemotherapeutic agents, infectious diseases remain a leading cause of mortality worldwide. Bacteria coevolving with infectious microorganisms have been driven to develop protection against environmental bioactive chemicals, and to resist their own antibiotics and defense compounds. Such resistance in pathogenic microorganisms provides protection against chemotherapeutic intervention and can lead to infections that are notoriously difficult to manage. Here, we introduce briefly the molecular mechanisms of action for common antibiotic classes and the structural determinants of bacterial resistance to antibiotics. Bacterial resistance to antibiotics interfering with cell wall biosynthesis is discussed with examples from β-lactams and glycopeptides. The molecular determinants of bacterial tolerance to protein biosynthesis inhibitors are analyzed with examples from aminoglycosides, marcolides and tetracyclines. Fluoroquinolone tolerance is described in connection with DNA regulation in the presence of inhibitors. The action of β-lactam and glycopeptides antibiotics, which target cell wall biosynthesis, is evaded through target modifications and antibiotic deactivation. β-Lactamases can be deployed to deactivate β-lactam antibiotics, while penicillin-binding proteins with altered binding sites provide an example of modification of the antibiotic target. Alteration of peptidoglycan peptide termini helps protect cell wall synthesis from glycopeptides, while reduced cross-linking and thickened cell wall obstruct glycopeptides access at the cell wall periphery. Specific molecular changes within ribosomal structures can foil aminoglycoside, tetracycline or macrolide attack. Further protection can be achieved by positioning proteins to protect protein synthesizing machinery from the action of tetracyclines. Important resistance mechanisms protecting bacterial protein synthesis include enzymatic deactivation of aminoglycosides and efflux systems expelling incoming macrolides, aminoglycosides and tetracyclines. Renewed efforts to explore alternative strategies for infection management have brought to the top of scientific agenda defense peptides, lanthionine antibiotics, phage therapies and other antimicrobial techni-

New Strategies Combating Bacterial Infection. Edited by Iqbal Ahmad and Farrukh Aqil
Copyright © 2009 WILEY-VCH Verlag GmbH & Co. KGaA, Weinheim
ISBN: 978-3-527-32206-0

ques, which have not seen wide use against clinically significant pathogens and resistance to which remains uncommon.

1.1
Introduction

'It's time to close the book on infectious diseases, declare the war against pestilence won, and shift national resources to such chronic problems as cancer and heart disease.' These were the optimistic words of William H. Stewart, the United States Surgeon-General, in 1967. However, even during these years of antimicrobial optimism researchers still recognized that microbes would take the opportunity to reproduce about every 30 min as a fresh chance to mutate, migrate, adapt to new hosts and environments, and resist hostile agents that threatened them [1]. The honeymoon period of the 1950s and 1960s for antibiotic development was over, and even William H. Stewart had to admit later that his observation had been 'dead wrong' [2].

Despite the enormous success of antibiotics as chemotherapeutic agents, infectious diseases remain a leading cause of mortality worldwide. Antibiotics are compounds that are (literally) 'against life' – naturally produced by soil microorganisms and airborne molds in response to threats such as overcrowding [3]. Bacteria coevolving with these microorganisms have been driven to develop protection against environmental bioactive chemicals, and to resist their own antibiotics and defense compounds. Such resistance in pathogenic microorganisms provides protection against chemotherapeutic intervention and can lead to infections that are notoriously difficult to manage. The widespread use of antibiotics in recent history has placed significant selective pressure on bacteria in favor of less susceptible strains.

Prior to Fleming's discovery of penicillin in 1929, an antibacterial, pyocyanase, was isolated by Emmerich and Low in 1899 from *Pseudomonas pyocyanase*, and had been used therapeutically against infections caused by staphylococci. In the 1920s, Lieske, Gratia and Dath demonstrated that soil microorganisms like actinomycetes could, in most instances, produce antibacterial substances. Fleming was the first to recognize the potential of penicillin even though he never produced therapeutically usable amounts of penicillins. His discoveries opened the door for Chain, Florey and colleagues to obtain clinically usable amounts of penicillin and large-scale production commenced by 1940. Even at this early stage, the ability of some strains of bacteria to develop penicillin resistance had become evident. By 1946 around 15% of strains of *Staphylococcus aureus* isolated in London hospitals were resistant to penicillin G; 1947 saw this figure approach 40% and by 1948 it was 60% [4].

The discovery of the 6-aminopenicillanic (6-APA) basic nucleus of penicillin molecules in 1959 meant chemical synthesis could be employed to prepare virtually any penicillin structure. The 1960s and 1970s saw the introduction of semisynthetic penicillins, for example ampicillin and amoxicillin, which had good activity against both Gram-positive and Gram-negative bacteria, and methicillin and oxacillin, which were potent against penicillinase-producing *S. aureus*. However, penicillins had their limitations. Most were only active against Gram-positive bacteria, some people developed life-threatening allergic reactions to them and resistance was developing rapidly. A solution came in 1945 from Giusepe Brotze who identified a mold growing near a sewage outfall as *Cephalosporium acremonium*, and showed it inhibited the growth of typhoid bacilli and other bacteria [3]. Thirteen years later Abraham and Newton elucidated the structure of the agent responsible. It contained a β-lactam ring fused to a six-membered ring containing a sulfur atom rather than the five-membered ring seen in penicillins. The Eli Lilly Company successfully removed the side-chain of cephalosporin C to produce a bare cephalosporin nucleus (7-aminocephalosporanic acid).

Strains of *S. aureus* and *Streptococcus pneumoniae* resistant to antibiotics can cause serious infections in hospitals and the community. Significant media attention in recent years has highlighted the problem and has coined the term 'superbugs' in reference, chiefly, to methicillin-resistant *S. aureus* (MRSA) and resistant enterococci. Although reported to a lesser extent, *Staphylococcus epidermidis* and *Staphylococcus saprophyticus* can colonize medical implants and cause serious antibiotic-resistant infections. Antibiotic-resistant commensal bacteria do not present a health problem in healthy individuals and only become problematic when they colonize atypical sites where they may cause persistent infections. In one example, *S. aureus* was a commensal in around 30% of the general population [4], while resistant hospital-acquired strains can lead to significant therapeutic challenges in immunocompromised patients.

In the 1960s, 10% of *S. aureus* strains produced penicillin-destroying enzymes (penicillinases); today the figure approaches 100%. Despite the introduction of methicillin in 1959 to tackle the increasing problem of penicillin resistance, it took only 3 years for MRSA to appear in 1961. Once resistance is genetically encoded it can spread rapidly within a population of bacterial species, or even to another bacterial species through transduction, transformation, conjugation or transposition.

Transduction involves phage replication that accidentally includes replication of resistance-encoding bacterial DNA. When newly replicated phages infect another bacterium the resistance genes can recombine into the DNA of the new host. Transformation involves recombination with the bacterial chromosome of new resistance genes present on absorbed fragments of exogenous bacterial DNA. Conjugation allows plasmid DNA conferring resistance to be transferred via the sex pilus to another bacterium and is a major mechanism of gene transfer in enterococci. Transposons confer resistance to antibiotics by transferring genes between different plasmids or from plasmids to chromosomes.

Social awareness of antibiotic-resistant hospital acquired infections, most notably with *Staphylococcus* and *Clostridium*, is at all times high. Each year in England and Wales about 5000 lives are lost to these infections, and the annual costs to the National Health Service amount to £1 billion. With resistance to antibiotics rapidly spreading, progress in search of novel chemotherapeutic agents is slow. The United kingdom has one of the highest levels of incidence of MRSA in Europe [5]. In 1993 there were 216 deaths where *Staphylococcus* infection was the final underlying cause of death. This figure rose to 546 deaths in 1998 [6]. Sophisticated resistant gene transfer mechanisms are partly responsible; however, it is our usage of antibiotics that places evolutionary pressure on bacteria and selects those conferring resistance. This is exacerbated by prescribing at subtherapeutic levels and poor patient compliance, and declining antibiotic research and development has not helped. Hospital antibiotic usage demands attention because the large numbers of immunocompromised patients in hospitals creates a suitable environment for the development of antibiotic-resistant bacteria. This environment is primed further by the fact that only around 30% of all hospital antibiotics are used for definitive therapy, where susceptibility patterns for the infection-associated pathogen are known [7].

Here, we introduce briefly the molecular mechanisms of antibiotic action, and discuss the molecular and structural determinants of bacterial resistance to antibiotics. Bacterial resistance to antibiotics interfering with cell wall biosynthesis is discussed with examples from β-lactams and glycopeptides. The molecular determinants of bacterial tolerance to protein biosynthesis inhibitors are analyzed with examples from aminoglycosides, marcolides and tetracyclines. Finally, fluoroquinolone tolerance is described in connection with DNA regulation in the presence of inhibitors.

1.2
Molecular Mechanisms of Resistance

Understanding the molecular mechanisms underlying antibiotic resistance requires an understanding of bacterial structure and function. Structural and metabolic differences between bacterial and mammalian cells make it possible to selectively kill bacteria with antibiotics. For example, the presence of a cell wall in bacterial, but not mammalian, cells can be exploited by antibiotics. Gram-negative bacteria have an inner cell wall made up of a thin layer of peptidoglycan and an outer wall referred to as the outer membrane owing to its resemblance to the cytoplasmic membrane. The periplasm lies between the outer cell wall and the inner cell wall. The Gram-positive cell has no outer membrane and therefore no periplasm. Instead, the cell wall is made up of a thick layer of peptidoglycan and accessory polymers. Mechanical strength in the cell wall is crucial as it must withstand osmotic pressures in excess of 5–15 atm without rupturing [8].

β-Lactam and glycopeptide antibiotics target cell wall biosynthesis, to which there are three stages [8]. First, UDP-*N*-acetylglucosamine (GlcNAc) is converted to UDP-*N*-acetylmuramyl-pentapeptide in the cytoplasm [9]. UDP-GlcNAc is made by linking glucose-1-phosphate (from glucose or glucosamine) to UDP (from pyrimidine biosynthesis starting with glutamate). MurC, D and E sequentially add L-Ala-D-γ-Gln and Lys (Gram-positives) or meso-diaminopimelate (DAP) (Gram-negatives) in ATP-dependent amide-forming steps to create UDP-muramyl-tripeptide. MurF adds a preformed D-Ala-D-Ala in the fourth amide-forming step to create the UDP-muramyl-L-Ala-γ-D-Gln-L-Lys-D-Ala-D-Ala pentapeptide. Transpeptidase catalyzed cross-linking between the third amino acid (L-Lys in Gram-positives, meso-DAP in Gram-negatives) and the penultimate amino acid (D-Ala) of adjacent peptidoglycan chains, with loss of the terminal D-Ala, gives the cell wall rigidity.

The second stage involves transferring muramyl-pentapeptide from UDP to a C55 isoprenol-P carrier. The MraY-catalyzed movement of the muramyl-pentapeptide to the membrane interface makes lipid I. The peptidoglycan chain consists of alternating N-acetylmuramic acid and GlcNAc, linked via β1–4 links in a reaction catalyzed by MurG. This generates the disaccharide-pentapeptide attached to bactoprenol-PP, which is lipid II. Lipid II flips the disaccharide-pentapeptide across the plasma membrane to its outer face. Here, transglycosylation involves transglycosylases catalyzing binding of the disaccharide pentapeptide to the 4-OH group of GlcNAc of the existing peptidoglycan strand.

Aminoglycoside, tetracycline and macrolide antibiotics target protein synthesis. The 70S bacterial ribosome consists of a 50S and a 30S subunit. The 50S subunit contains one molecule of 55S RNA and one molecule of 23S RNA plus 32 proteins. The 30S subunit contains one molecule of 16S RNA and 21 proteins. Bacterial mRNA binds to the 30S ribosomal subunit, attracting N-formylmethionyl-tRNA (fMet-tRNA) to its AUG initiator codon and forming the 30S initiation complex. The 70S initiation complex is completed by adding the 50S subunit. fMet-tRNA binds to the peptidyl donor (P)-site, which is adjacent to the aminoacyl acceptor (A)-site. Vacant at first, the A-site is where aminoacyl-tRNA bearing the appropriate anticodon and its specific amino acid will bind to. Affinity of aminoacyl-tRNA for the A-site is low and elongation factor (EF)-Tu, which hydrolyzes GTP, is required to increase affinity of the 70S complex for aminoacyl-tRNA. Aminoacyl-tRNA becomes bound to the A-site in the form of a ternary complex: EF-Tu·aminoacyl-tRNA·GTP. Codon–anticodon recognition at the A-site is associated with hydrolysis of GTP and results in the release of EF-Tu·GDP. Following this release is formation of a peptide bond catalyzed by peptidyltransferase which is located in the 50S subunit. Peptide bond formation involves the transfer of N-formylmethionine (or peptidyl residue in subsequent chain elongation cycles) from tRNA in the P-site to aminoacyl-tRNA in the A-site, where it is joined to the new amino acid. Deacylated tRNA is lost to the exit (E)-site, the newly formed peptidyl tRNA will move from the A-site to the P-site, and the mRNA and ribosome move to incorporate the next codon into the A-site. The peptidyltransferase-catalyzed transfer of the N-formylmethionine to the new amino acid is called translocation and is repeated for subsequent tRNA complexes. However, the maximum rate of translocation is dependent on EF-G, which hydrolyzes GTP to promote translocation.

Bacterial DNA is replicated in a stepwise manner along the DNA circle, progressing along a continuously advancing point, the replication fork [9]. Replication involves separation of the double-stranded DNA and alignment of nucleotides with their complementary bases on each single strand. However, separation of strands wound in a circular helix generates loops called positive supercoiled twists in the single strands. If this was unresolved, positive superhelicity would increase until the rising torsional strain prevented any further unwinding of the DNA at the replication fork [10]. DNA gyrase and topoisomerase IV work in a coordinated fashion to restore the proper conformational structure of DNA and safeguard against the occurrence of high-level replication-induced helical stress.

Discovering new antibiotics requires a comprehensive understanding of the molecular mechanisms and structural determinants of resistance, and how new drugs may be affected by existing mechanisms and ones which we can anticipate emerging [11]. Destroying or inactivating the antibiotic, pumping out the antibiotic and modifying the antibiotic target are important mechanisms conferring bacterial resistance, and will be reviewed in the context of specific antibiotics.

1.3
β-Lactams

Penicillins, cephalosporins and other β-lactam antibiotics contain a four-membered amide ring and target cross-linking in the final stage of cell wall biosynthesis (transpeptidation). The four-membered ring is structurally similar to the terminal acyl-D-Ala-D-Ala unit of the nascent peptidoglycan (Figure 1.1). Both contain the CO–N bond known to be cleaved by penicillin-binding proteins (PBPs) found in the cell membranes of bacteria. There are usually four or more PBPs with transglyco-sylation and transpeptidation activity. They are named numerically relative to their molecular size (PBP1 is the heaviest). High-molecular-mass PBPs have an essential role in peptidoglycan synthesis, whereas low-molecular-mass PBPs have only a minor role. Normally, PBPs utilize the D-Ala-D-Ala moiety of the pentapeptide as a substrate, forming an acyl-enzyme intermediate with release of the terminal D-Ala during transpeptidation. β-Lactam antibiotics act as pseudosubstrates and subvert this process by acylating the active sites of the PBPs [12]. Transpeptidation activity is essentially inhibited because deacylation of the penicilloylated PBPs is so slow.

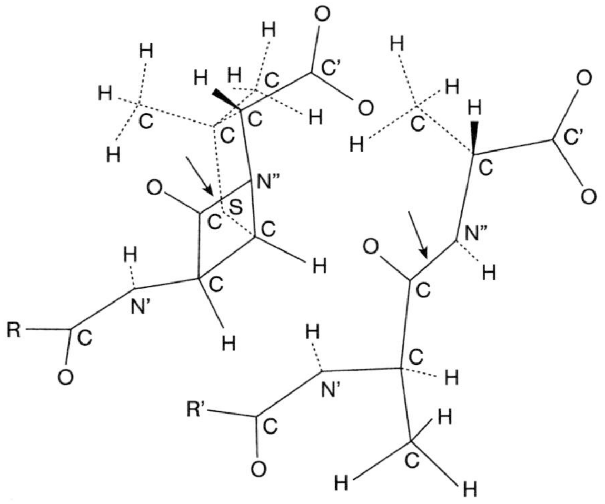

Figure 1.1 Dreiding stereomodels of penicillin (upper left) and of the acyl-D-Ala-D-Ala end of the nascent peptidoglycan (lower right). Arrows indicate CO–N bond common to both structures and the heavy lines indicate the portion of the penicillin molecule believed to resemble the peptide backbone of acyl-D-Ala-D-Ala [10].

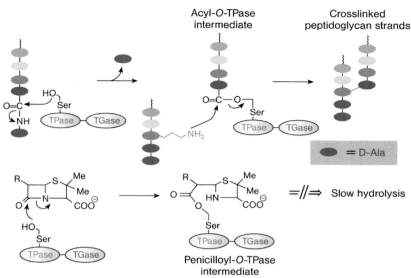

Figure 1.2 Inhibition of transpeptidase activity by penicillins [12].

Fifty-five percent all antibiotics are β-lactams [13], so resistance in this class is clinically important. β-Lactam hydrolysis by β-lactamases was observed by Abraham and Chain even before penicillin was introduced into the clinical setting. There are four classes of β-lactamases – classes A, C and D, which utilize an active-site serine, and class C, which requires divalent metal cations (Zn^{2+}) for catalysis [17]. The four-membered, strained lactam ring is the chemically activated functionality in the drugs which acylates and irreversibly modifies the cell wall cross-linking PBPs (Figure 1.2).

β-Lactamases (penicillinases) open the β-lactam ring by hydrolyzing the CO–N bond that is normally involved in the acylation of PBPs. The resultant ring-opened structure (penicilloic acid) has no activity as a pseudosubstrate of PBPs. Deactivation occurs in the periplasm, so that penicillin is deactivated before it reaches its cell wall target (Figure 1.3). Gram-negative bacteria produce β-lactamases within their cytoplasm and then secrete them into their periplasm, whereas Gram-positive bacteria secrete β-lactamases into the external environment. β-Lactamases confer high levels of resistance as a single β-lactamase molecule can hydrolyze 10^3 penicillin molecules in a second. Thus, if 10^5 enzymes are secreted per resistant cell, 100 million molecules of penicillin are destroyed every second [7].

Amino acid sequences of PBP active sites contain an active serine in the sequence Ala/Gly-Ser-X-X-Lys, of which the Ser-X-X-Lys sequence is conserved in all PBPs. Moreover, PBP1–3, 5 and 6 of *Escherichia coli* have been shown binding to this conserved serine [10]. Physiologically this serine acts as a nucleophile in the PBP active site and reacts with acyl-D-Ala-D-Ala to yield an acyl-D-alanyl-enzyme intermediate. In penicillin-sensitive bacteria, penicillin binds to this active-site serine forming an acetylated penicilloyl-enzyme intermediate, preventing any

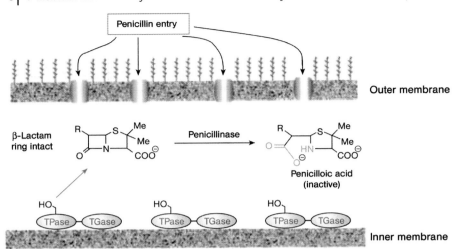

Figure 1.3 Chemical modification of penicillin by penicillinases [7].

normal cross-linking activity. Consistent with their ability to hydrolyze the CO–N bond, some β-lactamases also contain an active-site serine in the Ser-X-X-Lys motif [15] (see also [16] for review). This serine binds penicillin and later releases it as the hydrolyzed penicillin product. Penicillin-bound PBPs are immobilized as the acyl-enzyme because hydrolysis is so slow; however, for active-site serine β-lactamases the acyl-enzyme hydrolysis is rapid. Active-site serine β-lactamases can be divided into three molecular classes (A, C and D) on the basis of their primary structures [14].

The similarity between DD-peptidases and serine β-lactamases with respect to their catalytic pathways has prompted comparison studies between the two molecules, and structural similarities have been shown between class A β-lactamase and the D-Ala-D-Ala penicillin target [17]. Moreover, tertiary structures of class A and C β-lactamases and DD-peptidase hint at an ancestral link between these three enzymes [18]. All have a β-lactam-binding site in their center at the left edge of the five-stranded β-sheet. The reactive serine lies at the N-terminal end of a central α-helix [18] (orange in Figure 1.4). If β-lactamases are ancestrally linked to DD-peptidases, what makes DD-peptidases the unfortunate target of β-lactams, while β-lactamases escape harm and are primed ready to search and destroy β-lactams? Clearly β-lactamases do not catalyze transpeptidation and likewise DD-peptidases do not hydrolyze penicillin as efficiently as β-lactamases. This slow rate of deacylation of penicilloylated PBPs is the basis for β-lactam activity.

Interactions exist between the B3 β-strand of DD-peptidase and the C-terminal carboxylic acid group, the carbonyl group of the C-terminal peptide bond and the amido group of the penultimate peptide bond [18]. If a β-lactam is substituted for the D-Ala-D-Ala peptide, only the first two of these interactions is present. A better interaction between the C6 β-acylamido group of the β-lactam and the β-strand could

(a)

(b)

(c)

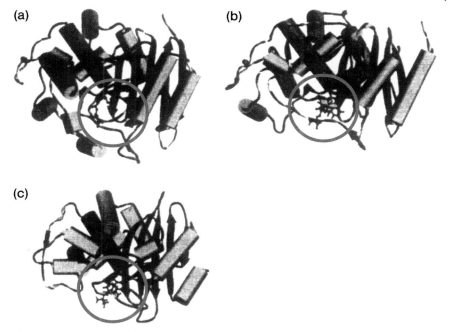

Figure 1.4 Molecular models of (a) D-Ala-D-Ala carboxypeptidase/transpeptidase from *Streptomyces* spp., (b) class C β-lactamase of *Enterobacter cloacae* and (c) class A β-lactamase of *Bacillus licheniformis*. Cylinders and arrows represent α-helices and β-strands, respectively. The reactive serine is at the N-terminus of the central helix H2. The Ω-loop lies at the lower center of each molecule. The red circle denotes the D-Ala-D-Ala-binding site in (a), and the β-lactam-binding site in (b) and (c) [18].

be achieved if the strand was pointed outward to improve hydrogen bonding between the acylamide and the carbonyl group of the β-strand [18]. Class A β-lactamase structure is optimized for hydrolysis of β-lactams and has completely lost its activity on D-Ala-D-Ala peptides. In comparison to class C the B3 β-strand of class A β-lactamases has an accentuated tilt which favors binding and hydrolysis of β-lactams. The Ω-loop protrudes more deeply into the active site allowing the backbone carbonyl group of Asn170 to hydrogen bond with the amide residue on the B3 β-strand. This increases the tilt of B3 β-strand. In contrast to DD-peptidases, class A and C β-lactamases possess a water molecule in the oxyanion pocket which increases the electrophilicity of the oxyanion hole. The absence of a water molecule in the oxyanion hole means DD-peptidases are less able to stabilize the oxyanion β-lactam tetrahedral intermediate [18].

Serine β-lactamases possess four major catalytic elements (Table 1.1), and the active-site serines are Ser70, Ser64 and Ser115 in class A, C and D β-lactamases, respectively [19]. The Ω-loop is also conserved in all serine β-lactamases, suggesting a role in the enzymes function. However, despite this, the exact role of other specific amino acid residues remains unclear.

Table 1.1 Major catalytic elements in class A, C and D β-lactamases [19].

	Group A	Group C	Group D
First element	S70–X–X–K	S64–X–X–K	S67–X–X–K
Second element	S130–X–N	Y150–A–N	S115–X–V
Third element	K234–T/S–G	K314–T–G	K205–T/S–G

The active site of the lactamases can be regarded as consisting of two ensembles of amino acids – a recognition ensemble and a catalytic ensemble. In addition to the active serine, the catalytic ensemble consists of a general base, which receives a proton from the serine side-chain. This activates the serine for nucleophilic addition to the oxygen on the carbonyl group on the β-lactam and allows the nitrogen to accept a proton from an unknown donor. There is also an oxyanion hole consisting of two amino acids that stabilize the tetrahedral oxyanion intermediate through hydrogen bonding and a cationic recognition site for the carboxylate group on the β-lactam [20]. The catalytic ensemble is most probably highly conserved and linked with the major catalytic elements in Table 1.1. The recognition ensemble is suggested to be more variable, consisting of the remaining amino acids and dependent on the specific selective pressures exerted on the bacteria.

Carbapenems are a group of broad-spectrum antibiotics defined by the substitution of the typical sulfur atom at position one with a carbon atom, this structure resists inactivation by most β-lactamases. However, class B β-lactamases can significantly hydrolyze carbapenems [16, 21]. Class B β-lactamases require divalent cations, primarily zinc, and there are three subgroups (B1–B3). The metallo-β-lactamases (MBLs) possess a distinct set of amino acids that define the architecture of their active site which is largely superimposible among the different MBLs. Most MBLs contain the principle binding motif His-X-His-X-Asp [21], although B2 enzymes possess an asparagine instead of a histidine at the first position. B2 enzymes also differ from other MBLs, by only containing a single zinc ion, while other MBLs contain two zinc ions in their active site. Similarly to the serine β-lactamases, the general consensus is that MBL hydrolysis of the β-lactam ring is via nucleophilic attack where their active site orients and polarizes the β-lactam bond to facilitate attack by zinc-activated water/hydroxides [20, 21].

Cefotaxime, cefotriaxone and ceftazidime are extended-spectrum β-lactamase (ESBL) antibiotics introduced in response to growing β-lactamase resistance. Plasmid-encoded class A TEM-1 β-lactamase is a common cause of resistance to penicillins and cephalosporins, but not ESBL antibiotics. However, variants of this class have been found with a G238S substitution displaying increased hydrolysis of ESBL antibiotics. A direct hydrogen bond between the hydroxyl side-chain group of Ser238 and the oxime group of ESBLs is necessary but not sufficient for hydrolytic activity of cefotaxime and ceftazidime [22]. Additionally, an intramolecular hydrogen bond, possibly from Ser238 to the main chain CO group of Asn170 of the Ω-loop, is thought to stabilize the Ω-loop in a new location which enables the enzyme to broaden its substrate profile to include ESBL antibiotics [22]. Fortunately, in terms of β-lactamase resistance, bacteria have not yet completely undermined our efforts. β-Lactamase inhibitors (e.g. clavulanic acid, sulbactam and tazobactam) are proteins that can be used in combination with β-lactam antibiotics. Despite little by way of antimicrobial activity, the β-lactamase inhibitors bind with higher affinity to the β-lactamases than the β-lactams and can be used to distract the β-lactamases while the β-lactams work without the threat of enzymatic inactivation. However, such a simple strategy only requires the bacteria to learn how to overwhelm the β-lactamase inhibitors with enzymes for them to gain the advantage.

Methicillin is an antistaphylococcal β-lactamase-stable penicillin introduced to overcome β-lactamase resistance. Methicillin resistance is caused by an acquired gene (*mecA*) which results in the synthesis of a fifth PBP (PBP2a), in addition to the intrinsic 1–4 PBPs [23]. Strains of *S. aureus* that acquire the *mecA* gene are MRSA [13]. PBP2a from *S. pneumoniae* exhibits a much lower affinity for β-lactam antibiotics, especially penicillins, when compared with other PBPs [24]. Similarly to other PBPs, PBP2a undergoes acylation with its peptidoglycan substrate at an active-site serine (Ser403) [13].

The stability of the acyl-PBP intermediate (PBP–penicillin) is what conveys the antibiotic action. A very stable intermediate essentially inhibits the transpeptidation activity of PBPs for a long time. Studying the kinetics of this reaction sequence has revealed two aspects that may cumulatively give rise to PBP2a resistance. The first is a significantly reduced rate constant for SauPBP2a acylation (k_2) compared to penicillin-sensitive PBPs. This makes formation of the PBP–penicillin intermediate very unlikely. Moreover SauPBP2a has an elevated dissociation constant (k_d) for the noncovalent preacylation Michaelis complex with the antibiotic, meaning encounters between the antibiotic and PBP2a are not favorable [13]. Interestingly, a resistant form of PBP2x (R-PBP2x) has an overall antibiotic binding efficiency (k_2/k_d) that is over 1000-fold slower than penicillin-sensitive PBP2x. Kinetic studies also noted differences between the deacylation rate constants (k_3) of penicilloyl-R-PBP2x and penicilloyl-S-PBP2x, equating to an increase in the deacylation rate of over 70-fold for R-PBP2x [25].

Elucidation of the structure of PBP2a provides some explanation for the kinetics. PBP2a from *S. aureus* (SauPBP2a) has an N-terminal extension and transmembrane anchor, a non-penicillin-binding domain (nPD) and a C-terminal transpeptidase domain (see Figure 1.5). The nPD domain positions the transpeptidase domain more than 100 Å away from the transmembrane anchor, suggesting a structural role in giving the transpeptidase domain substantial reach from the cell membrane [26]. The transpeptidase of SauPBP2a shares a similar overall fold with other transpeptidases and the serine β-lactamases. In the apo conformation of the helix α2 N-terminus, Ser403 is in a poor position for nucleophilic attack. Cα, Cβ and Oγ from Ser403 have to move slightly upon acylation. For acylation to occur under normal circumstances, a twisted apo conformation of strand β3 is required to accommodate the helix α2 N-terminus due to steric clash between Ser598 backbone carbonyl in the acyl-PBP complex and the Ser403 Cβ in the apo structure. Twisting of the β3 strand is also required for binding of nitrocefin, due to steric clash between the nitrocefin carboxylate and the Gly599 Cα in the apo structure [26]. Lim

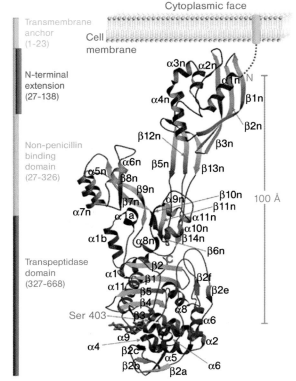

Figure 1.5 Structure of SauPBP2a. The bilobed N-terminal nPB domain is orange and the N-terminal extension is colored green. The transpeptidase domain is blue, with the position of the active site indicated by the red nitrocefin adduct (shown in stick rendering) [26].

and Strynadka describe a closed active-site structure for SauPBP2a, where formation of the Michaelis complex requires a conformation different to the conformation required for acylation. This transition from Michaelis complex to acyl-PBP complex (k_2) is so energetically costly that it does not happen, which is why SauPBP2a is resistant to β-lactams [26].

PBP2a resists modification by β-lactam antibiotics while remaining a competent catalyst in transpeptidation. Coupled with its closed active site [26], this has led to the suggestion that physiological substrates of PBPs interact with the surface of PBP and facilitate opening of the active site, making entry of peptidoglycan possible. Evidence for this is the increased nitrocefin (chromogenic cephalosporin) acylation of PBP2a (k_2) that accompanies increasing concentrations of peptidoglycan fragments.

Moreover, it was proved that the protein undergoes a conformational change consistent with a decrease in helicity [27]. β-Lactam antibiotics are small in comparison to the peptidoglycan substrate and so are not capable of interacting with

SauPBP2a in the same way, explaining why they do not gain access to the active site. The low affinity of PBP2a to β-lactams suggests that it is a naturally resistant form of PBP and therefore was probably never a target of β-lactams [24]. The theory that PBP2a takes over the biosynthesis of the cell wall from the four sensitive native PBPs is unlikely because PBP2a resistance is dependent on continued transglycosylation by native PBP2 [27, 28]. This cooperation may make this mechanism of resistance vulnerable to novel attempts at overcoming resistance in the future.

PBP2x and PBP2b are essential for bacterial growth [30] and are therefore targets of β-lactams [31]. The resistant strain of PBP2x (R-PBP2x) in *S. pneumoniae* has a transpeptidase domain (residues 266–616) carrying numerous substitutions compared to homologous sequences from β-lactam-sensitive streptococci (S-PBP2x) [32]. In recent years there have been exhaustive attempts to identify these mutations and their influence on the resistance profile of PBP2x. Thr338 does not contact the antibiotic directly; however, positioned just after the active-site Ser337, it is the most frequent mutation seen in clinically resistant pneumococci [32]. PBP2x from resistant strains (R-PBP2x) contain a T338A mutation. Imposing this mutation on S-PBP2x reduces the acylation efficiency by a factor of 2 [33] because it abolishes a crucial hydrogen bond between the hydroxyl group of the threonine residue and a buried water molecule [34]. Mutation T338A can coexist in some highly resistant strains with mutation M339F. Replacing methionine with phenylalanine introduces a bulkier side-chain which strains the active-site structure and reorientates the hydroxyl group of Ser337 [26]. It is likely that this distorted catalytic center contributes to the reduction in acylation efficiency seen in clinical strains containing the double mutation compared to those sensitive strains.

Figure 1.6 illustrates the closeness and parallel positioning of cefuroxime in relation to the β3 strand, and also indicates residues T338, T550 and Q552. Replacing Gln552 of S-PBP2x from the R6 strain with glutamic acid, which is commonly found in R-PBP2x, reduces the acylation efficiency by over two-thirds for both penicillin and cefotaxime [32]. Glutamic acid introduces a negative charge into what is normally a

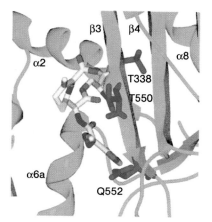

Figure 1.6 View of cefuroxime bound in the active site of S-PBP2x [32].

globally positive active site. This conflicts with the global negativity of β-lactams and reduces the acylation efficiency [32]. Reversion of this mutation confirms its role in the expression of resistance [35]. Crystal structures reveal an overall similarity between the structures of antibiotic-sensitive and antibiotic-resistant strains except in their active-site regions (Figure 1.7a) [34]. Although not disordered, strand β3 (Figure 1.7b), most notably segment Ser548 to Thr550, is displaced by 0.5 Å in resistant strains, while strands β4 and β5 superimpose very well [34].

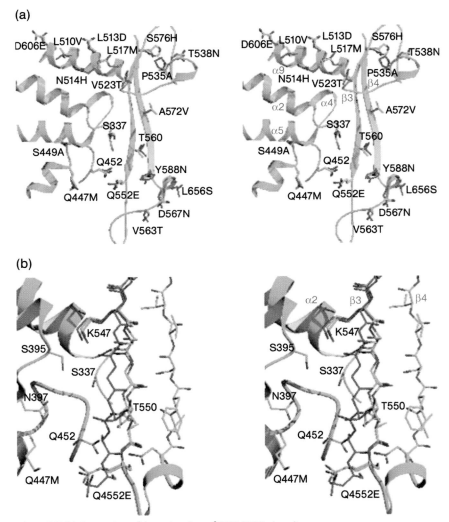

Figure 1.7 (a) Stereo view of the active sites of 5259-PBP2x (cyan) and R6-PBP2x (magenta). (b) View of the displacement of the β3 strands in PBP2a from *S. aureus* (magenta), R6-PBP2x (yellow) and 5259-PBP2x (cyan) [34].

Mutations S389L and N514H with PBP2x of *S. pneumoniae* have two distinct effects on the active site. First, there is steric clash between them which moves the catalytic residue Ser395 to an unfavorable position for enzymatic function. Here, only four of the six stabilizing interactions normally seen between Ser395 and cefotaxime in susceptible strains are formed [36], and these two mutations result in an open active site. The importance of an open active site in the expression of resistance is undetermined; however, it maybe beneficial to resistant strains of *S. pneumoniae* which produce abnormal indirectly cross-linked cell walls.

The *murMN* operon contains *murM* and *murN* genes, and produces branched muropeptides [37]. Instead of the linear-stem peptides (L-Ala-D-iGln-L-Lys-D-Ala) seen in susceptible strains, resistant strains have branched-stem peptides carrying Ala-Ser or Ala-Ala dipeptides on the ε-amino group of the stem peptide lysine residues [25]. Producing branched cell wall precursors is important in the expression of penicillin resistance in *S. pneumoniae* because interrupting the *murMN* operon causes virtually complete inhibition of the expression of penicillin resistance [32]. Although speculative, the theory that branched cell wall precursors are more successful than antibiotics in competition for active-site occupancy is conceivable. It is not implausible to suggest that this is because an open active site is more accessible to branched muropeptides.

Figure 1.7(a and b) shows the close proximity of Thr550 to the active-site Ser337. Alanine is often substituted for threonine at this position in resistant strains and reduces cefotaxime acylation efficiency by almost 20-fold. Interestingly this mutation is neutral for penicillin [32]. Preserving the hydroxyl group, while altering the steric property of the side-chain of position 550, is consistent with a T550S mutation. While this mutant is unable to discriminate between penicillin G and cefotaxime, and shows acylation efficiency values very close to those of Q552E mutant, it is not found clinically.

1.4
Glycopeptides

Glycopeptide antibiotics interfere with the transpeptidation stage of late cell wall biosynthesis. Vancomycin is a clinically used glycopeptide, which targets the terminal Lys (Dap)-D-Ala-D-Ala residues in the mature peptidoglycan intermediate, lipid II, and in unbranched linear peptidoglycan strands. It forms five hydrogen bonds between amides of the cross-linked heptapeptides and the D-Ala-D-Ala dipeptide terminus of each uncross-linked peptidoglycan pentapeptide side-chain (Figure 1.8). The D-Ala-D-Ala terminus is present in lipid I and lipid II; however, vancomycin cannot enter the bacterial cell due to its size and hydrophobicity, and only lipid II and the UDP-MurNAc-pentapeptide are affected [38]. Thus formed, the vancomycin–peptide complex becomes unavailable as substrate for transpeptidases and transglycosylases.

First produced in 1958, vancomycin is traditionally a last resort antibiotic for treating patients who are gravely ill or infected by organisms resistant to other

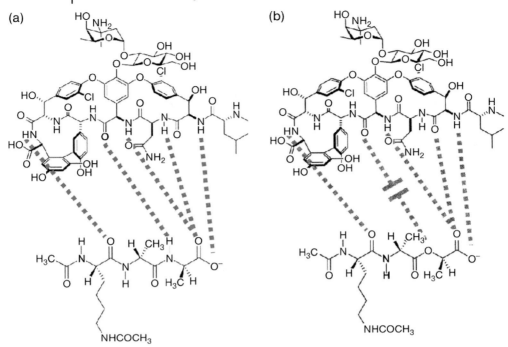

Figure 1.8 (a) Complex formation between vancomycin and *N*-acyl-D-Ala4-D-Ala5. Key hydrogen bonds stabilizing the complex are shown (after (8)). (b) Complex formation between vancomycin and D-Ala-D-Lac does not occur as one of the required hydrogen bonds cannot form [39, 40].

antibiotics such as β-lactams. Due to their unique mode of action, it was considered that development of resistance to glycopeptides was unlikely [8]. That year, nearly 30 years after the introduction of vancomycin, the first vancomycin-resistant enterococci (VRE) in Europe were reported [41]. Since then, VRE have become a major cause of nosocomial infections. Vancomycin-dependent enterococci which require vancomycin for their growth have emerged in patients treated for long periods with vancomycin [42]. Inactivation of the *ddl* gene for D-Ala:D-Ala ligase is responsible and mutants lacking D-Ala:D-Ala ligase are incapable of forming peptidoglycan precursors with complete D-Ala-D-Ala terminals. Such deficiency leads to utilization of an alternative pathway, in which bacteria synthesize peptidoglycan precursors terminating on D-Ala-D-Lac and are not susceptible to vancomycin [43].

Resistance to vancomycin in enterococci is achieved by altering their peptidoglycan precursors so that glycopeptides can no longer bind. Out of the four VanA–D VRE phenotypes, VanA and VanB are the two most clinically relevant. VanA, VanB and VanD have a substitution of the C-terminal D-Ala to D-Lac, whereas VanC and VanE have a D-Ser substituted in this position. D-Lac suppresses a hydrogen bond crucial for antibiotic binding, whereas D-Ser does not alter the hydrogen bonds, but is responsible for a conformational change which slightly reduces affinity for vancomycin [41]. VanA is characterized by inducible, high-level resistance to both vancomycin and teicoplanin, whereas VanB is resistant to vancomycin but susceptible to teicoplanin [42]. The five *van* genes, *vanRSHAX*, are necessary for both phenotypes.

The resistance mechanism for VRE was first elucidated by Walsh and Courvalin in the 1990s [39]. VanR and VanS proteins form a two component regulator system [40]. VanR senses vancomycin outside the bacterial cell, while VanS activates the transcription of the *VanHAX* genes. VanHAX proteins reprogram the peptidoglycan termini from the *N*-acyl-D-Ala-D-Ala (glycopeptide target) to *N*-acyl-D-Ala-D-Lac. VanH reduces pyruvate to D-lactate, D-Ala-D-Lac is made by a *VanA*-encoded ligase and VanX is a D-Ala-D-Ala dipeptidase which removes the D-Ala-D-Ala intermediate. Additionally, VanY is a DD-carboxypeptidase that cleaves the D-Ala terminal peptide from any normal peptide and VanZ modestly increases the minimum inhibitory concentration (MIC) for teicoplanin, but not for vancomycin, through an unknown mechanism [45]. As a result, D-Ala-D-Lac accumulates sufficiently within the cytoplasm to be added to the UDP-muramyl-tripeptide in a MurF-catalyzed reaction. UDP muramyl-L-Ala-D-γ-Glu-L-Lys-D-Ala-D-Lac is generated, which can then be incorporated into the growing peptidoglycan chain. Organization of the *vanA* operon is similar to that of the *vanB* and *vanD* operons. The mechanisms of VanA-, VanB- and VanD-type resistance are identical, and there is a high degree of similarity in the amino acid sequences of their Van proteins [43].

Figure 1.8 highlights the differences between D-Ala-D-Lac and D-Ala-D-Ala. Resistance is conferred by a 1000-fold decrease in binding affinity between D-Ala-D-Lac and glycopeptides [46]. One of the five hydrogen bonds does not form because of an absent amide group which is replaced by an ester oxygen whose lone pair of electrons exhibit repulsion towards an electronegative oxygen on vancomycin. Many hypotheses have attempted to explain why the VanB phenotype does not confer teicoplanin resistance. One suggests a lipid chain unique to teicoplanin anchors it to the membrane, which enhances antibiotic binding because both antibiotic and target are attached by membrane anchors to the same template [47]. This renders teicoplanin inaccessible to the sensor kinase, so it does not induce resistance [43]. An alternative hypothesis suggests VanS senses peptidoglycan intermediates or degradation products that result from blocking cell wall biosynthesis [48]. Localization of teicoplanin to the membrane means teicoplanin is in a good position to inhibit transglycosylation by binding to lipid II, whereas vancomycin would preferentially act on the nascent peptidoglycan chain to inhibit transpeptidation [8, 49, 50]. The phenotypic difference between VanA and VanB could be a result of differences in gene expression [43] as the specificity of induction in the VanB phenotype is a characteristic of VanS$_B$ [51].

Vancomycin usage has increased with the spread of MRSA, which exerts a greater selective pressure on staphylococcal bacteria and facilitates the development of resistance. Vancomycin intermediate *S. aureus* (VISA) and glycopepetide intermediate *S. aureus* (GISA) strains have been reported in many countries, and the first clinical case of vancomycin-resistant *S. aureus* (VRSA) was documented in the United States in 2002 [52]. VRSA describes *S. aureus* with a MIC of 32 mg/l or less, VISA have a MIC of 8–16 mg/l and hetero-VRSA (hVRSA) have a MIC of 4 mg/l or less, but which show a population heterogeneity similar to Mu3 (the archetypal hVRSA) when subjected to a full population analysis profile [7]. In one model, vancomycin-susceptible mutants spontaneously emerge due to the slow growth of VRSA strains. These susceptible cultures maintain subpopulations with vancomycin resistance (hVRSA) [53], which produce VRSA mutants when selected for by vancomycin [54]. The model suggests vancomycin and β-lactams maybe involved in vancomycin-sensitive *S. aureus* (VSSA) to hVRSA conversion [53]. Therefore, long-term treatment of MRSA with β-lactams is a possible risk factor for vancomycin resistance.

Enterococci are part of the normal flora of the alimentary canal and are opportunistic, with the capacity to acquire and spread antimicrobial resistant factors [55]. *vanA* genes have been transferred via plasmid-mediated conjugation from enterococci to staphylococci *in vitro* [56], although the mechanism of vancomycin resistance in *S. aureus* appears to be novel. All strains are negative for the *van* genes; however, a thickened cell wall with reduced levels of peptidoglycan cross-linking is common [7, 57]. Decreased cross-linking increases the number of D-Ala-D-Ala side-chains to which glycopeptides bind, therefore VRSA (e.g. Mu50) has an increased binding of vancomycin molecules [57].

Vancomycin-resistant Mu50 cells have an increased proportion of nonamidated muropeptides containing D-glutamate instead of D-glutamine [35, 57]. Pentapeptides containing D-glutamate are poorer substrates for transpeptidases and so consequently fewer cross-links are formed [53]. Additionally, experiments suggest that nonamidated murein monomers may have higher affinity for binding to vancomycin than amidated murein monomers [57]. Cui *et al.* conclude that a unit weight of purified peptidoglycan with a high proportion of nonamidated muropeptides consumes more vancomycin molecules than a unit weight of peptidoglycan with low nonamidated muropeptide content [58]. While nonamidated murein monomers seem to influence vancomycin susceptibility, there is evidence to suggest that absent or greatly reduced levels of PBP4 may also decrease the susceptibility of some strains of staphylococci to vancomycin [59–61]. The introduction of plasmid encoding PBP4 homologs into doubly D-cycloserine/vancomycin-resistant *S. aureus* restores sensitivity to both drugs [62]. PBP4 has transpeptidase activity and so its absence is likely to result in reduced cell wall cross-linking, similar to the effect of nonamidated muropeptides. It has been suggested that this mechanism is strain-specific as not all GISA strains show reduced PBP4 expression [61].

For bacteria, reduced cross-linking comes at a price. Resistance is only conferred if reduced cross-linking is accompanied by an increase in cell wall thickness [7]. Inevitably this increase in cell wall biosynthesis is accompanied by greater nutrient demands. For example, Mu50 cells incorporate 2.3 times more glucose molecules into the cell peptidoglycan compared to VSSA strains [58], suggesting this fitness cost may divert glucose away from other important energy-consuming processes [53]. However, this compromise enabled bacteria to produce many more D-Ala-D-Ala decoys, which positioned at the periphery of the peptidoglycan matrix can act by binding glycopeptides and clogging up their cell walls in an effort to block other glycopeptide molecules from penetrating their sites of cell wall biosynthesis.

1.5
Lantibiotics

Widespread resistance to conventional antibiotics has revived interest in the group of lanthionine-containing peptide antibiotics. These are subdivided into positively charged membrane-disrupting molecules [e.g. nisin (from *Lactococcus lactis*) and subtilin (from *Bacillus subtilis*)], type A and negatively charged, more hydrophobic antibiotics [e.g. mersacidin (from *Bacillus* spp. HIL Y-85,54728)], classified separately into type B [64, 65]. The best known example, nisin (Figure 1.9), has found use exclusively as a preservative in the food industry and clinical pathogens have not been subjected to sustained antibiotic pressure from molecules in this class. The first reports of an N-type inhibitory substance from *L. lactis* date back to 1928 [66]. Nisin is known to interact with lipid membranes [67, 68] and recently was shown to have multimode action against Gram-positive bacteria, including resistant pathogens MRSA [69] and VRE [70]. The action of nisin is mediated by pyrophosphate recognition [71] of mature cell wall intermediates lipid II and undecaprenyl pyrophosphate, which takes place on the outer leaflet of bacterial plasma membranes. There, lipid II can be engaged in binary membrane-lytic complexes by lantibiotics nisin [72] and subtilin [73]. The increase in membrane permeability leads to deregulation of cell division and cell shape regulation, minicell formation, and bacterial death [74]. Through formation of complexes with the membrane-associated precursors, lantibiotics also inhibit cell wall biosynthesis by removing lipid II and undecaprenyl pyrophosphate from the biosynthetic pathway.

Structural details of target recognition, obtained by nuclear magnetic resonance (NMR) in dimethylsulfoxide from nisin and modified lipid II [75], reveal pyrophosphate engagement, which involves residues from rings A and B. Solid-state NMR studies of nisin/lipid II in membranes [71] and enzymatically digested subtilin/lipid II, also in membranes [73], reveal the key involvement of the N-terminal amino group in nisin and the essential role of residue Trp1 in subtilin to antibiotic function.

The high activity of nisin and subtilin against Gram-positive organisms requires effective self-protection of the producer organisms, which belong to the target group. Resistance is coencoded with lantibiotic production and in nisin it is dependent on the expression of *nisI*, as well as the genes *nisE*, *nisF* and *nisG*. The *nisI*-encoded protein reduces nisin binding to membranes, while the proteins produced from *nisE*, *nisF* and *nisG* expression appear to be homologous to ATP-binding cassette (ABC)

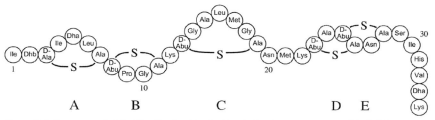

Figure 1.9 Structure of nisin. The five lanthionine/methyllanthionine rings are labeled A–E.

multidrug pumps [76]. Subtilin resistance is encoded by analogous genes in *Bacillus*, *spaIFEG*. [77]. However, the mechanism of self-immunity relies on highly specific antibiotic recognition, and cross-resistance of *L. lactis* to subtilin and of *B. subtilis* to nisin is not observed [78].

1.6
Aminoglycosides

Aminoglycosides are a large group of naturally occurring or semisynthetic polycationic compounds that inhibit translation in bacteria. Their structure consists of amino sugars linked via glycosidic bonds to amino-substituted cyclic alcohols, aminocyclitols [79].

Translation in bacteria requires specific interactions between aminoacyl-tRNA anticodons and mRNA codons. This occurs in a highly conserved rRNA sequence of the 30S ribosomal subunit. Aminoglycosides target a major groove in the model A-site of bacterial 30S rRNA. Binding within a pocket created by an AA base pair and a single bulged adenine, they induce codon misreading [80]. The most common mechanism of aminogylcoside resistance is enzymatic modification. However, bacteria can also reduce their susceptibility through target modification of the ribosomal-binding site or defects within uptake and efflux processes.

Specific nucleotides in the aminoacyl-tRNA site (A-site) region of *E. coli* 16S rRNA are protected by paromomycin from dimethyl sulfate (DMS), suggesting that this is a site of action for aminoglycosides [80]. These include the C1407–G1494 base pair, A1408, A1493 and U1495. Base pairing in the lower stem and asymmetry of the internal loop resulting from the presence of a nucleotide at position 1492 are also required for specific binding [81]. The base pair 1409–1491 forms the floor of the antibiotic-binding pocket and mutations affecting the secondary structure of this base pair confer resistance to aminoglycosides such as paromomycin [82]. Streptomycin also protects specific bases within the A-site region of *E. coli* 16S rRNA [83]. Moreover, mutations corresponding to this highly conserved region in *M. tuberculosis* confer resistance to streptomycin [84]. Target modification also extends beyond the-binding site of streptomycin. High-level streptomycin resistance is frequently due to mutations within the 530 loop region [85]. Although it may be possible for such a conformational change to result so that aminoglycoside binding is affected, it is more likely that the resulting conformational change prevents bound streptomycin-induced codon misreading. Mutations within ribosomal protein S12 also confer aminoglycoside resistance [86]. This highlights the importance of the 530 loop in aminoglycoside resistance, as ribosomal protein S12 helps stabilize this region [83, 87]. The evidence strongly implicates the A-site in aminoglycoside binding and shows that it is amenable to target modification in order to express resistance.

Structural differences between bacterial and human ribosomes concerning base pairs involved in antibiotic binding allow aminoglycosides to target selectively bacterial cells without harming human cells. The structure of the A-site of *E. coli* 16S ribosomal RNA reveals important interactions between the A-site RNA and rings I and II of

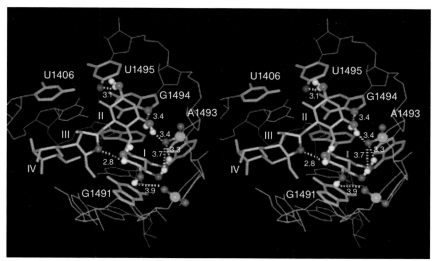

Figure 1.10 Stereo projection of paromomycin in the 16S
rRNA-binding site showing specific contacts between rings I and II
of paromomycin and the A-site RNA. The RNA is blue;
paromomycin is tan; the view is into the major groove of the RNA
core [80].

paromomycin that can be seen in Figure 1.10 [80]. The exocyclic amine and hydroxyl
groups of rings I and II make specific contacts that stabilize the antibiotic–RNA
complex and are primary targets for aminoglycoside-modifying enzymes conferring
resistance. Rings III and IV contribute weakly to specific antibiotic binding and
function [80].

Ribosomal RNA methylase enzymes are expressed in many aminoglycoside-
producing actinomycetes to protect them from antibiotics of their own or from
other microorganisms. For example *Streptomyces tenjimariensis* methylates A1408
and confers high-level resistance to aminoglycosides [88, 89]. In the base pair
A1408–A1493, essential in aminoglycoside binding to the 16S rRNA, A1408 is the
hydrogen acceptor and its loss leads to kanamycin resistance [80]. This target
modification is also apparent in *Micromonospora purpurea*, which methylates
G1405 and confers resistance to gentamicin through the resulting steric clash with
ring III of the antibiotic [88–90]. Interestingly, clinical isolates of *Pseudomonas
aeruginosa* and *Serratia marcescens* have also been found with genes encoding 16S
rRNA methylase activity [91, 92]. Sharing considerable primary sequence similarity
to aminoglycoside producers suggests possible gene transfer from actinomycetes to
Gram-negative pathogens [93].

Antibiotic deactivation is a major contributor to aminoglycoside resistance.
Enzymatic acetylation of the 3-amino group of paromomycin disrupts hydrogen
bonding with the target, which involves the amino group and the N7 of G1494, as well
as specific electrostatic and hydrogen-bonding contacts with the A1493 phosphate [80]. Ring I of paromomycin fits tightly in a pocket within the bacterial 16S

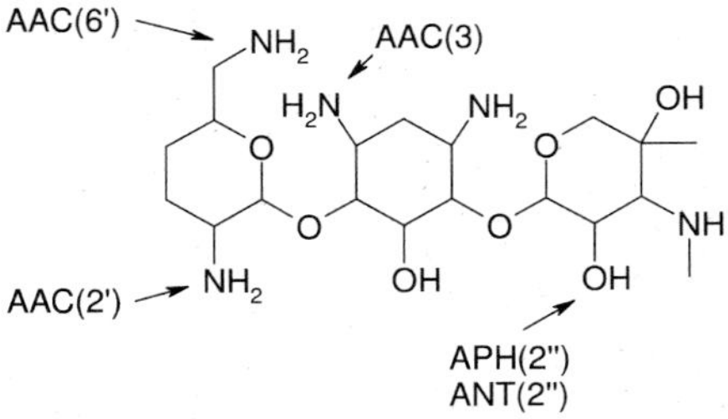

Figure 1.11 Structure of gentamycini and sites of enzymatic attack following [95].

rRNA and changes to the exocyclic groups of this ring interfere with binding. Acetylation of the 6′-amino also group gives rise to aminoglycoside resistance, and phosphorylation and adenylation of the 3′- and 4′-OH groups, respectively, leads to steric and electrostatic penalties to complex formation [80].

Enzymatic modifications of hydroxyl or amino groups on the aminocyclitol residues lead to reduced antibiotic activity. Modified aminoglycosides have diminished bacterial 16S rRNA A-site binding and lose their ability to inhibit protein synthesis [94]. Aminoglycoside *N*-acetyltransferases (AACs) acetylate the amino groups (*N*-acetylation) of the antibiotic and are dependent on acetyl-CoA. *O*-Phosphotransferases (APHs) phosphorylate the hydroxyl groups (*O*-phosphorylation). *O*-Nucleotidyltransferases (ANTs) add AMP moieties to hydroxyl groups (*O*-adenylation). The function of APHs and ANTs is ATP-independent. The position of the group attacked and the ring that carries it are indicated by the number of the enzyme (Figure 1.11).

Acquisition of Gcn5-related *N*-acetyltransferases (GNATs) provides bacteria with resistance to gentamicin, tobramycin and netilmicin. The crystal structure of plasmid-encoded AAC(3)-Ia from *Serratia marcescens* (SmAAC) has been determined [96] and the chromosomal gene *aac(6′)-Ii* from *Enterococcus faecium* has been characterized [97]. The four motifs C, D, A and B are characteristic of GNATs. Motif A seems to be critical for activity of NATs because site-directed mutations to Arg101, Gly104 and Gly106 of the invariant segment Arg/Gln-X-X-Gly-X-Gly/Ala, resulted in human spermidine/spermine NAT having no measurable effect [98]. AAC(3)-Ia contains all of the motifs (C, D, A and B) common to GNATs.

Cofactor binding and acetyl transfer catalyzed by 3-NAT I from *P. aeruginosa* (PsAAT) suggests acetyl-CoA binds to the enzyme, followed by aminoglycoside binding [99]. The acetyl group is only attached to CoA via a relatively weak thioester linkage and can easily be transferred to the aminoglycoside, while the acetylated product is released followed by CoA. Structural similarities between SmAAT and PsAAT hint at a common mechanism of acetyl transfer [96]. The acetyl group from acetyl-CoA lies above the side-chain of Gln145. Gln145 forms part of the floor of the active site, placing the acetyl group within close proximity of the aminoglycoside [96] (Figure 1.12). The thumb-like structure of motif B is a β-hairpin projection made up of S5 and S6. Asp147, Asp150 and Asp151 from the S5/S6 loop create one wall of a narrow canyon-like feature with negative electrostatic potential that extends away from the acetyl-CoA-binding surface toward the convex face of the enzyme. Asp53 from the acidic H1/H2 loop forms the

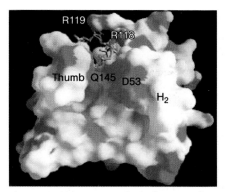

Figure 1.12 Active-site surface of SmAAT, showing CoA in the cofactor-binding notch and the negatively charged gentamicin-binding slot [96].

other wall [96]. The importance of negatively charged residues in the aminoglycoside-binding site shown here is supported by similar findings from work on the molecular structure of kanamycin nucleotidyltransferase [100].

APHs transfer a γ-phosphate group from ATP to the 3′- and/or 5′-hydroxyl group of aminoglycosides such as amikacin. APH(3′)-IIIa is a 264-residue enzyme existing as a monomer or covalent dimer linked via disulphide bridges, and is carried by opportunistic enterococci and staphylococci [101]. The monomers within the dimer are arranged in a head-to-tail/tail-to-head fashion. The disulphide bond joins Cys19 located in a β-sheet (β1) of one monomer to Cys156, which is part of a loop region between the helices αA and αB of the other monomer and vice versa. It is proposed that there is movement around this linkage, which is supported by the fact that there are few other significant bonds between the two monomers, except for hydrogen bonding between Asp150 and the side-chains of Arg5 and Trp85 of the partner molecule. The binding sites of each monomer face each other but are more likely to be independent due to the 20 Å distance between the two [101].

According to the crystal structure of the APH(3′)-IIIa monomer [101], there is a 94-residue N-terminal lobe and a 157-residue C-terminal lobe, tethered by a 12-residue stretch containing a short β-strand and α-helix. Between the two lobes is a deep cleft containing the ATP-binding site. Akin to the other aminoglycoside modifying enzymes mentioned here, APH(3′)-IIIa operates a Theorell–Chance kinetic mechanism [102] where phosphorylation follows binding of ATP and binding of the aminoglycoside. Release of the modified drug is followed by the rate-limiting dissociation of ADP. There is a dramatic similarity in structure between APH(3′)-IIIa and protein kinases [101], which suggests a mechanistic and evolutionary link between these antibiotic-resistance enzymes and Ser/Thr/Tyr protein kinases [11]. This relationship is strengthened by the report that aminoglycoside kinases can act as serine protein kinases [103].

Asp166 in cAPK has been implicated as the catalytic base required for the deprotonation of the substrate hydroxyl group for efficient attack at the γ-phosphate

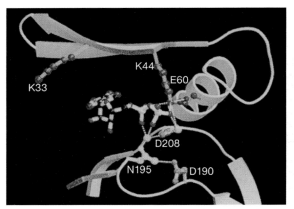

Figure 1.13 Nucleotide-binding site of APH(3′)-IIIa located in the cleft between the N- and C-terminal lobes. Conserved residues are shown interacting with ADP and the two magnesium ions. Although in the binding pocket, Lys33 does not interact directly with the nucleotide [101]

of ATP. The structure of the nucleotide-binding site of APH(3′)-IIIa (Figure 1.13) suggests a similar role for Asp190 because it is in the correct position to interact with the incoming hydroxyl. Site-directed mutagenesis of Asp190 causes a minimum 650-fold decease in k_{cat} [101] confirming this role. Affinity labeling demonstrates that the conserved Lys44 is close to the triphosphate-binding pocket [104]. Figure 1.13 shows Lys44 located on β-strand 3 and positioned directly over the ATP triphosphate-binding site, where it interacts with α- and β-phosphates. Site-directed mutagenesis also supports the role of Lys44 in ATP binding [101].

Le Goffic *et al.* originally isolated kanamycin nucleotidyltransferase (KNT) from *S. aureus* and showed it could catalyze the transfer of a nucleoside monophosphate group from a nucleotide to the 4′-hydroxyl group of kanamycin [104]. The enzyme utilizes ATP, GTP or UTP and can inactivate a wide range of aminoglycosides [100]. Contrary to nondenaturing gel electrophoresis findings [105], crystallographic studies suggest that KTNase is a dimer, the interface between the monomers of which is formed by electrostatic interactions. KNTase possibly operates a Theorell–Chance kinetic mechanism like nucleotidyltransferase 2″-I. Direct in-line displacement of pyrophosphate from P1 (nucleotide) by C2″-OH (aminoglycoside) is believed to produce nucleotidylaminoglycoside in a single step, accompanied by inversion of the phosphorous [106]. Enzyme turnover is controlled by the rate-limiting step of product release after pyrophosphate release [100].

Figure 1.14(a) shows the pronounced basket-like cleft formed by the two subunits of the dimmer, which accommodates different aminoglycosides. The N-terminal domain is delineated by Met1 to Glu127 and characterized by a five-stranded mixed β-pleated sheet, whereas the C-terminal domain contains five α-helices and is formed by Ala128 to Phe253 [79]. The subunits wrap round each other to form the aminoglycoside-binding site. The adenine portion of the nucleotide is involved in

(a)

(b)

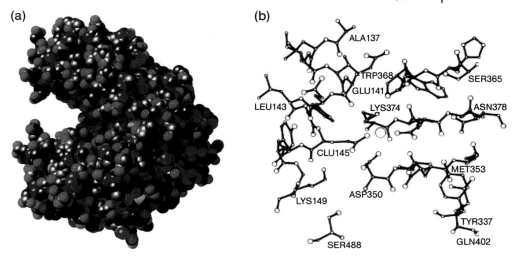

Figure 1.14 (a) Space-filling representation of all the atoms in the KNTase model. (b) Close-up view of the putative zinc-binding site [100].

few specific interactions with the protein, which explains why ATP, GTP and UTP are all substrates for the enzyme [79]. Close-up views of the zinc-binding site (Figure 1.14b) show it surrounded by a ring of seven negatively charged amino acid residues: Glu141, Glu142 and Glu145 from subunit 1, and Glu52, Glu67 Glu76 and Asp50 from subunit 2. This crown of negativity would bind kanamycin and other positively charged aminoglycosides, implicating the zinc-binding region as the active site.

KNTase bound with kanamycin and the nonhydrolyzable nucleotide analog AMPCPP also suggests a group of seven negatively charged amino acid residues form a portion of the binding site [79]. Consistent with other research on aminoglycosides [80], the amino sugar of kanamycin interacts most extensively with KNTase. It is wedged against the adenine ring of AMPCPP, while the third amino sugar has few specific interactions with the protein [79]. The exocylic hydroxyl and amino groups on this amino sugar can form hydrogen bonds with the side-chains of Glu67, Glu76 and Lys74 from subunit I, and Glu141 and Glu145 from subunit II. Ser94 and Glu141 from subunit II interact with the aminocyclitol moiety. Of particular interest is Glu145, whose carboxylate group is within hydrogen-bonding distance of the 4′-hydroxyl group. A resistance mechanism involving nucleophilic attack of the α-phosphorous on the nucleotide by activated kanamycin has been proposed [79]. Glu145 acts as a general base abstracting a proton from the 4′-hydroxyl group, which activates kanamycin for subsequent attack at the α-phosphate of the nucleotide. Both the nucleotide and the antibiotic are in the proper orientation for a single in-line displacement reaction. Additionally, Lys149 is within close proximity of the α-phosphoryl oxygens and would increase the electrophilic character of the phosphorous center, making it more susceptible for nucleophilic attack [107]. This mechanism is limited by the 5.0 Å distance for nucleophilic attack, which is too long. However, Pedersen *et al.* emphasize that the mechanism is based on a nucleotide analogue and enzyme catalysis may induce conformational changes that reduce this distance [79].

Aminoglycoside uptake involves three consecutive steps [107]. The first is adsorption of the cationic aminoglycoside to the surface, facilitated by electrostatic interactions with the negatively charged lipopolysaccharides in outer membranes of Gram-negative bacteria. The second and third steps are dependent on transmembrane potentials generated by the respiratory chain. Thus, anaerobic bacteria are intrinsically resistant [108] and strains with mutations in ATP synthetases have reduced susceptibility [107]. Some bacteria seem to have also developed permeability mechanisms linked to uptake and efflux in order to counteract the accumulation of aminoglycosides intracellularly. Models for this have been proposed; however, the exact mechanics of this have yet to be elucidated.

Multidrug efflux is another component of aminoglycoside resistance. An important structural example is AcrB from *E. coli* reveals a multiprotein complex involving AcrB, AcrA and TolC, which enable drug translocation across the plasma membrane followed by transport via TolC across the periplasm and the outer membrane [109, 110]. In this model, AcrB facilitates proton counterflow-driven drug transport from the plasma membrane inner leaflet into a protein cavity on the outer side of the membrane, as well as sequestering of drug form the membrane outer leaflet. From there, TolC provides a route to the bacterial exterior.

While AcrB does not efflux aminoglycosides [93], the model based on AcrB (Figure 1.15) can be applied to AcrD-mediated aminoglycoside efflux. Localized in the cytoplasm, resistance nodulation cell division (RND) proteins use membrane proton-motive force as their energy source. In Gram-negative bacteria they interact with a membrane fusion protein located in the periplasmic space and an outer membrane protein (OMP) to form a continuous tripartite channel. Trimeric TolC, the OMP that interacts with AcrB, forms a barrel composed of 12 outer membrane-spanning β-strands and 12 α-helices extending into the periplasmic space for over 100 Å. The internal cavity is open to the external environment to provide solvent access. Each monomer of AcrB contains 12 transmembrane domains (TMDs) and two large periplasmic domains. Substrates bind to the ring-like arrangement of TMDs connected to the periplasmic funnel [93]. There is an opening between the two periplasmic domains connecting the periplasm with the central channel. These 'vestibules' can channel substrates selected from the outer leaflet of the cell membrane or from the periplasmic space into the efflux apparatus for exportation out of the cell.

There are more acidic residues at the entrance of the AcrD vestibules compared to vestibules of AcrB [111]. These extra acidic residues may give AcrD the capacity to efflux aminoglycosides which are attracted towards the phospholipid through electrostatic interactions. Aminoglycoside hypersusceptibility accompanying the disruption or deletion of genes encoding AcrB in *E. coli* [112] implicates RND protein efflux in aminoglycoside resistance. Resistance to aminoglycosides, due to efflux, can

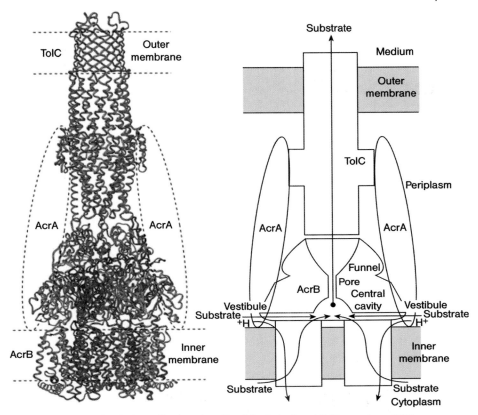

Figure 1.15 Model of multidrug efflux based on *E. coli* AcrB, AcrA and TolC complexes [110].

be mediated by the major facilitator superfamily (MFS) proteins. This has been shown in mycobacteria [113, 114], where resistance is encoded by *tap* and *P55* genes. Genes and proteins homologous to *P55*/P55 have been detected in many mycobacteria, including *M. tuberculosis* [93].

1.7
Macrolides

Macrolides are another class of antibiotics, structurally distinct from the aminoglycosides, that inhibit bacterial protein synthesis. Macrolides are characterized by a 14-, 15- or 16-membered macrolactone ring. The structure of erythromycin (Figure 1.16) consists of a 14-membered macrolactone ring with a cladinose sugar and a desosamine sugar attached. The 16-membered ring macrolides have two sugars attached through an amino sugar. Telithromycin lacks an α-ʟ-caladinose at position 3 on the erythronolide A-ring and is a semisynthetic macrolide from the ketolide class, which are active against bacteria resistant to erythromycin A [95].

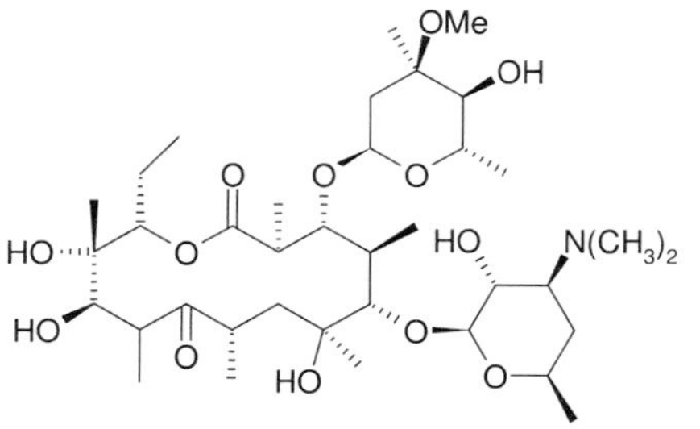

Figure 1.16 Structure of erythromycin.

While macrolides are the focus here, lincosamides and type B streptogramins will be referred to regarding their overlapping functions (the MLS_B phenotype) and their common 50S target. Erythromycin halts protein synthesis after formation of the initiation complex, probably by interfering with the translocation reaction [95]. Erythromycin, clarithromycin and roxithromycin bind to the entrance of the tunnel which channels nascent peptides away from the peptidyltransferase center [115]. While binding of macrolides may not block peptidyltransferase activity, erythromycin bound to the tunnel would interfere with channeling of nascent peptides (Figure 1.17).

Widespread use of macrolides has led to the emergence of resistance in *S. aureus* and *Streptococcus pyogenes*. Impermeability and efflux has provided Gram-negative bacilli with intrinsic macrolide resistance, except the azalides [95]. The three types

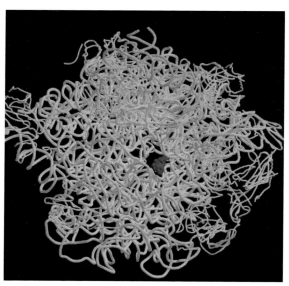

Figure 1.17 Top view of the *Deinococcus radiodurans* 50S subunit showing erythromycin (red) bound at the entrance of the tunnel [115].

of macrolide resistance are target modification conferring the MLS_B phenotype, active efflux and macrolide-inactivating enzymes. The lactone ring and the desosamine sugar are the two reactive components of macrolides which form hydrogen bonds with the peptidyltransferase cavity of 23S rRNA [115]. The 6-OH group is within hydrogen-bonding distance of N6 of A2062Ec, and the 11-OH group and 12-OH group hydrogen bond with O4 of U2609Ec. The 2'-OH group of the desosamine sugar forms hydrogen bonds with N6 and N1 of A2058Ec and N6 of A2059Ec [113]. Dimethylation of the N6 amino group of A2058 located in domain V of 23S rRNA is consistent with this, and confers resistance to erythromycin and its derivatives (except 16-membered macrolides). This post-transcriptional modification is mediated by adenine-specific N-methyltransferase activity and is encoded for by *erm* (erythromycin ribosome methylation) genes. Dimethylation of N6 adds a bulky substituent to the residue, causing steric hindrance for binding and abrogating the hydrogen bonding with the 2'-OH group of the macrolide [115]. A substitution of guanine for adenine at 2058 has been described in staphylococci and streptococci [95]. This change confers resistance because it would disrupt hydrogen bonding therefore prevent macrolides from binding to the 23S rRNA. Acquisition of mutant A2059G gives rise to macrolide and lincosamide resistance despite increasing the generation time of R6. This ML phenotype is not conferred by any other existing determinants known [116].

Ribosomal proteins L4 and L22 have been implicated in resistance [117] and mutations in which have been reported in clinical isolates of *S. pneumoniae* and *S. pyogenes* [95]. However, the distances between erythromycin and L4 and L22 are too great for meaningful interactions [115]. Multiple changes in reactivity to chemical probes indicate that both mutations disrupt the conformation of specific residues in domains II, III and V. Domain V includes residue A2058; however, neither mutation shows any effects at or within its vicinity [118]. The MLS_B phenotype is associated with mutations or methylation of A2058, so mutations L4 and L22 do not give rise to MLS. Rather, they confer resistance to erythromycin, spiramycin and tylosin, but not lincosamides [118]. Research suggests mutations in L4 confer an MS_B phenotype implicating a 3-amino-acid substitution and an 18-bp insertion [116]. Supporting ribosomal proteins binding at multiple rRNA sites, perturbations of the 23S rRNA structure have been observed with mutations within ribosomal proteins L4 and L22. The resultant 23S rRNA is resistant to macrolides.

Efflux is clinically relevant in *S. pneumoniae*, *S. pyogenes* [119, 120] and most Gram-negative bacteria. Acquired *mefE* and *mefA* encode the membrane protein Mef, an active efflux pump accounting for over 50% of the resistance seen in *S. pneumoniae* and *S. pyogenes* [95]. *MefE* and *mefA* are 90% identical, and are now referred to collectively as *mef*(A) [121]. *mef*(A) confers resistance to macrolides, but not lincosamides (the M-R phenotype). The macrolide efflux pump, AcrAB-TolC of *H. influenzae* and *E. coli* is an example of an intrinsic efflux pump that acts in synergy with slow penetration of outer cell walls to give the high MICs seen in Gram-negative bacteria [122].

The ribosomes of *E. coli* K12 strain expressing *mefA* display no evidence of methylated 23S rRNA and are more sensitive to macrolides that macrolide-susceptible strains [123], confirming a novel gene unrelated to *erm* genes [120]. Erythromycin inactivation and *msrA*-encoded efflux have been ruled out as potential candidates for the M-R phenotype seen in streptococci [123]. Moreover, discernible regions of MefA show no homology to the Walker motifs A and B which characterize ABC transporters. Similarities in tertiary structure exist between MefA and Tet antiporters of Gram-negative bacteria, consistent with it being a member of the MFS.

As a 12-transmembrane secondary transporter with H^+ antiporter activity, MefA couples drug efflux to a downhill electrochemical gradient of protons. MefA has a simple 12-transmembrane segmented structure possible resulting from duplication of a gene encoding a six-transmembrane segment protein [124]. The N-terminal halves of MFS proteins exhibit greater sequence similarity than the C-terminal halves and so form the site of variable ligand binding, whereas the C-terminal is probably involved in energization of the protein [125, 126]. Macrolide recognition is explained using a model based on the C-terminal half of the BmrR protein (BRC) [127]. BRC (Figure 1.18) consisting of eight β-strands and three α-helices is only active as a dimer. Binding of the ligand is governed by a buried charged acidic residue, and the α2 helix periodically undergoes a change to expose the binding site and facilitate binding of the ligand [128]. Binding of electroneutral ligands is governed by the size of the opening to the binding site as well as hydrophobic forces. Once bound, hydrophobic and aromatic residues stabilize the ligand until a second conformational change, presumably linked to proton translocation, results in expulsion of the ligand [126].

Extrusion of macrolides can be explained using the hydrophobic 'vacuum cleaner model' or the 'flippase model'. The 'vacuum cleaner model' [124] proposes macrolides move freely into the lipid phase of the membrane and on reaching MefA are actively expelled out of the cell. The 'flippase model' proposes that on reaching the protein within the membrane, the drug is 'flipped' to the outer layer. Structural characteristics of multidrug-resistant proteins favor the vacuum cleaner model [124]. The 12 TMDs of MefA would form a central aqueous pore open to the environment, closed to the cytosol and lined with aromatic amino acids which could facilitate the

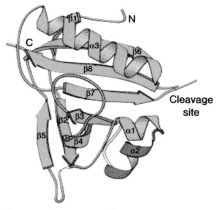

Figure 1.18 Monomer of BRC. The essential acid residue is indicated as Glu 134 in β7 [128].

transport of hydrophobic substrates [129]. In contrast to MefA, MsrA/B in staphylococci acts as an ATP-dependent efflux pump conferring the MS phenotype. *S. aureus* cells expressing MsrA accumulated significantly less erythromycin than cells without MsrA [120]. MsrA efflux is abolished by arsenate and strongly inhibited by dinitrophenol [130] consistent with its dependence on ATP. MsrA/B-mediated efflux is relatively uncommon compared to methylase-mediated resistance in staphylococci [131].

As a member of the ABC family, the structure of MsrA consists of two TMDs. These typically comprise of six transmembrane-spanning α-helices and two nucleotide-binding domains (NBDs), located on the inner surface of the bacterial cell membrane as sites of ATP hydrolysis (Figure 1.19). Deletion of 42 C-terminal codons from a fragment of *S. epidermidis msrA* results in loss of the MS resistance phenotype demonstrating the importance of the C-terminal domain [132, 133]. All NBDs possess a Walker A motif (P-loop) and a Walker B motif which hydrogen bond extensively with the nucleotide. The Walker B motif provides the catalytic base. The ABC signature motif, H-loop, Q-loop and stacking aromatic are only characteristic of transport ABCs and coordinate the bound nucleotide or nucleophile water [134]. NBDs couple conformational changes induced by ATP binding, hydrolysis and

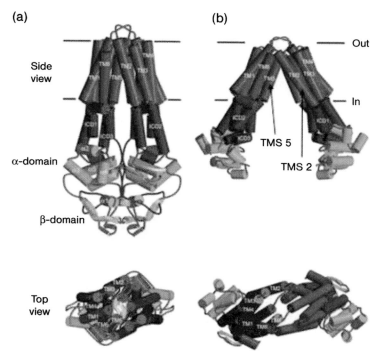

Figure 1.19 Structure of (a) MsbA from *Vibrio cholerae* and (b) MsbA from *E. coli*, showing the location of the transmembrane domains (red), the intracellular domain (blue) and the nucleotide-binding domain (cyan) [126].

release to the transport process [134] in order to energize transport of macrolides and type B streptogramins. Figure 1.19 shows the tilt of the monomers of a typical MDR, away from the normal of the membrane forming a cone-shaped chamber that excludes access from the outer leaflet of the membrane so a ligand can only bind from the inner leaflet [126].

An ATP switch model has been proposed as the first coherent model for ABC transport [134]. Binding of ligands such as macrolides induces a conformational change in the TMD which is then relayed via intracellular domains to the NBD. The conformational change here enhances the binding of ATP and lowers the activation energy required for 'closed' dimer formation. ATP binding is associated with a 'closed' NBD dimer conformation, whereas dissociation of the NPD dimer to an 'open' conformation is caused by ATP hydrolysis. This acts as a switch for conformational changes back in the TMDs that mediate transport of the macrolide. The 'closed' dimer conformation may expose the macrolide to unfavorable positively charged residues lining the inner leaflet half of the TMD chamber. This prompts the macrolide to 'flip' to the more favorable, hydrophobic environment of the outer leaflet half of the chamber. This flip induces a conformational change in TMDs, especially TMS-2 and -5 and the NBDs which would lead to expulsion of the macrolide [126]. ATP hydrolysis appears to initiate resetting of the transporter to it basal 'open' conformation [134].

Inactivating enzymes are important resistance mechanisms in *Nocardia* spp., and include esterases which hydrolyze the lactone ring and enzymes which fix a glucose or a phosphate at the 2′-OH group of D-desosamine [95]. Until recently there had been no reports of esterase activity in Gram-positive organisms; however, strain 01A1032 of *S. aureus* exhibits esterase activity of 14- and 16-membered macrolides, and also the ability to efflux azithromycin [133]. Ribosomes from strain 01A1032 demonstrated 100% inhibition by erythromycin and azithromycin, ruling out any activity of Erm methylases and ribosomal mutations. Polymerase chain reaction (PCR) results gave a PCR product consistent with an *msrA*-like gene, which implicates an ABC protein in mediating the efflux resistance mechanism. It remains unclear if efflux and enzymatic inactivation are part of one mechanism, act synergistically or are mutually exclusive [133].

1.8
Tetracyclines

Tetracyclines are polycyclic broad-spectrum antibiotics with activity against both Gram-positive and Gram-negative organisms. The main structural feature of a tetracycline molecule is a linear fused nucleus of four rings (Figure 1.20). In the 1950s and 1960s tetracycline was one of the most widely used antibiotics [135] due to its broad spectrum of activity, few side-effects, low toxicity, low production costs and the possibility for oral administration. Widespread resistance to tetracycline now limits its use, with it being discontinued as first-line therapy for the treatment of sexually transmitted diseases, after the appearance of tetracycline-resistant *Neisseria*

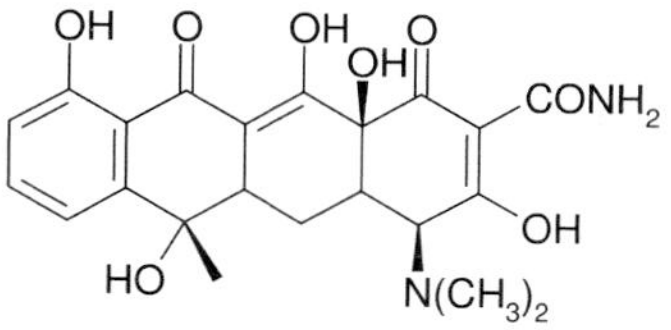

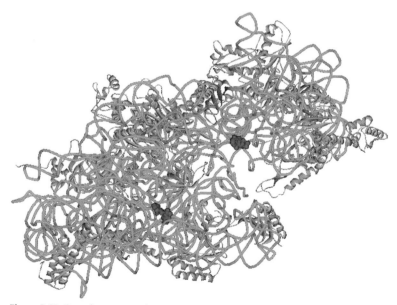

Figure 1.20 Structure of tetracycline.

gonorrhoeae [136]. Tetracyclines target both mammalian (80S) and bacterial (70S) ribosomes, although 70S ribosomes are more sensitive to tetracycline, which is also found in higher concentrations in bacterial cells compared to mammalian cells. Tetracycline disrupts the codon–anticodon interaction between mRNA and tRNA by inhibiting the binding of aminoacyl-tRNA to the A-site. The initial stages of ternary complex (EF-Tu·GTPaa·tRNA) binding such as decoding and GTP hydrolysis are still possible; however, accommodation of the tRNA into the 50S A-site is blocked, preventing further extension of the peptide chain. The ribosome becomes locked in a nonproductive and energetically expensive cycle of ternary complex binding and release (Figure 1.21) [137].

Tetracycline binds to a single site on the 30S ribosomal subunit, which includes a region of 16S rRNA containing base A892. The anticodon of bound aminoacyl-tRNA is positioned very close to a 16S rRNA region which includes a base at 1400. This suggests that tertiary folding of 16 rRNA brings bases 892 and 1400 very close to each other and that bound tetracyclines block aminoacyl-tRNA binding by disturbing this

Figure 1.21 Crystal structure of tetracycline (red) bound to the 30S ribosomal subunit from *Thermus thermophilus* [137] (PDB code: 1HNW). Visualized using Discovery Studio (Accelrys).

folding [10]. Even though ribosomal protein S7 is not associated with 16S rRNA, tetracycline binds strongly to both the 30S ribosomal subunit and S7 ribosomal protein [138]. It is suggested that binding of tetracycline to 16S rRNA causes such a distortion in RNA structure that ribosomal protein S7 is presented to the 892–1400 domain [10].

Crystallographic study reveals two tetracycline-binding sites [137]. The primary site (20 Å wide and 7 Å deep) is formed by an irregular minor groove of H34 (RNA residues 1196–1200:1053–1056) in combination with residues 964–967 from the H31 stem loop. Hydrophobic interactions are apparent between bases 1054 and 1196 of H34 and the fused-ring system of tetracycline. However, tetracycline interacts primarily through hydrogen bonding between oxygen atoms of one side of the tetracycline molecule and the exposed sugar phosphate backbone oxygen atoms of H34. There is also a hydrogen bond to the oxygen phosphate of G966 from H31. Important salt bridges exist between a putative magnesium ion on the hydrophilic side of tetracycline and phosphate oxygen atoms of G1197 and G1198. Tetracycline bound to the primary site sterically interferes with aminoacyl-tRNA binding to the A-site [137]. The secondary binding site is associated with the H27 Switch region (residues 891–894:908–911) and H11 (residues 242–245) [132]. Binding of tetracycline here cannot directly interfere with tRNA binding; however, it could interfere with the transition between open and closed states of the 30S ribosomal subunit [139], which is important for the decoding reaction [140]. Pioletti *et al.* identified six (Tet-1–6) tetracycline-binding sites on the 30S subunit [141]. Tet-1 corresponds with the same residues from H34 and H31 as the primary site location and Tet-5 is associated with the H27 switch region of the secondary site [141]. The roles of the four other binding sites are less clear cut. Consistent with this is enhanced DMS modification of C1054 and the protection of A892 from this modification, by tetracycline [142].

Tetracycline uptake is energy dependent [143] and mediated by a change in pH rather than a transport protein [144]. Existing in protonated form (TH2) and magnesium-chelated form (THMg), TH2 is free to diffuse into the cell through the phospholipid membrane, whereas THMg is not. Intracellular pH is high compared to the pH outside the cell and so is trapped within the bacterial cell is a greater proportion of THMg [144]. Altering porin proteins such as OmpF can confer resistance by limiting tetracycline diffusion [135]; however, efflux and ribosomal protection are the most common resistant determinants [145].

There are tetracycline-specific and multidrug efflux pumps conferring tetracycline resistance [146, 147]. Tetracycline and a narrow spectrum of related antibiotics is exported in *E. coli* by TetB and other proteins encoded by the *tet* genes [145]. TetB alone normally does not give rise to resistance, but is usually associated with TetA or OtrA efflux pumps [148]. Tet proteins belong to the major facilitator superfamily and have a predicted six-plus-six transmembrane domain organization, which provides the structural basis underlying an electrically neutral proton tetracycline antiport system [149, 150]. A model for the 'tetracycline transport cycle' in resistant cells proposes tetracycline is probably effluxed as a metal–chelate complex coupled with proton influx [151].

(a)   (b)

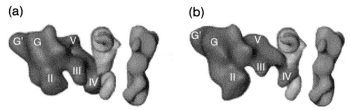

Figure 1.22 (a) Positioning of EF-G in GTP form (red) together with A-site (pink)- and P-site (green)-bound tRNAs.
(b) Positioning of Tet(O)·GTPγS (red), the P-site-bound fMet-tRNA (green) together with tRNA in a position equivalent to the observed position of the A-site-bound tRNA in the pretranslocational ribosome [152].

Ribosomal protection is conferred by ribosomal protection proteins (RPPs) that associate with the ribosome and make it insensitive to tetracycline inhibition. Tet(O) and Tet(M) are well-studied examples, sharing 75% amino acid sequence identity [153] and other members of RPPs [Tet(S), Tet(T), Tet(Q), TetB(P), Tet(W) and OtrA] function through similar mechanisms to Tet(O) and Tet(M) [139]. Tet(M) binds to ribosomes, shares considerable amino acid homology to EF-G and exhibits similar ribosome-dependent GTPase activity to this promoter of translocation [154]. Moreover, binding of Tet(O) or EF-G enhances the reactivity of A1408 to DMS modification [142]. Both Tet(M) and Tet(O) catalyze the release of tetracycline from the ribosome in a GTP-dependent manner.

The tip of domain IV of Tet(O) is positioned to interact with the noncovalent junction between the shoulder and head of the 30S subunit (Figure 1.22). This region is made up of helices 18 and 33/34 of 16S rRNA. H34 contributes to the Tet-1/primary tetracycline-binding site [152] and is involved in maintenance of translational fidelity, particularly in stop codon decoding and frame shifting [155]. The interaction between the tip of domain IV of Tet(O) and the junction linking the head and shoulder of the 30S subunit (Figure 1.23) is consistent with its role in chasing tetracyclines from the

(a)   (b)

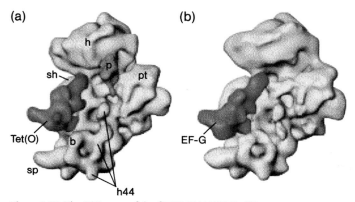

Figure 1.23 The 30S parts of the fMET-tRNA·70S·Tet(O) map. Landmarks: sh, shoulder; h, head [152].

primary binding site [142]. However, Tet(O) binding does not promote translocation and leaves the conformation of the ribosome virtually unchanged [148]. An interaction between domain III of Tet(O) and S17 ribosomal protein could also be relevant in Tet(O)-mediated resistance [152].

Binding of Tet(O) protects C1214 from DMS chemical modification and enhances the reactivity of A1408 [142]. Interestingly, C1214 can be found on helix 34 (h34) and is close but does not overlap the tetracycline-binding site. Located on h44, A1408 is distinct from the tetracycline-binding site. Enhanced DMS modification of C1054 upon tetracycline binding is inhibited by Tet(O); however, protection from DMS modification of A892 by tetracycline binding is unaffected by Tet(O) binding. This suggests Tet(O) only prevents tetracycline from binding to the primary site C1054 [142]. Protection of C1214 from chemical modification indicates the base is being shielded directly or it undergoes a conformational change resulting in decreased accessibility [142]. Tet(O) has close proximity with h34 (Figure 1.24) suggesting a direct interaction or indirect through other contacts in h34, however it does not approach h44. The enhancement of reactivity of A1408 indicates a Tet(O)-induced conformational change which has been suggested to involve ribosomal protein S12, whose core makes contact with the backbone of h44 around residues 1491 and 1492 [142].

Tet(O) intercalates into the elongation cycle of the post-translocational ribosome (POST complex) [156]. The PRE complex is characterized by deacyl-tRNAf^{MET} in the P-site and AcPhe-tRNA^{Phe} in the A-site, whereas the POST complex has an empty A-site. Tet(O) and EF-Tu cannot simultaneously bind to the ribosome because they occupy overlapping sites. Instead, Tet(O) binding was found to stabilize a conformation of the GAR (L11 region, i.e. H42/43/44) which left the ribosome in a conformation favorable for interaction with EF-TU [156].

Tetracycline binds to the A-site on the post-translocational ribosome where it sterically interferes with the accommodation of aminoacyl-tRNA. A conformational

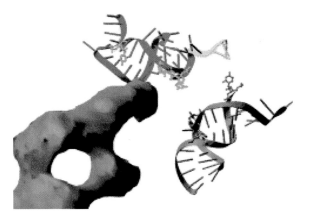

Figure 1.24 Interaction of domain IV of Tet(O) (red density) with the region around the primary tetracycline-binding site. Blue ribbon: helix 34; red ribbon: helix 44 [142].

change in the decoding site follows tetracycline binding without release of tRNA bound to the E-site accompanies. Tet(O) present in low abundance in the cell recognizes the blocked ribosome by virtue of its open A-site, prolonged pausing in the POST state and possibly the drug induced conformational change [156]. The interaction between Tet(O) and the ribosome triggers release of tetracycline prior to GTP hydrolysis [157] and induces rearrangements in the A-site [142]. Tet(O) hydrolyzes GTP leaving the ribosome in a conformation which is compatible with EF-TU binding. The conformational changes prolong at the A-site after Tet(O) release enhancing the ability of the ternary complex to compete with tetracycline [156].

1.9
Fluoroquinolones

The development of quinolones began with the introduction of nalidixic acid in 1962, they are synthetic compounds based on the 4-quinolone nucleus (Figure 1.25). First-generation quinolones include cinoxacin and oxolinic acid, which despite a limited spectra, had potent Gram-negative activity. Second-generation quinolones were fluorinated (fluoroquinolones), and included ciprofloxacin which was introduced in 1987 and was the first widely use quinolone. Fluoroquinolones have excellent Gram-negative activity and can be administered orally [9]. The late 1990s saw the introduction of the third generation of quinolones (e.g. levofloxacin and moxifloxacin) with increased Gram-positive activity [158]. Quinolones inhibit DNA replication by targeting DNA gyrase and topoisomerase IV. The proposed model of quinolone action on DNA gyrase in Figure 1.26 shows quinolones binding to the enzyme–DNA complex through significant hydrogen bonds between the 3-carboxy and 4-oxy groups common to quinolones and the unpaired bases of the unwound single strand. This binding prevents DNA gyrase from relegating the cleaved DNA and double-strand breaks accumulate, setting off the SOS repair system which ultimately leads to bacterial cell death. In response to this threat, bacteria have developed resistance mechanism, which include altering the target enzymes (DNA gyrase and topoisomerase IV), changing cell wall permeability and quinolone efflux mechanisms [159].

Quinolones do not bind directly to either the individual subunits of type II topoisomerases or the complete tetramers. Instead, they act on type II topoisomerases

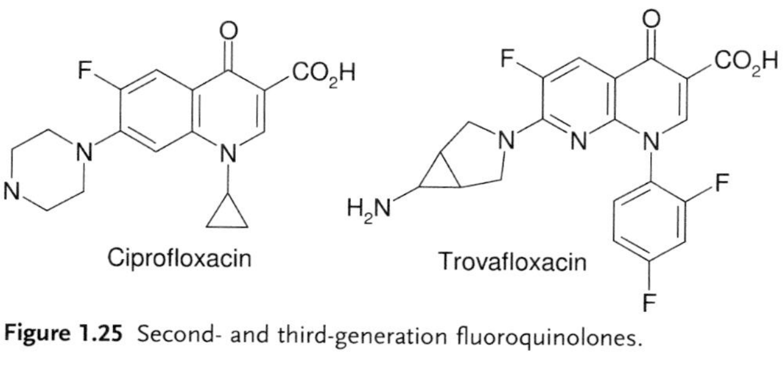

Figure 1.25 Second- and third-generation fluoroquinolones.

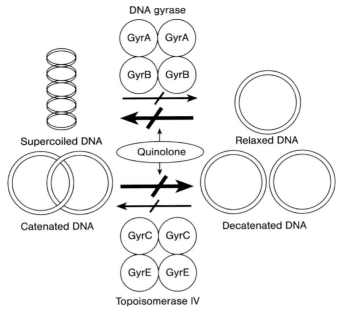

Figure 1.26 Quinolones block DNA gyrase and topoisomerase activity by stabilizing the enzyme–DNA complex. This inhibits the movement along the DNA of other proteins such as DNA and RNA polymerase [165].

by trapping or stabilizing an enzyme reaction intermediate in which both DNA strands are covalently linked to the breakage reunion subunits (GyrA from DNA gyrase or GrlA from topoisomerase IV) [160, 161]. Stabilization of this cleavage complex initiates a series of events leading to cell death [162]. Mutations in type II topoisomerases are often found in a discrete sequence of genes called the quinolone-resistant determinant region (QRDR) [163]. For example, determination of the nucleotide sequence for the QRDR of *gyrA* from codon 55 to 138 revealed that three (C5, C14 and L1) out of 12 of the clinical isolates of *Bacteroides fragilis* tested demonstrated the same Ser82Phe mutation in GyrA. All three of these clinical isolates showed the highest MIC of 4 µg/ml out of the clinical strains tested [164].

Mutations in type II topoisomerases are not just restricted to the QRDRs and the spontaneous *S. aureus* mutant, MT52244c9 is evidence for this. MT5224c9 was obtained in one step by plating MT5 *gyrB142* on ciprofloxacin. Compared to its parental strain, MT5224c9 showed an approximate 4-fold increase in MIC of ciprofloxacin yet an approximate 8-fold decrease in MIC of the courmarin novobiocin. The mutation is localized to *Sma*I chromosomal fragment A by a linkage to a *Tn*551 insertion in this fragment, which encodes topoisomerase IV *grlB* and *grlA* genes [166, 167]. The mutation in question is Asn470Asp and is actually confined to the GrlB subunit of topoisomerase IV [162], which hydrolyzes ATP for the energy-dependent introduction of negative supercoils into the DNA during replication.

Mutations in *grlA* and *gyrA* genes have also been shown to confer resistance to ciprofloxacin. Substitution of Ser80 with Phe or Tyr in the *grlA* gene was found to be the principle mutation (39.8%), while substitution of Ser84 with Leu in the *gyrA* gene was the principle mutation (35.2%) in *S. aureus* isolates [168].

Interestingly, the Asn470Asp mutation is located just outside the region of amino acids 451–458 which is highly conserved between GrlB and GyrB [162]. An homologous mutation (Asn493) in the yeast species *Saccharomyces cerevisiae* has been located through crystal structures to B′α3 helix, which is some distance from the active-site tyrosine involved in DNA strand breaking [169]. This raises the question as to how such a mutation can confer resistance. Despite the absence of structural data on the topoisomerase–DNA–quinolone complex, Fournier and Hooper suggest that resistance is unlikely to result from the fact that the Asn470Asp mutation introduces a negative charge because this would actually increase its affinity for the positively charged piperazinyl group on ciprofloxacin. In light of the fact that mutant enzymes exhibit reduced relaxation of supercoiled DNA [166], it is further suggested that the *grlB* mutation reduces the intrinsic catalytic efficiency of the enzyme. This could be true if quinolones only bind to specific steps in the catalytic cycle of the enzyme–DNA complex. Mutation could induce conformational changes which slow down the catalytic cycle, meaning that there would be fewer opportunities where the enzyme–DNA complexes are in an appropriate conformation for quinolones to bind to [162].

The mechanism shared by proteins which belong to the MFS has already been dealt with in detail in the macrolide chapter. Of course there are many other proteins which are part of this family. Fluoroquinolone resistance, for example, is associated with over expression of one such multidrug efflux protein, NorA [168]. Ciprofloxacin-resistant strains of *Mycobacterium smegmatis* have been shown to have an increased initial phosphate uptake [170]. Uptake of phosphate is mediated by the phosphate-specific transporter Pst which possesses a nucleotide-binding subunit (PstB) reported to exhibit ATPase activity [171]. This is consistent with Pst as a member of the ABC family of transport proteins. Monitoring phosphate uptake and fluoroquinolone sensitivity in strains with a disrupted *pstB* gene (WT^d) showed that there was a remarkable reduction in phosphate scavenging ability. Also compared to wild-type cells, the sensitivity of WT^d cells to ciprofloxacin, ofloxacin and sparfloxacin was increased by approximately 2-fold [170].

High-level fluoroquinolone resistance in *Coxiella burnetii* has been associated with two distinct nucleotide mutations in the *gyrA* gene [172]. Interestingly, a pH-related mechanism appears to contribute to perfloxacin resistance seen in *C. burnetii* [173]. Spyridaki *et al.* found that at pH 7.2 susceptible strains (SCB2 and SCB4), demonstrated a higher penetrability of pefloxacin than perfloxacin-resistant strains of *C. burnetii* (RCB2 and RCB4). The suggestion that an energy-dependent process such as active efflux may play a role in the reduced accumulation of perfloxacin was dismissed after the difference in accumulation of pefloxacin was not abolished by the addition of carbonyl cyanide *m*-chlorophenylhydrazone. However, the study showed that in an acid environment of pH 4.5, all strains demonstrated reduced intracellular concentration of pefloxacin. This provides some evidence to suggest a pH-related mechanism maybe involved in reducing the pefloxacin concentration in *C. burnetii* [173].

1.10
Conclusions

The effectiveness and remarkable success of antibiotics in controlling bacterial infections put them forward amongst the most successful examples of chemotherapeutic intervention. Between 1900 and 1980 the death rate from infectious diseases fell from 797 per 100 000 to 36 per 100 000 of the population [7]. However, the very success of antibiotic chemotherapy brought about the widespread concerns surrounding their use. The 'golden age' of antibiotic therapy, which characterized the 1950s and 1960s, has come to an end. In fact, within the last 40 years we have struggled in a ceaseless war with resistance where bacteria have adapted quickly to new antibiotic developments, threatening a return to the 'dark ages' of the preantibiotic era.

Infections resistant to antibiotic treatment present a major and increasing problem to hospital and community care. Resistance to all antibiotic classes has been reported and is easily transferred within mixed bacterial populations. Over the past 70 years a vast body of research has been dedicated to the elucidation of the molecular mechanisms of antibiotic action and bacterial resistance to antibiotics. Some of the molecular and structural determinants underpinning resistance to major antibiotic classes have been outlined here, with special attention to target modification, antibiotic deactivation and drug efflux.

The action of β-lactam and glycopeptides antibiotics, which target cell wall biosynthesis, is evaded through target modifications and antibiotic deactivation. β-Lactamases can be deployed to deactivate β-lactam antibiotics, while penicillin-binding proteins with altered binding sites provide an example of modification of the antibiotic target. Alteration of peptidoglycan peptide termini helps protect cell wall synthesis from glycopeptides, while reduced cross-linking coupled to a grossly thickened wall essentially stalls glycopeptides at the cell wall periphery. Specific molecular changes within ribosomal structures can foil aminoglycoside, tetracycline or macrolide attack. Further protection can be achieved by positioning proteins to protect protein synthesizing machinery from the action of tetracyclines. Important resistance mechanisms protecting bacterial protein synthesis include enzymatic deactivation of aminoglycosides and efflux systems expelling incoming macrolides, aminoglycosides and tetracyclines.

The dependence of our society on the use of antimicrobial chemotherapy to control infections maintains the evolutionary pressure on bacteria and drives the development of antibiotic resistance. This emphasizes the urgent necessity of a sustained effort to explore new antibiotic targets and mechanisms of action. Renewed efforts to explore alternative strategies for infection management have brought to the top of scientific agenda defense peptides, lanthionine antibiotics, phage therapies and other antimicrobial techniques, which have not seen wide use against clinically significant pathogens and resistance to which remains uncommon.

References

1 Courvalin, P. (2004) in *The Pfizer Journal*, Impact Communications, pp 6–12.

2 Maas, S. (2003) *Minnesota Medicine*, **86**, 8–11.

3 Mann, J. (1999) Fighting bacteria: the mould that grew on sewage: evolution of the cephalosporins, in *The Elusive Magic Bullet: The Search for the Perfect Drug*, Oxford University Press, Oxford. pp. 57–58.

4 Department of Health, *A Simple Guide to MRSA*, http://www.dh.gov.uk/assetRoot/ 04/11/34/79/04113479.pdf (accessed 20 October 2005).

5 Tiemersma, E.W., Bronzwaer, S.L., Lyytikainen, O., Degener, J.E., Schrijnemakers, P., Bruinsma, N., Monen, J., Witte, W., Grundman, H. and European Antimicrobial Resistance Surveillance System Participants (2004) *Emerging Infectious Diseases*, **10**, 1627–1634.

6 Crowcroft, N.S. and Catchpole, M. (2002) *British Medical Journal*, **325**, 1390–1391.

7 Toleman, M.A., Bennett, P.M. and Walsh, T.R. (2006) *Microbiology and Molecular Biology Reviews*, **70**, 296–316.

8 Kahne, D., Leimkuhler, C., Lu, W. and Walsh, C. (2005) *Chemical Reviews*, **105**, 425–448.

9 Bearden, D.T. and Danziger, L.H. (2001) *Pharmacotherapy*, **21**, 224S–232.

10 Russell, A.D. and Chopra, I. (1996) *Understanding Antibacterial Action and Resistance*, 2nd edn, Ellis Horwood, London.

11 von Wright, G.D. (2003) *Current Opinion in Chemical Biology*, **7**, 563–569.

12 Walsh, C. (2000) *Nature*, **406**, 775–881.

13 Fuda, C., Suvorov, M., Vakulenko, S.B. and Mobashery, S. (2004) *Journal of Biological Chemistry*, **279**, 40802–40806.

14 Ambler, R.P. (1980) *Philosophical Transactions of the Royal Society B: Biological Sciences*, **289**, 321–331.

15 Couture, F., Lachapelle, J. and Levesque, R.C. (1992) *Molecular Microbiology*, **45**, 1693–1705.

16 Walther-Rasmussen, J. and Hoiby, N. (2006) *Journal of Antimicrobial Chemotherapy*, **57**, 373–383.

17 Samraoui, B., Sutton, B.J., Todd, R.J., Artymiuk, P.J., Waley, S.G. and Phillips, D.C. (1986) *Nature*, **320**, 378–380.

18 Knox, J.R., Moews, P.C. and Frere, J.M. (1996) *Chemistry and Biology*, **9**, 937–947.

19 Majiduddin, F.K., Materon, I.C. and Palzkill, T.G. (2002) *International Journal of Medical Microbiology*, **292**, 127–137.

20 Page, M.I. and Laws, A.P. (1998) *Chemical Communications*, **1998**, 1609–1617.

21 Walsh, T.R., Toleman, M.A., Poirel, L. and Nordmann, P. (2005) *Clinical Microbiology Reviews*, **18**, 306–325.

22 Cantu, C. 3rd and Palzkill, T. (1998) *Journal of Biological Chemistry*, **273**, 26603–26609.

23 Chambers, H.F. (1997) *Clinical Microbiology Reviews*, **10**, 781–791.

24 Zhao, G., Meier, T.I., Hoskins, J. and McAllister, K.A. (2000) *Antimicrobial Agents and Chemotherapy*, **44**, 1745–1748.

25 Lu, W.P., Kincaid, E., Sun, Y.P. and Bauer, M.D. (2001) *Journal of Biological Chemistry*, **276**, 31494–31501.

26 Lim, D. and Strynadka, N.C. (2002) *Nature Structural Biology*, **9**, 870–876.

27 Fuda, C., Hesek, D., Lee, M., Morio, K., Nowak, T. and Mobashery, S. (2005) *Journal of the American Chemical Society*, **127**, 2056–2057.

28 Leski, T.A. and Tomasz, A. (2005) *Journal of Bacteriology*, **141**, 1815–1824.

29 Pinho, M.G., Filipe, S.R., De Lencastre, H. and Tomasz, A. (2001) *Journal of Bacteriology*, **183**, 6525–6531.

30 Kell, C.M., Sharma, U.K., Dowson, C.G., Town, C., Balganesh, T.S. and Spratt, B.G. (1993) *FEMS Microbiology Letters*, **106**, 171–175.

31 Grebe, T. and Hakenbeck, R. (1996) *Antimicrobial Agents and Chemotherapy*, **40**, 829–834.

32 Mouz, N., Di Guilmi, A.M., Gordon, E., Hakenbeck, R., Dideberg, O. and Vernet, T. (1999) *Journal of Biological Chemistry*, **274**, 19175–19180.

33 Mouz, N., Gordon, E., Di Guilmi, A.M., Petit, I., Petillot, Y., Dupont, Y., Hakenbeck, R., Vernet, T. and Dideberg, O. (1998) *Proceedings of the National Academy of Sciences of the United States of America*, **95**, 13403–13406.

34 Pernot, L., Chesnel, L., Le Gouellec, A., Croize, J., Vernet, T., Dideberg, O. and Dessen, A. (2004) *Journal of Biological Chemistry*, **279**, 16463–16470.

35 Boyle-Vavra, S., Labischinski, H., Ebert, C.C., Ehlert, K. and Daum, R.S. (2001) *Antimicrobial Agents and Chemotherapy*, **45**, 280–287.

36 Pares, S., Mouz, N., Petillot, Y., Hakenbeck, R. and Dideberg, O. (1996) *Nature Structural Biology*, **3**, 284–289.

37 Filipe, S.R., Severina, E. and Tomasz, A. (2001) *Microbial Drug Resistance – Mechanisms, Epidemiology, and Disease*, **7**, 303–316.

38 Bordet, C. and Perkins, H.R. (1970) *Biochemical Journal*, **119**, 877–883.

39 Bugg, T.D.H., Wright, G.D., Dutkamalen, S., Arthur, M., Courvalin, P. and Walsh, C.T. (1999) *Biochemistry*, **30**, 10408–10415.

40 Walsh, C.T., Fisher, S.L., Park, I.S., Prahalad, M. and Wu, Z. (1996) *Chemistry and Biology*, **3**, 21–28.

41 Perl, T.M. (1999) *American Journal of Medicine*, **106** (5A), 26S–37.

42 Van Bambeke, F., Chauvel, M., Reynolds, P.E., Fraimow, H.S. and Courvalin, P. (1999) *Antimicrobial Agents and Chemotherapy*, **43**, 41–47.

43 Gholizadeh, Y. and Courvalin, P. (2000) *International Journal of Antimicrobial Agents*, **16**, S11–S17.

44 Murray, B.E. (1998) *Emerging Infectious Diseases*, **4**, 37–47.

45 Cetinkaya, Y., Falk, P. and Mayhall, C.G. (2000) *Clinical Microbiology Reviews*, **13**, 686–707.

46 McComas, C.C., Crowley, B.M. and Boger, D.L. (2003) *Journal of the American Chemical Society*, **125**, 9314–9315.

47 Cooper, M.A. and Williams, D.H. (1999) *Chemistry and Biology*, **6**, 891–899.

48 Huang, W.M., Zhang, Z.L., Han, X.J., Wang, J.G., Tang, J.L., Dong, S.J. and Wang, E.K. (2002) *Biophysical Chemistry*, **99**, 271–279.

49 Ge, M., Chen, Z., Onishi, H.R., Kohler, J., Silver, L.L., Kerns, R., Fukuzawa, S., Thompson, C. and Kahne, D. (1999) *Science*, **284**, 507–511.

50 Kerns, R., Dong, S.D., Fukuzawa, S., Carbeck, J., Kohler, J., Silver, L. and Kahne, D. (2000) *Journal of the American Chemical Society*, **122**, 12608–12609.

51 Gold, H.S. (2001) *Clinical Infectious Diseases*, **33**, 210–219.

52 Sievert, D.M., Boulton, M.L., Stoltman, G., Johnson, D., Stobierski, M.G., Downes, F.P., Somsel, P.A., Rudrik, J.T., Brown, W., Hafeez, W., Lundstrom, T., Flanagan, E., Johnson, R., Mitchell, J. and Chang, S. (2002) *Journal of the American Medical Association*, **288**, 824–825.

53 Cui, L., Ma, X., Sato, K., Okuma, K., Tenover, F.C., Mamizuka, E.M., Gemmell, C.G., Kim, M.N., Ploy, M.C., El-Solh, N., Ferraz, V. and Hiramatsu, K. (2003) *Journal of Clinical Microbiology*, **41**, 5–14.

54 Hiramatsu, K., Aritaka, N., Hanaki, H., Kawasaki, S., Hosoda, Y., Hori, S., Fukuchi, Y. and Kobayashi, I. (1997) *Lancet*, **350**, 1670–1673.

55 Gilmore, M., Clewell, D., Courvalin, P., Dunny, G., Murray, B. and Rice, L. (2002) *The Enterococci: Pathogenesis Molecular Biology and Antibiotic Resistance*, ASM Press, Washington, DC.

56 Noble, W.C., Virani, Z. and Cree, R.G. (1992) *FEMS Microbiology Letters*, **72**, 195–198.

57 Hanaki, H., Labischinski, H., Inaba, Y., Kondo, N., Murakami, H. and Hiramatsu, K. (1998) *Journal of Antimicrobial Chemotherapy*, **42**, 315–320.

58 Cui, L., Murakami, H., Kuwahara-Arai, K., Hanaki, H. and Hiramatsu, K. (2000) *Antimicrobial Agents and Chemotherapy*, **44**, 2276–2285.

59 Sieradzki, K. and Tomasz, A. (2003) *Journal of Bacteriology*, **185**, 7103–7110.

60 Finan, J.E., Archer, G.L., Pucci, M.J. and Climo, M.W. (2001) *Antimicrobial Agents and Chemotherapy*, **45**, 3070–3075.

61 Sieradzki, K. and Tomasz, A. (1999) *Journal of Bacteriology*, **181**, 7566–7570.

62 Peteroy, M., Severin, A., Zhao, F., Rosner, D., Lopatin, U., Scherman, H., Belanger, A., Harvey, B., Hatfull, G.F., Brennan, P.J. and Connell, N.D. (2000) *Antimicrobial Agents and Chemotherapy*, **44**, 1701–1704.

63 Wootton, M., Bennett, P.M., Macgowan, A.P. and Alsh, T.R. (2005) *Antimicrobial Agents and Chemotherapy*, **49**, 3598–3599.

64 Sahl, H.G. (1991) *Nisin and Novel Lantibiotics*, ESCOM Science Publishers, Leiden.

65 Brotz, H. and Sahl, H.G. (2000) *Journal of Antimicrobial Chemotherapy*, **46**, 1–6.

66 Rogers, L.A. and Whittier, E.O. (1928) *Journal of Bacteriology*, **16**, 211–229.

67 Bonev, B.B., Chan, W.C., Bycroft, B.W., Roberts, G.C.K. and Watts, A. (2000) *Biochemistry*, **39**, 11425–11433.

68 van Kraaij, C., Breukink, E., Noordermeer, M.A., Demel, R.A., Siezen, R.J., Kuipers, O.P. and de Kruijff, B. (1998) *Biochemistry*, **37**, 16033–16040.

69 Brumfitt, W., Salton, M.R.J. and Hamilton-Miller, J.M.T. (2002) *Journal of Antimicrobial Chemotherapy*, **50**, 731–734.

70 Bartoloni, A., Mantella, A., Goldstein, B.P., Dei, R., Benedetti, M., Sbaragli, S. and Paradisi, F. (2004) *Journal of Chemotherapy*, **16**, 119–121.

71 Bonev, B.B., Breukink, E., Swiezewska, E., De Kruijff, B. and Watts, A. (2004) *FASEB Journal*, **18**, 1862–1869.

72 Breukink, E., Wiedemann, I., van Kraaij, C., Kuipers, O.P., Sahl, H.G. and de Kruijff, B. (1999) *Science*, **286**, 2361–2364.

73 Parisot, J.L., Carey, S., Breukink, E., Chan, W.C., Narbad, A. and Bonev, B. (2008) *Antimicrobial Agents and Chemotherapy*, **52**, 612–618.

74 Hyde, A.J., Parisot, J., McNichol, A. and Bonev, B.B. (2006) *Proceedings of the National Academy of Sciences of the United States of America*, **103**, 19896–19901.

75 Hsu, S.T.D., Breukink, E., Tischenko, E., Lutters, M.A.G., de Kruijff, B., Kaptein, R., Bonvin, A. and van-Nuland, N.A.J. (2004) *Nature Structural and Molecular Biology*, **11**, 963–967.

76 Siegers, K. and Entian, K.D. (1995) *Applied and Environmental Microbiology*, **61**, 1082–1089.

77 Stein, T., Heinzmann, S., Dusterhus, S., Borchert, S. and Entian, K.D. (2005) *Journal of Bacteriology*, **187**, 822–828.

78 Stein, T., Heinzmann, S., Solovieva, I. and Entian, K.D. (2003) *Journal of Biological Chemistry*, **278**, 89–94.

79 Pedersen, L.C., Benning, M.M. and Holden, H.M. (1995) *Biochemistry*, **34**, 13305–13311.

80 Fourmy, D., Recht, M.I., Blanchard, S.C. and Puglisi, J.D. (1996) *Science*, **274**, 1367–1371.

81 Karimi, R. and Ehrenberg, M. (1994) *European Journal of Biochemistry*, **226**, 355–360.

82 De Stasio, E.A., Moazed, D., Noller, H.F. and Dahlberg, A.E. (1989) *EMBO Journal*, **8**, 1213–1216.

83 Springer, B., Kidan, Y.G., Prammananan, T., Ellrott, K., Bottger, E.C. and Sander, P. (2001) *Antimicrobial Agents and Chemotherapy*, **45**, 2877–2884.

84 Musser, J.M. (1995) *Clinical Microbiology Reviews*, **8**, 496–514.

85 Kotra, L.P., Haddad, J. and Mobashery, S. (2000) *Antimicrobial Agents and Chemotherapy*, **44**, 3249–3256.

86 Finken, M., Kirschner, P., Meier, A., Wrede, A. and Bottger, E.C. (1993) *Molecular Microbiology*, **9**, 1239–1246.

87 Stern, S., Powers, T., Changchien, L.M. and Noller, H.F. (1988) *Journal of Molecular Biology*, **201**, 683–695.

88 Cundliffe, E. (1989) *Annual Review of Microbiology*, **43**, 207–233.

89 Skeggs, P.A., Thompson, J. and Cundliffe, E. (1985) *Molecular and General Genetics*, **200**, 415–421.

90 Thompson, J., Skeggs, P.A. and Cundliffe, E. (1985) *Molecular and General Genetics*, **201**, 168–173.

91 Yokoyama, K., Doi, Y., Yamane, K., Kurokawa, H., Shibata, N., Shibayama, K., Yagi, T., Kato, H. and Arakawa, Y. (2003) *Lancet*, **362**, 1888–1893.

92 Nicolas, G., Auger, J., Beaudoin, M., Halle, F., Morency, H., LaPointe, G. and Lavoie, M.C. (2004) *Journal of Microbiological Methods*, **59**, 351–361.

93 Magnet, S. and Blanchard, J.S. (2005) *Chemical Reviews*, **105**, 477–498.

94 Llano-Sotelo, B., Azucena, E.F., Jr, Kotra, L.P., Mobashery, S. and Chow, C.S. (2002) *Chemistry and Biology*, **9**, 455–463.

95 Finch, R.G., Greenwood, D., Norrby, S.R. and Whitley, R.J. (2003) *Antibiotic and Chemotherapy: Anti-infective Agents and Their Use in Therapy*, Chruchill Livingstone, Edinburgh.

96 Wolf, E., Vassilev, A., Makino, Y., Sali, A., Nakatani, Y. and Burley, S.K. (1998) *Cell*, **94**, 439–449.

97 Mawson, A.J. and Costar, K. (1993) *Letters In Applied Microbiology*, **17**, 256–258.

98 Lu, L., Berkey, K.A. and Casero, R.A., Jr. (1996) *Journal of Biological Chemistry*, **271**, 18920–18924.

99 Williams, J.W. and Northrop, D.B. (1979) *Journal of Antibiotics*, **32**, 1147–1154.

100 Sakon, J., Liao, H.H., Kanikula, A.M., Benning, M.M., Rayment, I. and Holden, H.M. (1993) *Biochemistry*, **32**, 11977–11984.

101 Robichon, D., Gouin, E., Debarbouille, M., Cossart, P., Cenatiempo, Y. and Hechard, Y. (1997) *Journal of Bacteriology*, **179**, 7591–7594.

102 McKay, G.A. and Wright, G.D. (1995) *Journal of Biological Chemistry*, **270**, 24686–24692.

103 Daigle, D.M., McKay, G.A., Thompson, P.R. and Wright, G.D. (1999) *Chemistry and Biology*, **6**, 11–18.

104 Le Goffic, F., Baca, B., Soussy, C.J., Dublanchet, A. and Duval, J. (1976) *Annals of Microbiology (Paris)*, **127**, 391–399.

105 Sadaie, Y., Burtis, K.C. and Doi, R.H. (1980) *Journal of Bacteriology*, **141**, 1178–1182.

106 Van Pelt, J.E., Iyengar, R. and Frey, P.A. (1986) *Journal of Biological Chemistry*, **261**, 15995–15999.

107 Taber, H.W., Mueller, J.P., Miller, P.F. and Arrow, A.S. (1987) *Microbiological Reviews*, **51**, 439–457.

108 Bryan, L.E., Kowand, S.K., Evan Den Elzen, H.M. (1979) *Antimicrobial Agents and Chemotherapy*, **15**, 7–13.

109 Murakami, S., Nakashima, R., Yamashita, E., Matsumoto, T. and Yamaguchi, A. (2006) *Nature*, **443**, 173–179.

110 Murakami, S., Nakashima, R., Yamashita, E. and Yamaguchi, A. (2002) *Nature*, **419**, 587–593.

111 Yu, E.W., Aires, J.R. and Nikaido, H. (2003) *Journal of Bacteriology*, **185**, 5657–5664.

112 Rosenberg, E.Y., Ma, D. and Nikaido, H. (2000) *Journal of Bacteriology*, **182**, 1754–1756.

113 Silva, P.E., Bigi, F., de la Paz Santangelo, M., Romano, M.I., Martin, C., Cataldi, A. and Ainsa, J.A. (2001) *Antimicrobial Agents and Chemotherapy*, **45**, 800–804.

114 Ainsa, J.A., Blokpoel, M.C., Otal, I., Young, D.B., De Smet, K.A. and Martin, C. (1998) *Journal of Bacteriology*, **180**, 5836–5843.

115 Schlunzen, F., Zarivach, R., Harms, J., Bashan, A., Tocilj, A., Albrecht, R., Yonath, A. and Franceschi, F. (2001) *Nature*, **413**, 814–821.

116 Tait-Kamradt, A., Davies, T., Appelbaum, P.C., Depardieu, F., Courvalin, P., Petitpas, J., Wondrack, L., Walker, A., Jacobs, M.R. and Sutcliffe, J. (2000) *Antimicrobial Agents and Chemotherapy*, **44**, 3395–3401.

117 Wittmann, H.G., Stoffler, G., Apirion, D., Rosen, L., Tanaka, K., Tamaki, M., Takata, R., Dehio, S., Otaka, E. and Osawa, S. (1973) *Molecular and General Genetics*, **127**, 175–189.

118 Gregory, S.T. and Dahlberg, A.E. (1999) *Journal of Molecular Biology*, **289**, 827–834.

119 Tait-Kamradt, A., Clancy, J., Cronan, M., Dib-Hajj, F., Wondrack, L., Yuan, W. and Sutcliffe, J. (1997) *Antimicrobial Agents and Chemotherapy*, **41**, 2251–2255.

120 Clancy, J., Petitpas, J., Dib-Hajj, F., Yuan, W., Cronan, M., Kamath, A.V., Bergeron, J. and Retsema, J.A. (1996) *Molecular Microbiology*, **22**, 867–879.

121 Roberts, M.C., Sutcliffe, J., Courvalin, P., Jensen, L.B., Rood, J. and Seppala, H. (1999) *Antimicrobial Agents and Chemotherapy*, **43**, 2823–2830.

122 Nikaido, H. (1998) *Clinical Infectious Diseases*, **27**, S32–S41.

123 Sutcliffe, J., Tait-Kamradt, A. and Wondrack, L. (1996) *Antimicrobial Agents and Chemotherapy*, **40**, 1817–1824.

124 Van Bambeke, F., Balzi, E. and Tulkens, P.M. (2000) *Biochemical Pharmacology*, **60**, 457–470.

125 Griffith, J.K., Baker, M.E., Rouch, D.A., Page, M.G.P., Skurray, R.A., Paulsen, I.T., Chater, K.F., Baldwin, S.A. and Henderson, P.J.F. (1992) *Current Opinion in Cell Biology*, **4**, 684–695.

126 Langton, K.P., Henderson, P.J. and Herbert, R.B. (2005) *Natural Product Reports*, **22**, 439–451.

127 Zheleznova, E.E., Markham, P., Edgar, R., Bibi, E., Neyfakh, A.A. and Brennan, R.G. (2000) *Trends in Biochemical Sciences*, **25**, 39–43.

128 Zheleznova, E.E., Markham, P.N., Neyfakh, A.A. and Brennan, R.G. (1999) *Cell*, **96**, 353–362.

129 Pawagi, A.B., Wang, J., Silverman, M., Reithmeier, R.A. and Deber, C.M. (1994) *Journal of Molecular Biology*, **235**, 554–564.

130 Ross, J.I., Eady, E.A., Cove, J.H., Cunliffe, W.J., Baumberg, S. and Wootton, J.C. (1990) *Molecular Microbiology*, **4**, 1207–1214.

131 Zhong, P. and Shortridge, VD. (2000) *Drug Resistence Updates*, **3**, 325–329.

132 Ross, J.I., Eady, E.A., Cove, J.H. and Baumberg, S. (1995) *Gene*, **153**, 93–98.

133 Wondrack, L., Massa, M., Yang, B.V. and Sutcliffe, J. (1996) *Antimicrobial Agents and Chemotherapy*, **40**, 992–998.

134 Higgins, C.F. and Linton, K.J. (2004) *Nature Structural and Molecular Biology*, **11**, 918–926.

135 Speer, B.S., Shoemaker, N.B. and Salyers, A.A. (1992) *Clinical Microbiology Reviews*, **5**, 387–399.

136 Moran, J.S. and Zenilman, J.M. (1990) *Reviews of Infectious Diseases*, **12** (Suppl. 6), S633–S644.

137 Brodersen, D.E., Clemons, W.M. Jr, Carter, A.P., Morgan-Warren, R.J., Wimberly, B.T. and Ramakrishnan, V. (2000) *Cell*, **103**, 1143–1154.

138 Goldman, R.A., Hasan, T., Hall, C.C., Strycharz, W.A. and Cooperman, B.S. (1983) *Biochemistry*, **22**, 359–368

139 Connell, S.R., Tracz, D.M., Nierhaus, K.H. and Taylor, D.E. (2003) *Antimicrobial Agents and Chemotherapy*, **47**, 3675–3681.

140 Ogle, J.M., Murphy, F.V., Tarry, M.J. and Ramakrishnan, V. (2002) *Cell*, **111**, 721–732.

141 Pioletti, M., Schlunzen, F., Harms, J., Zarivach, R., Gluhmann, M., Avila, H., Bashan, A., Bartels, H., Auerbach, T., Jacobi, C., Hartsch, T., Yonath, A. and Franceschi, F. (2001) *EMBO Journal*, **20**, 1829–1839.

142 Connell, S.R., Trieber, C.A., Stelzl, U., Einfeldt, E., Taylor, D.E. and Nierhaus, K.H. (2002) *Molecular Microbiology*, **45**, 1463–1472.

143 McMurry, L.M., Park, B.H., Burdett, V. and Levy, S.B. (1987) *Antimicrobial Agents and Chemotherapy*, **31**, 1648–1650.

144 Yamaguchi, A. Ohmori, H., Kaneko-Ohdera, M., Nomura, T. and Sawai, T. (1991) *Antimicrobial Agents and Chemotherapy*, **31**, 1648–1650.

145 Chopra, I. and Roberts, M. (2001) *Microbiology and Molecular Biology Reviews*, **65**, 232–260.

146 Paulsen, I.T., Brown, M.H. and Skurray, R.A. (1996) *Microbiological Reviews*, **60**, 575–608.

147 Levy, S.B. (1992) *Antimicrobial Agents and Chemotherapy*, **36**, 695–703.

148 Lyras, D. and Rood, J.I. (1996) *Antimicrobial Agents and Chemotherapy*, **40**, 2500–2504.

149 Schnappinger, D. and Hillen, W. (1996) *Archives of Microbiology*, **165**, 359–369.

150 Yamaguchi, A., Ono, N., Akasaka, T., Noumi, T. and Sawai, T. (1990) *Journal of Biological Chemistry*, **265**, 15525–15530.

151 Nikaido, H. and Thanassi, D.G. (1993) *Antimicrobial Agents and Chemotherapy*, **37**, 1393–1399.

152 Spahn, C.M., Blaha, G., Agrawal, R.K., Penczek, P., Grassucci, R.A., Trieber, C.A., Connell, S.R., Taylor, D.E., Nierhaus, K.H. and Frank, J. (2001) *Molecular Cell*, **7**, 1037–1045.

153 Leblanc, D.J., Lee, L.N., Titmas, B.M., Smith, C.J. and Tenover, F.C. (1988) *Journal of Bacteriology*, **170**, 3618–3626.

154 Burdett, V. (1991) *Journal of Biological Chemistry*, **266**, 2872–2877.

155 Moine, H. and Dahlberg, A.E. (1994) *Journal of Molecular Biology*, **243**, 402–412.

156 Connell, S.R., Trieber, C.A., Dinos, G.P., Einfeldt, E., Taylor, D.E. and Nierhaus, K.H. (2003) *EMBO Journal*, **22**, 945–953.

157 Trieber, C.A., Burkhardt, N., Nierhaus, K.H. and Taylor, D.E. (1998) *Biological Chemistry*, **379**, 847–855.

158 O'Donnell, J.A. and Gelone, S.P. (2000) *Infectious Disease Clinics of North America*, **14**, 489–513.

159 Ball, P. (1994) *Infection*, **22**, S140–S147.

160 Fournier, B. and Hooper, D.C. (1998) *Antimicrobial Agents and Chemotherapy*, **42**, 121–128.

161 Drlica, K. and Hooper, D.C. (1993) in *Quinolone antimicrobial agents*, eds. Hooper, D.C. and Wolfson, J.S. (American Society for Microbiology, Washington, D.C.), pp. 19–40.

162 Fournier, B. and Hooper, D.C. (1998) *Antimicrobial Agents and Chemotherapy*, **42**, 121–128.

163 Deplano, A., Zekhnini, A., Allali, N., Couturier, M. and Struelens, M.J. (1997) *Antimicrobial Agents and Chemotherapy*, **41**, 2023–2025.

164 Bachoual, R., Dubreuil, L., Soussy, C.J. and Tankovic, J. (2000) *Antimicrobial Agents and Chemotherapy*, **44**, 1842–1845.

165 Hooper, D.C. (1999) *Drug Resistance Updates*, **2**, 38–55.

166 Trucksis, M., Wolfson, J.S. and Hooper, D.C. (1991) *Journal of Bacteriology*, **173**, 5854–5860.

167 Ng, E.Y., Trucksis, M. and Hooper, D.C. (1996) *Antimicrobial Agents and Chemotherapy*, **40**, 1881–1888.

168 Tanaka, M., Wang, T., Onodera, Y., Uchida, Y. and Sato, K. (2000) *Journal of Infection and Chemotherapy*, **3**, 131–139.

169 Berger, J.M., Gamblin, S.J., Harrison, S.C. and Wang, J.C. (1996) *Nature*, **379**, 225–232.

170 Bhatt, K., Banerjee, S.K. and Chakraborti, P.K. (2000) *European Journal of Biochemistry*, **267**, 4028–4032.

171 Chan, F.Y. and Torriani, A. (1996) *Journal of Bacteriology*, **178**, 3974–3977.

172 Musso, D., Drancourt, M., Osscini, S. and Raoult, D. (1996) *Antimicrobial Agents and Chemotherapy*, **40**, 870–873.

173 Spyridaki, I., Psaroulaki, A., Kokkinakis, E., Gikas, A. and Tselentis, Y. (2002) *Journal of Antimicrobial Chemotherapy*, **49**, 379–382.

2
Novel Approaches to Combat Drug-Resistant Bacteria

Iqbal Ahmad, Maryam Zahin, Farrukh Aqil, Mohd Sajjad Ahmad Khan, and Shamim Ahmad

Abstract

There is an increasing concern that the need for new antibacterial agents is greater than ever because of the emergence of multidrug resistance in common bacterial pathogens and the rapid emergence of new infections. Despite the critical need for new antibacterial agents, development is declining. Therefore, solutions for encouraging and facilitating the development of new agents are needed. It is believed that sustained success in the long-term battle against bacterial pathogens will require new strategies and new chemical scaffolds that target other cellular processes. In recent years, new antibacterial drug targets have been explored and validated. Some of the vulnerable targets are bacterial quorum sensing, interference with virulence and pathogenicity, drug-resistance mechanisms (efflux pump) and targets based on plasmid physiology. It is expected and believed by many workers that unexplored microbial diversity may provide novel and effective antibacterial drugs, if improved screening strategies are attempted and intelligent test systems are applied. This chapter will discuss recent strategies to obtain novel antibacterials from microbial products to combat drug-resistant bacteria. Further, roles of alternative therapy (phage therapy) and antibiotic use policy are discussed as possible ways to minimize the problem of drug resistant bacteria.

2.1
Introduction

One of the greatest accomplishments of modern medicine had been the development of antimicrobial drugs for the treatment of infectious diseases. Alexander Fleming discovered the first antibiotic, penicillin, in 1928 and over half century of extensive research most acute bacterial infections had been treated effectively with antibacterial drugs. The remarkable success of antibacterial drugs had given an impression in the late 1960s and early 1970s that infectious diseases had been congruent. However,

New Strategies Combating Bacterial Infection. Edited by Iqbal Ahmad and Farrukh Aqil
Copyright © 2009 WILEY-VCH Verlag GmbH & Co. KGaA, Weinheim
ISBN: 978-3-527-32206-0

40 years later, infectious diseases still remain the third leading cause of death in the United States [1] and the second leading cause of death worldwide [2]. This is mainly due to the development of drug resistance, emergence of new infectious agents and nonavailability of suitable drugs for many infectious diseases. Indiscriminate and excessive use of antibiotics in medical as well as nonmedical settings has resulted in the selection and development of antibiotic resistance among clinical isolates. Bacterial pathogens have developed several mechanisms to overcome antibiotic pressure and now thrive even in the presence of multidrugs. Bacteria may develop resistance by mutation and by acquisition of new genes through bacterial genetic exchange mechanisms (conjugation, transformation and transduction). The horizontal transfer of resistance plasmids among Gram-negative and other bacteria has helped in the fast dissemination of resistance genes not only to pathogenic, but also other commensal and nonpathogenic bacteria as well. Many pathogenic bacteria now have gained the status of problematic multidrug-resistant (MDR) bacteria. [3]. The incidence of antibiotic-resistant bacteria such as methicillin-resistant *Staphylococcus aureus* (MRSA), vancomycin-resistant enterococci (VRE), *Mycobacterium tuberculosis* and extended-spectrum β-lactamase (ESβL)-producing enteric bacteria (*E. coil, Klebsiella pneumoniae*, etc.) all over the world has increased considerably. If this trend continues and the discovery of antibacterials with novel modes of action is not accelerated, there may be a stage when bacterial epidemics due to multidrug resistance might emerge and become much problematic. Therefore, there is an urgent need to continue our efforts both at the academic as well as the pharmaceutical industry level to develop new drugs and possible ways to combat drug-resistant pathogens.

An investigation by Spellberg *et al.* [4] has indicated a continuous decline in the discovery of new antibacterial agents by 56% over the past 20 years (1998–2002 versus 1983–1987). Projecting future development, new antibacterial agents constitute six out of 506 drugs discovered in development programs of the largest pharmaceutical and biotechnology companies of the world. This has clearly indicated a marked slowdown in development of new antibiotics.

The major pharmacy companies are still in the antibiotic development market, but it seems that they are loosing interest on these drugs. Since 1998, only nine antibiotics or new uses of old antibiotics have been approved by the United States Food and Drug Administration (FDA) and only six antibiotics are in phase 2 and phase 3 clinical trials. Pfizer is the current leader, having obtained FDA approval for three antibiotics since 1998 and two antibiotics in phase 2 and 3 trials. One possible reason for the slowdown in antibiotic research is advances made in health care that have prevented the occurrence of epidemics, and also the increased attention required by many other problems like cancer, arthritis and new viral diseases. Other reasons may be the introduction of new guidelines for a raised utility bar for new antibiotics to file patents. According to the Tufts Center for the Study of Drug Development, the cost of developing a new drug has risen from US$231 million in 1987 to about US$802 million in 2001. This increase is primarily due to the rising cost of clinical trials. Other factors affecting companies' decreased interest regarding antibiotic research and development progress include the difficulty in improving the efficacy of currently available antibiotics and the low return on investment compared with those in research of nonantibiotic drugs [5–7].

However, many pharmaceutical companies and academic institutions have continued their efforts in the search for novel antibacterial compounds from natural products using classical and molecular approaches. The need for additional antibiotics is becoming increasingly vital; the number of newly developed antibiotics has been on the wane. Indeed, seeking novel targets that are different from those affected by currently available antibiotics could provide a powerful tool for dealing with infections caused by antibiotic-resistant bacteria. On the other hand, the role of natural product-derived compounds in modern drug discovery and compounds in antibacterial clinical trials has been published [7–9]. Many authors have highlighted the need for exploring alternative approaches to develop antibacterial products – this is a worthwhile task and re-examining the potential of promising older methods might be of value. Phage therapy, use of probiotics and natural products such as herbal medicines and honey, and so on, may solve many problems in the management of drug-resistant bacterial infections alone or in combination with modern drugs. This chapter focuses on various approaches to obtain antibacterial compounds with novel modes of action or that could act in combination with older antibiotics. Other approaches to combat bacterial infections such as the use of plant essential oils and derived compounds, and natural products like honey are discussed in other chapters in this book.

2.2
Approaches to Antibacterial Drug Discovery and Combating the Problem of Drug Resistance

The emerging and sustained resistance to antibiotics and the poor pipeline of new antibacterial drugs is creating a major health issue worldwide. Bacterial pathogens are increasingly becoming resistant to even the most recently approved antibiotics. The interest of larger pharmaceutical companies in developing new antibiotics is declining due to both inherent difficulties and economic reasons. However, academic institutions and a few pharmaceutical companies are still actively engaged in new antibiotic discovery.

There are five distinct classical approaches to obtain new microbial metabolites as described by Crueger and Crueger [10].

(i) Screening for the production of new metabolites with new isolates and/or new test method. This is probably the only way to obtain completely new classes of substances.

(ii) Chemical synthesis or chemical modification of microbial substances. Researchers in pharmaceutical companies as well as academic institute continue their efforts to synthesizing new chemical entities new chemical entity (NCEs) in the search for new antibiotics, specifically against MRSA. Several examples are available at the initial level. However, in many cases desirable or ideal characteristics of their chemotherapeutic agent may not be reached [11, 12].

(iii) Biotransformation, which produces a change in a chemical molecule by means of a microbial or enzymatic reaction

(iv) Interspecific protoplast fusion. This approach exploits the recombination of genetic information from rather closely related producer strains.

(v) Gene technology, in which genes may be transferred between unrelated strains which are products of known substances. Alternatively, transfer may be to non-producers which contain 'silent' genes leading to the generation of modified or even new substances. This approach has been productive for many therapeutic compounds and the use of this technology in the development of novel antibiotics is to be exploited.

In recent years many new approaches have been emphasized by academic and pharmaceutical laboratories to combat the problem of drug resistance. These approaches include (i) development of new antibiotics from less-explored microbial and other natural products, (ii) developing antiresistance strategies such as inhibition of efflux pumps and β-lactamases, and elimination of resistance plasmids, (iii) synergistic interaction of antibiotics with other nonantibiotics or natural products, and (iv) anti-infective strategies like developing antipathogenic drugs, and exploiting novel targets based on genomic and proteomic information of the pathogens [3, 13]. We will discuss some of the current trends in these areas of investigation.

2.2.1
Search for New Antibiotics from Microbial Sources

The success of a screening program depends on both the kinds of organisms used and the methods for detection of activity. According to an estimate, the choice of strains has a 30–40% influence on outcome; the test procedure has a 60–70% influence. Of the natural antibiotics discovered so far, 67% come from microorganisms and the majority of these were isolated from actinomycetes, some from fungi and other bacteria. Of the vast diversity of microorganisms, only about 1% of the world microorganisms are extensively studied. Therefore, cultivable microorganisms from various niches have great promise to yield novel microbial and other therapeutic properties [10].

Recent advances in technology have sparked a revival in the discovery of natural product antibiotics, at least in academic institutions and certain pharmaceutical companies specially targeting old and new microbial sources such as streptomyces and other actinomycetes, cyanobacteria and uncultured bacteria. Many leading scientific groups believe that there is a vast potential for novel antibiotics from streptomyces only [14, 15]. It has been estimated that only 1–3% of all streptomyces antibiotics have been discovered. To find the remaining 97–99% will require a combination of high-throughput screening and modern technologies. Similarly, other microbes of extreme environments and different ecologies are to be assessed by a holistic approach to screening to obtain novel compounds against MDR bacteria. To access the greater diversity of microbes, there are both conventional culturing and novel culturing methods. Since only a fraction of the microbial population can be cultured by classical methods, efforts to expand the range of bacteria that can be tapped for antibiotic research are being facilitated by several strategies such as expanded conventional culturing approaches, novel culture methods, heterologous DNA-based methods, metagenomics, comparative genomics and combinatorial biosynthesis. A brief introduction of these strategies is given below as described by Clardy *et al.* [7].

2.2.1.1 Conventional Culturing Methods

Many investigators have emphasized the need for exploring more culturable microorganisms in the search of novel antibiotic compounds with improved screening strategies. Recently Baltz [15] has estimated the frequency of antibiotic production by actinomycetes and depicted that 10^4 strains would include 2500 antibiotic producers. It was further estimated that less than one part in 10^{12} of the Earth's surface has been screened for actinomycetes [15]. Enrichment techniques may be utilized for slow-growing actinomycetes. Development of intelligent test systems is the key to the discovery of antibiotics with novel modes of action and to avoid the rediscovery of known compounds. Baltz [16] noted that screening strains of *Escherichia coli* K-12 engineered to harbor 15 antibiotic resistance genes to exclude the common antibiotics produced by actinomycetes can be an effective strategy. Similarly, other potential microorganisms should also be subjected to the screening programs. Microbes such as extremophiles, endophytic and symbiotic bacteria, and those occupying unexplored special niches should be taken into consideration.

2.2.1.2 Novel Culturing Environments

The potential and natural power of microbes to provide novel bioactive compounds may be realized with the fact that the majority of the microbial world is unexplored and only 1% of the bacterial population is culturable in nature, from which we have derived natural products so far. Thus, concerted efforts are needed to develop techniques to increase the culturability of other new bacteria. According to an estimate, approximately one-third of the bacterial divisions have no cultured representatives and are known only through rRNA sequences [17–19]. New strategies to culture the uncultured bacteria have been reported [20]. The concept of the techniques is based on the establishment of a molecular milieu of a multispecies community while a single strain is cultured in a small diffusion chamber located with in this permissive environment. Under these conditions uncultured bacteria form colonies that are unable to form colonies on the artificial media. Similarly, other promising approaches to culture previously uncultured organisms have also been described [21]. It is believed that exploration of specific requirements of noncultivable organisms may lead to the development of suitable methods for the cultivation of microorganism.

2.2.1.3 Heterologous DNA-Based Approaches

Heterologous expression of foreign genes in a host that can be cultured more easily and is amenable to genetic modification is a powerful strategy for the discovery of the small-molecule repertoire of the bacterial world. For a successful outcome in this approach, it requires that biosynthetic, resistance and regulatory genes reside on a contiguous stretch of DNA that has an appropriate metabolic and genetic background. There are good number of examples indicating the success and future potential of this strategy. *E. coli* and *Pseudomonas putida* have been developed as engineered hosts expressing drug candidates and other small molecules [7]. The pantocins, a group of antibiotics produced by *Pantoea agglomerans* and discovered in the early 1980s, are an early example of this approach to identify new antibiotics. The characteristic of these DNA-based studies is the reliance on a phenotypic (antibiotic) assay, in contrast to the target-based assays [22, 23].

2.2.1.4 Metagenomics and Molecular Engineering

As the majorities of microbes are nonculturable, direct isolation of DNA from the environment and use of it in heterologous expression systems is an attractive strategy to explore new antibiotics/small molecules. In the last two decades, the techniques have been successfully developed and utilized for studying microbial diversity based on direct isolation of DNA from soil and further analysis [24]. Many attempts have been made by screening DNA libraries of soil DNA. Early examples have indicated discovery of terragin-A – a metal-chelating compounds which often shows antibiotic activity [25]. Brady and Clardy, [26] using an *E. coli* expression system, tested 7×10^5 metagenomic DNA clones from soil and used *Bacillus subtilis* as an indicator strain, and identified a series of compounds with antibiotic activity (e.g. *N*-acyl tyrosine). Other successful examples are the discovery of indirubin and related small molecules [27], and triarylcations [28] from metagenomics bacterial artificial chromosome libraries in *E. coli*. The impact of genomics and combinatorial synthesis on the discovery and manipulation of natural products has been recently reviewed by many experts, highlighting their power and limitations [29, 30].

Molecular engineering approaches producing new antibiotics have been in development for about 25 years. Advances in cloning and analysis of antibiotic gene clusters, engineering biosynthetic pathways in *E. coli*, transfer of engineered pathways from *E. coli* to *Streptomyces* expression hosts, and stable maintenance and expression of cloned genes have streamlined the process in recent years. Advances in understanding mechanisms and substrate specificities during assembly by polyketide synthase, nonribosomal peptide synthetases, glycosyltransferases and other enzymes have made molecular engineering design and outcome more predictable. Complex molecular scaffolds not amenable to synthesis by medicinal chemistry such as vancomycin, daptomycin and erythromycin are now tractable by molecular engineering [9].

Drugs which contain new antibacterial templates with novel mechanisms of action should have advantages over known antibacterials in the fight against MDR bacteria and the emergence of new pathogens. In a recent review [8], it has been shown that natural product templates of actinonin, pleuromutilin, ramoplanin and tiacumicin B are undergoing clinical evaluation. In addition to these, the new templates present in the recently discovered lead antibacterials arylomycin, GE23077, mannopeptimycin, muraymycin, caprazamycin, nonathiacin and ECO-0501 represent a potential new class of antibacterial agents. Microbial natural products still appear as the most promising source of the future antibiotics that scientific society is expecting. Recent advances in the development of various molecular targets, high-throughput screening strategies, and novel methods of cultivating microbes and modifying their metabolic and molecular targets are promising for antibiotic drug discovery from microbial and other natural products.

2.2.2
Role of Genomics in New Antibacterial Drug Discovery

The impressive progress in genome research in the last decade has influenced biological sciences research including antibiotic research and development. Since 1995, considerable development has been made in deciphering the genome of many bacterial pathogens. At present, the complete genome sequences of more than 200 bacteria are now publicly available, among which many are pathogenic bacteria (www.ncbi.nim.nih/gov/genome/complete.html). The use of genetically modified bacteria as well as holistic approaches such as transcriptome profiling has been proven of asset in validating the mode of action of screening hits from target assays. Recent literature has indicated a large number of potentially essential genes in different pathogenic bacteria including *E. coli* [31], *Haemophilus influenzae* [32], *Helicobacter pylori* [33], *Staphylococcus aureus* [34, 35] and *Streptococcus pneumoniae* [36], mainly by plasmid insertion mutagenesis, conditional mutants and transposon mutagenesis methods [37]. A gene is regarded as being essential when the bacterium cannot survive its genetic inactivation. Transposon mutagenesis, genetic footprinting using diverse hybridization and the polymerase chain reaction technique enabling mapping of insertion sites in genomes are used [31–33, 38–40].

Global expression profiling techniques hold the power to analyze bacterial physiology holistically as never before. Whereas the genome provides the static blueprint of all properties that a cell is able to develop, the transcriptome and proteome are highly dynamic and ideally suited to capture snapshots of bacterial physiology when challenged by environmental changes, including external stress or antibiotic treatment. Antibiotics with different modes of action induced discriminative expression profiles and the signatures obtained with well-known antibiotics are references for comparisons with the profile induced by a novel antibacterial agent of interest. Based on bacterial functional genomics, a number of bacterial targets have been identified in important Gram positive as well as negative pathogens. These targets belong to functional category of gene products such as translation followed by replication, cell wall biosynthesis and fatty acid biosynthesis. Several pharmaceutical companies have started genome-sequencing programs, and have embraced genomic information as the basis for a rational and target-directed antibacterial drug discovery process. Functional genomics techniques are very attractive and are implemented in the modern antibacterial drug discovery process. Knockout analysis and mutation studies aid in the selection and validation of potential novel targets by proving their essentiality for bacterial survival [37].

Functional genomics techniques are now in place at various stages of the early drug discovery process, and have been proven highly successful for *in vitro* target validation and determination of the mode of action of novel antibacterial agents. However, antibacterial drug discovery from functional genomics, transcriptome and proteome expression profiling is still in its infancy [41–43].

Readers are advised to see the excellent articles published recently on genomics, functional genomics, and their application in novel drug target identification, validation and drug discovery [37, 44–47].

It has been proposed by leading scientists in the area that there are many constraints and problems with genome-based drug discovery programs for antibacterials, and they suggested that application of functional genomics has enormous potential and also supports classical screening strategies. Many scientists believe that a strategy combining the traditional and genomic approaches, together with the battery of available novel techniques, would result in the discovery of the new classes of antibacterial that are needed to combat MDR bacteria from natural product [48].

2.2.3
Antimicrobial Peptides as New Anti-Infective Drugs

An alternative to the synthesis of NCEs is their identification by screening natural sources such as soil bacteria and extracts from plants and fungi. One boon from this screening process is the identification of natural antimicrobial peptides and a wide variety of organisms from both the plant and animal kingdoms. Insects, frogs and humans all secrete peptides at their mucosal and epithelial surfaces. These include magainin, cercopins, attacins and defensins [49]. Mammalian defensins are small cationic peptides that are thought to work by inducing pores in the bacterial cell wall [50]. Another important peptide is nisin, produced by *Lactococcus lactis*. It is thought that nisin, like vancomycin, binds to lipid II, a component of many bacterial cell membranes, and permeabilizes the membrane, causing dissipation of ion gradients and eventual cell death [13]. Recently excellent review articles have been published on peptide antibiotics and their therapeutic potential [9]. It has been reported that short cationic amphibolic peptides, with antimicrobial and/or immunomodulatory activities, provide a template for two separate classes of antimicrobial drugs: (i) direct-acting antimicrobials and (ii) indirect-acting therapeutic drugs. The role of cationic host defense peptides in modulating the innate responses and boosting infection resolving immunity while dampening potentially harmful proinflammatory responses gives their peptides the potential to become an entirely new therapeutic approach against bacterial infections [51].

2.3
Combination Drug Therapy

Interactions between antibacterial drugs of different classes *in vitro* have been investigated for a long time. The nature of interactions may be synergistic, antagonistic or neutral. Combination drug therapy has been recognized and used to combat drug-resistant pathogens such as β-lactamase-producing bacteria as much as against *M. tuberculosis*. Principle modes of combinations are described as follows [52].

(i) A second compound (an adjuvant) prevents the degradation or modification of the primary drug (an antibiotic) (e.g. protection of β-lactam antibiotic against β-lactamases by clavulanate, β-lactamase inhibitors).
(ii) A second compound (an adjuvant) allows the accumulation and retention of the primary drug (an antibiotic) by inhibiting efflux pumps.
(iii) A second compound (an adjuvant) inhibits the intrinsic repair pathway or tolerance mechanisms of cells to the primary drug (an antibiotic).
(iv) A second compound is itself an antibiotic that targets a similar or different pathway that is inhibited by the first drug.

The importance of synergistic effects on the efficacy of drug combinations has great potential to slow down the emergence of resistance. It is believed that drug combinations acting on multiple targets in pathogens would delay and decrease the ability of pathogens to accumulate simultaneous mutations that affect multiple targets [53–55]. This concept has been well documented for tuberculosis and leprosy [56, 57]. In 1986, the World Health Organization recommended multiple drug therapy to overcome resistance to *M. tuberculosis*.

A number of combination therapies have achieved commercial success in the treatment of bacterial infections. For example, a standard course of treatment for tuberculosis is a combination of isoniazid, rifampicin, pyrazinamide and ethambatol or streptomycin for 2 months (http://www.cdc.gov). The modifications of therapy should ideally be based on susceptibility testing of drugs and the susceptibility pattern from the isolates will guide the therapeutic design. Details of various drug combinations for the treatment of tuberculosis are illustrated in Chapter 3 of this book. Cottarel and Wiezbowski [52] suggested that the efficacy of combination therapies is partially dependent on the pathogen and type of infection.

Considering the progress made so far in synergistic studies, it is believed that there is an attractive opportunity to develop novel types of combination antibacterial therapies by combining known antibiotics. Some companies such as Combinato Rx (Boston, Massachusetts, USA) [58] and Biovertis (Vienna, Austria) initiated clinical development programs for dual-mode-of-action compounds.

MPex Pharmaceutical (San Diego, California, USA) is developing a compound that can interfere with the efflux pumps. In a Biovertis type of approach, a berberin is combined with MDR pump inhibitors [52].

Many other pharmaceuticals companies, both in developed and developing countries, are looking for effective combinational approaches to combat the problem of drug resistance.

Combination therapy is gaining momentum as systematic efforts to generate combination therapy products are increasing. It is well established for viral infection, tuberculosis and, to some extent, against fungal infection [59, 60]. It is expected that with the application of functional genomics, various novel targets will be identified, and by combination therapy these targets may be more effectively validated and used for disease control.

2.4
Strategies to Target Mechanisms of Drug Resistance

Various theoretical and some practical approaches have been described by different workers to target mechanisms of drug resistance. Due to the presence of a variety of mechanisms of antibiotic resistance among bacteria, it is difficult to overcome the long-term battle against drug resistance. However, the problem can be effectively managed by adapting multidimensional strategies. Various strategies to combat the drug resistance problem have been suggested by several workers [3, 13]. We described here some novel and alternative approaches for the management of drug resistance.

2.4.1
Bacterial Efflux Systems and Efflux Pump Inhibition

In order to limit the intracellular concentration of a toxic compound, bacteria can diminish its entry by developing a barrier of low permeability. This nonspecific phenomenon is intrinsic to Gram-negative bacteria and mycobacteria. A decrease in the intracellular concentration of an antibiotic could also be due to its extrusion outside the cell via an energy-dependent process called active efflux, mediated by efflux pumps. Such pumps are ubiquitous proteins, localized in the cytoplasmic membrane of all kinds of cells, from bacteria to eukaryotes.

Efflux (pumping of antimicrobials out of the bacterial cell) is a well-established mechanism of antimicrobial resistance in disease causing bacteria. Bacterial antimicrobial efflux systems fall into five classes: (i) the major facilitator (MF) superfamily (ii) , the ATP-binding cassette (ABC) family, (iii) the resistance nodulation division (RND) family, (iv) the small multidrug resistance family, and (v) the multidrug and toxic compound extrusion family [61, 62]. Antibiotic efflux systems can be drug or class specific, as for the MF family, Tet and Mef exporters of tetracyclins and macrolides, respectively, that are encoded by plasmids and other mobile genetic elements in both Gram-negative and Gram-positive organisms [63] or capable of accommodating a range of chemically distinct antimicrobials, as for the chromosomally encoded NorA like MF multidrug transporters prevalent in Gram-positive bacteria or RND multidrug transporters of Gram-negative bacteria [64]. These MF and RND transporters capture and expel a wide variety of structurally and functionally unrelated antibiotics of clinical significance, and so are important determinants of intrinsic and acquired multiple antibiotic resistance in a variety of Gram-negative pathogens [61]. Indeed, many classes of currently available antibiotics (e.g. macrolides, ketolides, oxazolidinones, streptogramines, lipopeptides, rifamycins and glycopeptides) do not have any clinically useful activity against Gram-negative bacteria because of the activity of RND pumps that are broadly distributed in these organisms. This is also an issue for drugs in clinical development and compounds at various stages of preclinical research [65].

To tackle bacterial resistance, it would be valuable to use inhibitors of resistance mechanisms, able to potentiate the activity of existing antibiotics. In this approach, the antibiotic is coadministered with an inhibitor that neutralizes the resistance and consequently the antibiotics are still useful even in resistant organisms. This strategy can be used when resistance involves antibiotic-inactivating enzymes and it has been validated with the successful use of β-lactamase inhibitors. A similar approach can be used for target-modifying enzymes and for efflux systems [66–68].

Efflux pumps can be considered as potentially effective antibacterial targets, and the identification and development of safe and effective bacterial efflux pump inhibitors is needed. Such molecules are expected to: (i) decrease the intrinsic resistance of bacteria to antibiotics, (ii) reverse acquired resistance even in highly resistant strains with multiple target mutations and (iii) reduce the frequency of emergence of resistant mutant strains [69]. Efflux inhibition is considered as a viable approach to deal with efflux-mediated resistance. The use of inhibitors with antimicrobials results in restoring of their activity despite of there being substrates for efflux. However, precise tailoring of pharmacokinetics of both inhibitor and antibiotic is difficult in such combination therapy.

A number of approaches, realized as well as hypothesized, for avoiding or inhibiting efflux have been discussed and reviewed in recent years [64, 70]. Poole and Lomovskaya classified various strategies to counter efflux-mediated resistance which include (i) bypass efflux or avoiding efflux, (ii) efflux inhibition by substrate-binding and site-specific inhibitors, (iii) efflux inhibition pump modulators (http://www.impexpharma.com), (iv) uncouple substrate, proton transport, block outer membrane factor components, (v) inhibition of pump assembly and (vi) inhibition of efflux gene expression [71]. New fluoroquinolones (FQs) such as garenoxacin, moxiflokacin and WCK771 are effective in treating *S. aureus* infections, partly due to the fact that these FQs are less accommodated by the membrane factor family NorA pumps than were older agents (norfloxacin and ciprofloxacin). A novel desFQ, DX-619 has been described that appears not to be impacted by any hitherto known efflux mechanism in *S. aureus* [72]. Other antibacterial agents such as the ketoliole subclass of macrolides (telithromycin and cethromycin) are also emerging as effective alternatives to macrolides and in treating streptococci with the MefA efflux mechanism [72–75]. It is believed that an approach based on developing agents which are less affected by efflux has merit.

Recent progress in efflux pump inhibitor research has yielded some effective compounds such as MC-2077,110 against *P. aeruginosa* RND pumps. Similarly, inhibitors like MC-002595 and MC-004124 [76, 77] have been developed. The edge of inhibitors targeting binding systems is expected because of the fact that it would be difficult to develop resistance to such compounds.

In addition to the rational design of efflux pump inhibitors, known efflux inhibitors are tested against different efflux systems. For example, a plant alkaloid, reserpine, potentiates the activity of FQs on MDR Gram-positive bacteria [78]. This compound is now also known to inhibit other systems in *L. lactis*. Similarly, the phenols epicatechin gallate and epigallocatechin gallate present in green tea inhibit P-glycoprotein [79–81] and cause Tet[r] reversal in staphylococci strains overexpressing the TetB or the LetK efflux pumps [82]. They are also known to potentiate the activity of norflox against NorA-overexpressing *S. aureus* strains. Natural products have been identified as inhibitors of the *S. aureus* NorA efflux pump, such as the prophyrin pheophorbide A and the flavonoligan 5′-methoxy hydrocarpin isolated from *Berberis* plants [83, 84]. Many other plant metabolites have also been described as potential inhibitors, as reviewed recently by Aqil *et al.* [3].

On the other hand, efflux inhibition in *P. aeruginosa* (RND family) will not be effective in eliminating β-lacatmases, which are more important determinants of resistance to these agents. Another difficulty may arise in the strains which possesses multiple RND systems able to confer resistance to a given antibiotic (e.g. *P. aeruginosa*) [61]. It is suggested by many workers that only broad-spectrum inhibitors may be able to interact with multiple transporters and can be considered for combination therapy with FQs. Inhibitors targeting RND pump functions might act at the substrate (i.e. antibiotic) binding sites and so act as competitive inhibitors of antimicrobial efflux.

Many questions and problems are to be addressed and resolved to make efflux inhibitors clinically effective. There is also a concern about the selectivity of some efflux pump inhibitors that have been shown to inhibit both eukaryotic and bacterial efflux systems. Nevertheless, an enhanced understanding of the molecular details of the efflux pump function and complex assembly and structure will be needed for discovering broad-spectrum efflux pump inhibitors.

Numerous studies have confirmed the feasibility of inhibiting efflux pump activity and so enhancing the drug susceptibility of efflux pump-mediated resistance. Various efflux pump inhibitors have been identified both from synthetic and natural sources; however, to date, besides their *in vivo* activity, none of them is in clinical use [64, 69].

2.5
Quorum Sensing Inhibition: A Novel Anti-Infective Drug Target

Quorum sensing (QS) is a means for bacterial communities to rapidly and coordinately change their genome expression pattern in response to environmental cues and population density. Bacteria that use QS produce and secrete certain signaling compounds called autoinducers, normally *N*-acyl homoserine lactones (AHLs). The bacteria also have receptors that can specifically detect the inducers. When an inducer binds to a receptor, it activates the transcription of certain genes, including those for autoinducer synthesis. When only a few other bacteria of the same kind are in the vicinity, diffusion reduces the concentration of the inducer in the surrounding medium to almost zero. Thus, the bacteria produce little inducer.

When a large number of bacteria of the same kind are in the vicinity, the concentration of inducer passes a threshold, whereupon more inducer is synthesized. This forms a positive feedback loop and the receptor becomes fully activated. Bacteria produce structurally diverse signal QS molecules. The best-characterized QS molecules in Gram-negative bacteria are AHLs and in Gram-positive bacteria are short peptides. The most-studied QS system is the LuxR–LuxI homologous systems and the cognate signal molecule, AHL. Similarly, many Gram-negative bacteria including *Bacillus, Streptococcus pneumoniae* and *S. aureus* use small peptides or modified peptides for signaling. Autoinducer-2 (AI-2) – a signaling molecule – is common to many bacteria [85].

Many bacterial traits such as production of virulence factors, secondary metabolites, biofilm formation, surface motility as well as gene transfer and so on are regulated by QS in several pathogenic and opportunistic bacteria [86–89].

QS signaling systems of bacterial pathogens are central regulators for the expression of virulence factors and represent highly attractive targets for the development of novel anti-infective therapeutics. The signal molecules and signaling systems in many pathogenic bacteria have been explored in last two decades [89].

The use of signal molecule-based drugs to attenuate bacterial pathogenicity rather than bacterial growth is attractive for several reasons. The compounds, both synthetic and natural, capable of interfering with QS systems have been termed antipathogenic drugs [88, 90]. Various types of QS systems have been explored and widely studied in Gram-negative bacteria. The concept of QS was first elucidated in the luminescent marine bacterium *Vibrio fischeri* [91]. Extensive work in this area by several groups, especially in the United States and Europe, has elucidated the QS system in many pathogenic bacteria such as *Pseudomonas aeruginosa, Burkholderia cepacia, Salmonella typhimurium, E. coli, Vibrio cholerae* and *Yersinia enterocolitica*. *P. aeruginosa* is one of the most-studied QS organisms because of its role as an emerging opportunistic human pathogen. Chronic respiratory infection by *P. aeruginosa* is a source of morbidity and mortality in nearly all individuals with cystic fibrosis [92]. Excellent review articles are available on QS organisms and QS is not the central focus of this chapter.

QS-inhibitory compounds might constitute a new generation of antimicrobial agents. In recent years, a number of biotechnology companies that aim to specifically develop anti-QS and anti-biofilm drugs have emerged, such as QSI (Lyngby, Denmark), Quorex Pharmaceuticals (Carlsbad, California, USA), Microbia (Cambridge, Massachusetts, USA) and SC (Martinsried, Germany) [93]. Several strategies aiming at the interruption of bacterial QS circuits are possible including:

(i) Inhibition of signal generation.
(ii) Inhibition of signal (AHL) dissemination and inhibition of AHL signal reception.

Other strategies include QS inhibition by higher organisms including phyto-compounds [90, 93].

Disruption of the signaling systems could check the bacteria from colonizing surfaces and causing disease. Such an inhibition acts through the specific interference with a signaling system, without affecting growth or killing the cell and hence without leading to the development of resistance, and it is currently the most sought requirement for new antimicrobial technologies. Using specific biosensor strains, anti-QS activity in plants, microbes and synthetic compounds could be tested and evaluated as described by many workers [94–97]. We have proposed a screening strategy for natural and microbial products screening for detecting anti-QS activity using suitable biosensor strains as used by Mclean *et al.* [94] (Figure 2.1).

2.5.1
Anti-QS Compound Effects *In Vitro*

Anti-QS compounds could be of great interest in the treatment of bacterial infections. The varying length of the acyl side-chain was found to be important (e.g. AHL with extended side-chains generally caused inhibition of LuxR homologs). Production of halogenated furanone compounds by Australian algae *Delisa pulchra* [98] prevents extensive surface growth by bacteria and higher fouling organisms. Furanones were reported to interact with bacterial AHL receptors and to promote their rate of proteolytic degradation [99], in contrast to the stabilizing effect of AHL signal binding. Halogenated furanones have been shown to inhibit several QS-controlled phenotypes including swarming, toxin production of *Vibrio harveyi* and bioluminescence of *Vibrio fischeri* [90].

AHL-degrading enzyme AiiA has been found in several *Bacillus* spp. including *B. thurengienesis*, *B. cereus* and *B. mycoides* isolated from the environment. This enzyme catalyzes the lactonolysis reaction and hence lowers the amount of bioactive AHL [100]. Furanone compounds and closely relative compounds downregulate the production and secretion of several virulence factors such as protease, chitinase and pyoverdin in *P. aeruginosa*. Biofilms developed in presence of furanone compounds become more susceptible to treatment with antibiotics and disinfectants [88, 101]. Production of enzymes that are able to degrade AHLs is not limited to *Bacillus*. Homologs of AiiA able to degrade AHLs have been identified in *Agrobacterium tumefaciens*, *Arthrobacter* sp. *K. pneumoniae*, *Comomonas* sp. and *Rhodococcus* sp. [102].

Park *et al.* [103] investigated that strain M664, identified as a *Streptomyces* sp., secretes an AHL degrading enzyme, AhlM reduced the concentration of AHL and decreased the production of virulence factors, including elastase, total protease and LasA in *P. aeruginosa*.

(5*Z*)-4-Bromo-5-(bromomethylene)-3-butylfuran-2(5*H*)-one (furanone) from red marine algae *Deliesea pulchra* was previously found to inhibit the growth, swarming and biofilm formation of Gram-positive bacteria. The QS disrupter furanone of *D. pulchra* is found to inhibit QS in *E. coli* via AI-2. Furanone appeared to alter AI-2 signaling posttranscriptionally [104].

Using the QS selector screen, Rasmussen *et al.* [95] identified a number of compounds that are able to block both LuxR- and LasR-based QS, which include P-Benzoquinone, 2,4,5-Tri-bromo-imidazole, Indole, 3-Nitro-benzen-sulfonamide and 4-Nitro-pyridine-*N*-oxide.

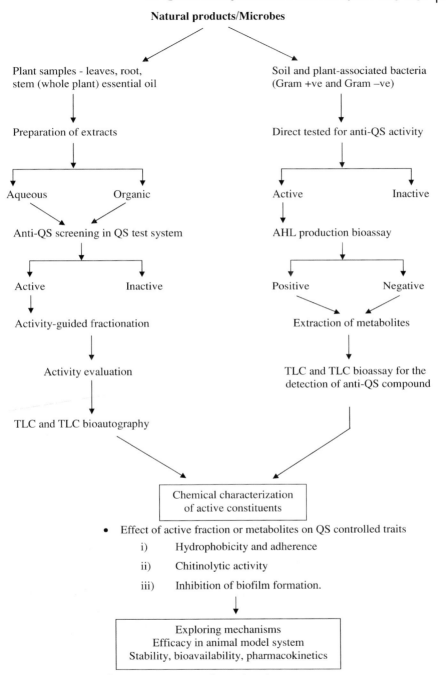

Figure 2.1 Flow chart for anti-QS screening of natural products.

Plants such as carrot, garlic, habanera and water lily produce compounds that interfere with bacterial QS [95]. Propolis produced by bees also contains QS-inhibitory activity. Garlic extracts containing at least three different compounds; pea seedling, crown vetch, soybean, rice and tomato have also been found to produce compounds able to mimic the activities of AHL and interfere with bacterial QS [105, 106]. Methanol extract of pea seedlings inhibited pigment production, extracellular protease activity and exochitinase activity in *Chromobacterium violaceum*. A 3-oxo-C_{12} homoserine lactone antagonist with a small difference in affinity for the receptor protein and ability to mediate degradation of the receptor would interfere efficiently with the QS system in the *P. aeruginosa* [107].

Qazi *et al.* [108] have showed that 3-oxo-C_{12} homoserine lactone inhibits the production of exotoxins as well as antagonizes QS at sub growth inhibitor concentration, but enhances protein A expression. The PA2385 gene in PAOl encodes an acylase. This protein has *in situ* quorum quenching activity which enables *P. aeruginosa* PAOl to modulate its own QS-dependent pathogenic potential (i.e. reduces production of virulence factor elastase and pyocyanin) [109].

Adonizio *et al.* [97] have screened 50 medicinal plants from Southern Florida. Of these, six plants showed QS-inhibitory activity: *Conocarpus erectus* L. (Combretaceae), *Chamaecyce hypericifolia* (L.) Mill spp. (Euphorbiaceae), *Callistemon virminalic* (Myrtaceae) *Bucida burceras* L. (Combretaceae), *Tetrazygia bicolor* (Melastomataceae) and *Quercus virginiana* Mill (Fagaceae). Choo *et al.* [110] also screened vanilla extract having QS inhibiting activity. Recently, Vattem *et al.* [111] tested several dietary phytochemicals for their anti-QS property as well as their effect on the swarming motility of *E. coli* and *P. aeruginosa*. Most of the tested herbs, spices and fruits showed varying levels of activities at sub minimum inhibitory concentrations.

We have made the first attempt in India to screen Ayurvedic medicinal plants, essential oils and certain food plants for their anti-QS activity, and reported on many plants. The active constituents are under characterization. Similarly, anti-QS activity has also been detected in seedling extracts of pea (Figure 2.2) and AHL degradation of *Bacillus* isolates [112, 113].

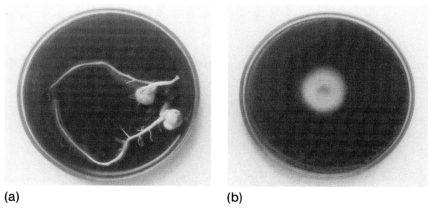

(a)   **(b)**

Figure 2.2 QS inhibition by (a) pea seedlings and (b) pea root exudates.

2.5.2
Efficacy of QS-Inhibitory Compounds in Animal Infection Models

Numerous compounds are able to interfere with the QS system in *in vitro* test systems. However, only a few have been tested in animal model systems, as recently described by Rassmusen and Givskov [90]. In some cases, QS-inhibitory compounds provide protection to infected animals and the pathogenicity of bacterium is affected. Whether QS-inhibitory compounds offer new hope in the continuing battle against MDR bacteria is not yet fully explored, but the results are promising. QS-inhibitory compounds inhibiting biofilm formation seems to be an attractive action. It is expected that future bacteria eradication or multidrug resistance control will depend on the synergistic action of multiple bioactive components.

2.6
Phage Therapy: An Alternative Therapy

Almost after 60 years of the introduction of antibiotics in chemotherapy, interest is reemerging in alternative ways of controlling bacterial infections due to the emergence of MDR pathogenic bacteria. Now one approach is reinventing phage therapy to control bacterial pathogens that have emerged as resistant to some, and in certain cases all, clinically approved antibiotics.

Lytic bacteriophages are bacterial viruses that were discovered independently by Fredrick Tworts and Felix d'Herelle in 1915 and 1917, respectively. It was immediately apparent that their potent bactericidal capacity should be harnessed to treat human infections. The industrialization of antibiotics in 1940s, however, changed the focus of anti-infective research and development in the west away from phage treatments. Now interest is again turning towards phages as adjunct therapies in controlling bacterial pathogens, specifically those that have become MDR [114].

A recent report by a special task force cochaired by the Centers for Disease Control, FDA and National Institutes of Health stated that 'The world may soon be faced with previously treatable diseases that have again become untreatable as in the pre antibiotic era' [115]. Lytic bacteriophages are very effective in lysing the bacteria that serve as their specific hosts. The mechanism of killing bacteria is different from those of antibiotics and fits nicely in the novel mode of action concept desired for all new antibacterial agents.

There are two possible ways of utilizing phages:

(i) Some research groups continue to develop whole phages as alternative treatments.
(ii) The isolation and optimization of purified phage components as antibacterials.

This has led to the novel opportunities in fighting against interactive infection.

Recently, an article published by Fischetti *et al.* [116] has highlighted the importance and limitations of reinventing phage therapy.

Tan *et al.* [13] have assessed the various strategies including phage therapy in the search for ways to cope with antibiotic resistance. The use of bacteriophage to fight MDR bacteria seems promising. Groups from Poland, Britain, the Soviet Union and the United States have published on the use of bacteriophages as an alternative to antibiotic therapy. The Polish and Soviet Union studies were based on subjective measurement of the clinical response of patients and demonstrated efficacy against many antibiotic-resistant pathogens, and could be used to treat mucosal, systemic and cutaneous infection in adults and children. On the other hand, the British study provides an objective assessment of the efficacy of bacteriophage therapy [117].

Therapeutic phages are not likely to be another magic bullet that totally solves the problem of antibiotic resistance. Phage-based therapeutics has its own specific limitations. There is ample opportunity of developing resistance against phages by bacterial pathogens. There are several other problems associated with developing phage therapy.

However, more concerted research, proper funding, evaluation and regulation strategies are needed. In 2005, an Editorial in *Drug discovery Today* (www.drugdiscoverytoday.com) focused on phage therapy as an attractive option for dealing with antibiotic-resistant bacterial infection. More recently, Fischetti *et al.* [117] reviewed phage therapy and indicated that although whole phages continue to generate interest as alternatives to antibiotics, the focus is shifting to the use of purified phage components as antibacterial agents.

2.7
Antibiotic Use Policy: Practical Management of the Problem of Antibiotic Resistance

Antibiotic resistance is at epidemic levels worldwide in both hospitals and the community. The reason for this epidemic is the inappropriate and excessive use of antibiotics. Other factors that contribute to this problem includes cross-infection in hospitals, rapid worldwide transport, immunosuppressants, increasingly invasive diagnostics and so on [118]. To control and minimize the resistance epidemics, multidimensional approaches are needed. One of the approaches is to frame antibiotic use policy based at the local, regional, national and international level, and should be implemented both in government and private hospitals as well as in the community.

Increasing antibiotic resistance is a global health problem. Guidance for antibiotic prescribing is a crucial part of any strategy to limit the development of resistance [119]. A resistance problem is present in both hospital and ambulatory care, and in these settings high usage of antibiotic is inappropriate. Therefore, an antibiotic policy should be a strategic goal in hospitals and in the community [118]. An antibiotic policy should review general issues regarding the use of antimicrobials in surgery, rather than deal with particular points of prophylaxis or treatment of specific surgical-site infections. The ever-increasing opportunities for the use of antimicrobials in patients with surgical infection pose many serious problems. Perhaps the most urgent is the excessive and inappropriate use of antibiotics, which may lead to the emergence of antimicrobial resistance.

Antibiotics should be used rationally in order to improve patient outcome, to contain the cost of treatment and, most importantly, to prevent the emergence of antibiotic resistance. Unfortunately, inappropriate use of antibiotics appears to be a universal phenomenon and numerous surveys from all around the world have shown high rates of inappropriate prescribing. To improve antibiotic usage, many hospitals have put in place antibiotic policies. Every institution should have an antibiotic policy that is an integral part of the hospital infection control program. The policy must be multifaceted, involving all levels of staff. Formulation of the policy should involve all stakeholders to ensure ownership and, most importantly, the policy must have the full support of the authorities. An antibiotic policy is more than just a set of guidelines. It includes education and feedback, strategies to improve prescribing, dealing with forces that influence prescribing, audit and research. Education and feedback are perhaps the most important elements of an antibiotic policy [120].

The guidelines should take into account local patterns of infections and resistance with an emphasis on major areas of deficiencies. Other factors that should be considered in formulating such guidelines include constraints in supporting facilities, the availability of the specific antibiotics and financial considerations. The microbiology laboratory plays an important role in the successful implementation of an antibiotics policy. Antibiotic resistance surveillance data should be disseminated on a regular and timely basis to those who need to know. Antibiotic cycling is a strategy that has received interest in recent times. It involves a scheduled rotation of one or different classes of antimicrobial agents with comparable spectra of activity in order to mitigate the effects of selective pressure [120–122].

2.8
Conclusions

The development of antibiotic resistance in bacteria is a never-ending process. Therefore, management of the emergence and spread of multidrug resistance in pathogenic bacteria requires coordinated attacks on many fronts. The problem of antibiotic resistance is complex, and thus needs a multidimensional and holistic approach of management.

Development of new antibiotics with novel modes of action from natural products including tapping unexplored microbial diversity, and other sources must continue in order to obtain new antibiotics against bacteria. However, antibiotic discovery strategies must also consider the antibiotic mechanisms of drug resistance of bacteria in order to develop new antibiotics against drug-resistant pathogens. Recent developments in high-throughput screening systems and exploration of novel molecular drug targets based on genomics and proteomics will definitely be of help to discover new antibiotics. On the other hand, the development of antipathogenic drugs based on molecular targets is an attractive strategy. The extended form of this strategy should also include the development of suitable vaccines. The utility of alternative strategies such as the use of probiotics, phage therapy and immunomodulators should be explored in the management of drug-resistant pathogens. The combinational approach seems to be the most effective approach at present to combat the current problem of drug resistance. On the other hand, antibiotic overuse must be stopped. Antibiotic use policy must be framed appropriately, and should be strictly implemented both in the hospital and community settings. Governments should encourage and provide generous funding to academic institutions to develop both new drug development as well as other aspects related to the development of alternative strategies to control bacterial infections and antibiotic use policy.

Acknowledgments

We are grateful to the University Grant Commission, New Delhi for financial assistance in the form of Major Research Project 33-208/2007(SR) on 'Quorum sensing inhibition by natural products', and also to Professor Robert J.C. McLean (Texas State University) for his kind cooperation and guidance.

References

1 Pinner, R.W., Teutsch, S.M., Simonsen, L., Klug, L.A., Graber, J.M., Clarke, M.J. and Berkelman, R.L. (1996) *Journal of the American Medical Association*, **275**, 189–193.

2 WHO (2002) *Death by Cause, Sex and Mortality Stratum in WHO Regions. Estimates for 2001. World Health Report 2002*, WHO, Geneva.

3 Aqil, F., Ahmad, I. and Owais, M. (2006), Targeted screening of bioactive plant extracts and phytocompounds against problematic groups of multidrug resistant bacteria in *Modern Phytomedicine: Turning Medicinal Plants into Drugs* (eds I. Ahmad, F. Aqil and M. Owais), Wiley-VCH Verlag GmbH, Weinheim, pp. 173–197.

4 Spellberg, B., Powers, J.H., Brass, E.P., Miller, L.G. and Edwards, J. Jr. (2004) *Clinical Infectious Diseases* **38**, 1279–1286.

5 Katz, M.L., Mueller, L.V., Polyakov, M. and Weinstock, S.F. (2006) *Nature Biotechnology*, **24**, 1529–1531.

6 http://csdd.tufts.edu/NewsEvents/RecentNews.asp?newsid=6 (2001).

7 Clardy, J., Fischbach, M.A. and Walsh, C.T. (2006) *Nature Biotechnology*, **24**, 1541–1550.

8 Butler, M.S. and Buss, A.D. (2006) *Biochemical Pharmacology*, **71**, 919–929.

9 Baltz, R.H. (2006) *Nature Biotechnology*, **24**, 1533–1540.

10 Crueger, W. and Crueger, A. (1990) *Biotechnology: A Textbook of Industrial Microbiology*, 2nd edn (English edition by T.D. Brock), Science Tech, Madison, WI.

11 Lloyd, A.W. (1998) *Drug Discovery Today*, **3**, 480.

12 Lloyd, A.W. (1998) *Drug Discovery Today*, **5**, 122.

13 Tan, Y.T., Tillet, D.J. and McKay, I.A. (2000) *Molecular Medicine Today*, **6**, 309–314.

14 Watve, M.G., Tickoo, R., Jogg, M.M. and Bhole, B.D. (2001) *Archives of Microbiology*, **176**, 386–390.

15 Baltz, R.H. (2005) *SIM News*, **55**, 186–196.

16 Baltz, R.H. (2006) *Journal of Industrial Microbiology and Biotechnology*, **33**, 507–513.

17 Torsvik, V., Goksoyr, J. and Daae, F.L. (1990) *Applied and Environmental Microbiology*, **56**, 782–787.

18 Hugenholtz, P., Goebel, B.M. and Pace, N.R. (1998) *Journal of Bacteriology*, **180**, 4765–4774.

19 Pace, N.R. (1997) *Science,* **276**, 734–740.

20 Kaeberlein, T., Lewis, K. and Epstein, S.S. (2002) *Science,* **296**, 1127–1129.

21 Joseph, S.J., Hugenholtz, P., Sangwan, P., Osborne, C.A. and Janssen, P.H. (2003) *Applied and Environmental Microbiology,* **69**, 7210–7215.

22 Beer, S.V. and Rundle, J.R. (1983) *Phytopathology,* **73**, 1346.

23 Wright, S.A., Zumoff, C.H., Schneider, L. and Beer, S.V. (2001) *Applied and Environmental Microbiology,* **67**, 284–292.

24 Saano, A. and Lindström, K. (1995), Isolation and identification of DNA from soil in *Methods in Applied Soil Microbiology and Biochemistry* (eds A. Kassem and N. Paolo), Academic Press, San Diego, CA, pp. 440–451.

25 Wang, G.Y.S., Graziani, E., Waters, B., Pan, W., Li, X., McDermott, J., Meurer, G., Saxena, G., Andersen, R.J. and Davies, J. (2000) *Organic Letters,* **2**, 2401–2404.

26 Brady, S. and Clardy, J. (2000) *Journal of the American Chemical Society,* **122**, 12903–12904.

27 MacNeil, I.A., Tiong, C.L., Minor, C., August, P.R., Grossman, T.H., Loiacono, K.A., Lynch, B.A., Phillips, T., Narula, S., Sundaramoorthi, R., Tyler, A., Aldredge, T., Long, H., Gilman, M., Holt, D. and Osburne, M.S. (2001) *Journal of Molecular Microbiology and Biotechnology,* **3**, 301–308.

28 Gillespie, D.E., Brady, S.F., Bettermann, A.D., Cianciotto, N.P., Liles, M.R., Rondon, M.R., Clardy, J., Goodman, R.M. and Handelsman, J. (2002) *Applied and Environmental Microbiology,* **68**, 4301–4306.

29 Bode, H.B. and Muller, R. (2005) *Angewandte Chemie (International Edition in English),* **44**, 6828–6846.

30 Menzella, H.G., Reid, R., Carney, J.R., Chandran, S.S., Reisinger, S.J., Patel, K.G., Hopwood, D.A. and Santi, D.V. (2005) *Nature Biotechnology,* **23**, 1171–1176.

31 Gerdes, S.Y., Scholle, M.D., Campbell, J.W., Balázsi, G., Ravasz, E., Daugherty, M.D., Somera, A.L., Kyrpides, N.C., Anderson, I., Gelfand, M.S., Bhattacharya, A., Kapatral, V., D'Souza, M., Baev, M.V., Grechkin, Y., Mseeh, F., Fonstein, M.Y., Overbeek, R., Barabási, A.L., Oltvai, Z.N. and Osterman, A.L. (2003) *Journal of Bacteriology,* **185**, 5673–5684.

32 Akerley, B.J., Rubin, E.J., Novick, V.L., Amaya, K., Judson, N. and Mekalanos, J.J. (2002) *Proceedings of the National Academy of Sciences of the United States of America* **99**, 966–971.

33 Salama, N.R., Shepherd, B. and Falkow, S. (2004) *Journal of Bacteriology,* **186**, 7926–7935.

34 Forsyth, R.A., Haselbeck, R.J., Ohlsen, K.L., Yamamoto, R.T., Xu, H., Trawick, J.D., Wall, D., Wang, L., Brown-Driver, V., Froelich, J.M., C, K.G., King, P., McCarthy, M., Malone, C., Misiner, B., Robbins, D., Tan, Z., Zhu, Z.Y., Carr, G., Mosca, D.A., Zamudio, C., Foulkes, J.G. and Zyskind, J.W. (2002) *Molecular Microbiology,* **43**, 1387–1400.

35 Ji, Y., Zhang, B., Van Horn, S.F., Warren, P., Woodnutt, G., Burnham, M.K.R. and Rosenberg, M. (2001) *Science,* **293**, 2266–2269.

36 Thanassi, J.A., Hartman-Neumann, S.L., Dougherty, T.J., Dougherty, B.A. and Pucci, M.J. (2002) *Nucleic Acids Research,* **30**, 3152–3162.

37 Freiberg, C. and Oesterhelt, H.B. (2005) *Drug Discovery Today: Targets,* **10**, 927–935.

38 Hutchison, C.A. III, Peterson, S.N., Gill, S.R., Cline, R.T., White, O., Fraser, C.M., Smith, H.O. and Venter, J.C. (1999) *Science,* **286**, 2165–2169.

39 Hare, R.S., Walker, S.S., Dorman, T.E., Greene, J.R., Guzman, L.M., Kenney, T.J., Sulavik, M.C., Baradaran, K., Houseweart, C., Yu, H., Foldes, Z., Motzer, A., Walbridge, M., Shimer, G.H. Jr. and Shaw, K.J. (2001) *Journal of Bacteriology,* **183**, 1694–1706.

40 Kang, Y., Durfee, T., Glasner, J.D., Qiu, Y., Frisch, D., Winterberg, K.M. and Blattner, F.R. (2004) *Journal of Bacteriology,* **186**, 4921–4930.

41 Fischer, H.P., Brunner, N.A., Wieland, B., Paquette, J., Macko, L., Ziegelbauer, K. and Freiberg, C. (2004) *Genome Research*, **14**, 90–98.

42 Bandow, J.E., Brötz, H., Leichert, L.I.O., Labischinski, H. and Hecker, M. (2003) *Antimicrobial Agents and Chemotherapy*, **47**, 948–955.

43 Hutter, B., Schaab, C., Albrecht, S., Borgmann, M., Brunner, N.A., Freiberg, C., Ziegelbauer, K., Rock, C.O., Ivanov, I. and Loferer, H. (2004) *Antimicrobial Agents and Chemotherapy*, **48**, 2838–2844.

44 Ricke, D.O., Wang, S., Cai, R. and Cohen, D. (2006) *Current Opinion in Chemical Biology*, **10**, 303–308.

45 Galperin, M.Y. and Koonin, E.V. (1999) *Current Opinion in Biotechnology*, **10**, 571–578.

46 Allsop, A.E. (1998) *Current Opinion in Biotechnology*, **9**, 637–642.

47 Burrack, L.S. and Higgins, D.E. (2007) *Current Opinion in Microbiology*, **10**, 4–9.

48 Walker, M.J.A., Barrett, T. and Guppy, L.J. (2004) *Drug Discovery Today: Targets*, **3**, 208–214.

49 Huttner, K.M. and Bevins, C.L. (1999) *Pediatric Research*, **45**, 785–794.

50 Breukink, E., Wiedemann, I., van Kraaij, C., Kuipers, O.P., Sahl, H.G. and de Kruijff, B. (1999) *Science*, **286**, 2361–2364.

51 Hancock, R.E.W. and Sahl, H.G. (2006) *Nature Biotechnology*, **24**, 1551–1557.

52 Cottarel, G. and Wierzbowski, J. (2007) *Trends in Biotechnology*, **25**, 547–555.

53 Coates, A., Hu, Y., Bax, R. and Clive, P. (2002) *Nature Reviews. Drug Discovery*, **1**, 895–910.

54 Taylor, P.W., Stapleton, P.D. and Luzioet, J.P. (2002) *Drug Discovery Today*, **7**, 1086–1091.

55 Walsh, C. (2003) *Nature Reviews Microbiology*, **1**, 65–70.

56 Saltini, C. (2006) *Respiratory Medicine*, **100**, 2085–2097.

57 Scollard, D.M., Adams, L.B., Gillis, T.P., Krahenbuhl, J.L., Truman, R.W. and Williams, D.L. (2006) *Clinical Microbiology Reviews*, **19**, 338–381.

58 Borisy, A.A., Elliott, P.J., Hurst, N.W., Lee, M.S., Lehar, J., Price, E.R., Serbedzija, G., Zimmermann, G.R., Foley, M.A., Stockwell, B.R. and Keith, C.T. (2003) *Proceedings of the National Academy of Sciences of the United States of America*, **100**, 7977–7982.

59 Jacobs, M.R., Jones, R.N. and Giordano, P.A. (2007) *Diagnostic Microbiology and Infectious Disease*, **57** (Suppl. 3), S55–S65.

60 Mukherjee, P.K., Sheehan, D.J., Hitchcock, C.A. and Ghannoum, M.A. (2005) *Clinical Microbiology Reviews*, **18**, 163–194.

61 Poole, K. (2005) *Journal of Antimicrobial Chemotherapy*, **56**, 20–51.

62 Putman, M., Veen, H.W. and Konings, W.K. (2000) *Microbiology and Molecular Biology Reviews*, **64**, 672–693.

63 Butaye, P., Cloeckaert, A. and Schwarz, S. (2003) *International Journal of Antimicrobial Agents*, **22**, 205–210.

64 Poole, K. and Lomovskaya, O. (2006) *Drug Discovery Today: Therapeutic Strategies*, **3**, 145–152.

65 Lomovskaya, O. and Bostian, K.A. (2006) *Biochemical Pharmacology*, **71**, 910–918.

66 Wright, G.D. (2007) *Chemistry and Biology*, **7**, R127–R132.

67 Miller, L.A., Ratnam, K. and Payne, D.J. (2001) *Current Opinion in Pharmacology*, **1**, 451–458.

68 Hajduk, P.J., Dinges, J., Schkeryantz, J.M., Janowick, D., Kaminski, M., Tufano, M., Augeri, D.J., Petros, A., Nienaber, V., Zhong, P., Hammond, R., Coen, M., Beutel, B., Katz, L. and Fesik, S.W. (1999) *Journal of Medicinal Chemistry*, **42**, 3852–3859.

69 Marquez, B. (2005) *Biochimie*, **87**, 1137–1147.

70 Lomovskaya, O., Lee, A., Hoshino, K., Ishida, H., Mistry, A., Warren, M.S., Boyer, E., Chamberland, S. and Lee, V.J. (1999) *Antimicrobial Agents and Chemotherapy*, **43**, 1340–1346.

71 Alekshun, M.N. and Levy, S.B. (1999) *Trends in Microbiology*, **7**, 410–413.

72 Strahilevitz, J., Truong-Bolduc, Q.C. and Hooper, D.C. (2005) *Antimicrobial Agents and Chemotherapy*, **49**, 5051–5057.

73 Ince, D., Zhang, X., Silver, L.C. and Hooper, D.C. (2002) *Antimicrobial Agents and Chemotherapy*, **46**, 3370–3380.

74 Kaatz, G.W., Moudgal, V.V. and Seoet, S.M. (2002) *Journal of Antimicrobial Chemotherapy*, **50**, 833–838.

75 Jacobs, M.R., Bajaksouzian, S., Windau, A., Appelbaum, P.C., Patel, M.V., Gupte, S.V., Bhagwat, S.S., De Souza, N.J. and Khorakiwala, H.F. (2004) *Antimicrobial Agents and Chemotherapy*, **48**, 3338–3342.

76 Renau, T.E., Leger, R., Filonova, L., Flamme, E.M., Wang, M., Yen, R., Madsen, D., Griffith, D., Chamberland, S. and Dudley, M.N. (2003) *Bioorganic and Medicinal Chemistry Letters*, **13**, 2755–2758.

77 Watkins, W.J., Landaverry, Y., Léger, R., Litman, R., Renau, T.E., Williams, N., Yen, R., Zhang, J.Z., Chamberland, S., Madsen, D., Griffith, D., Tembe, V., Huie, K. and Dudley, M.N. (2003) *Bioorganic and Medicinal Chemistry Letters*, **13**, 4241–4244.

78 Brenwald, N., Gill, M. and Wise, R. (1997) *Journal of Antimicrobial Chemotherapy*, **40**, 458–460.

79 Miller, J.H. (1995) *Antimicrobial Agents and Chemotherapy*, **39**, 2375–2377.

80 Miller, J.M.T.H. and Shah, S. (2000) *Journal of Antimicrobial Chemotherapy*, **46**, 852–853.

81 Jodoin, J., Demeule, M. and Beliveau, R. (2002) *Biochimica et Biophysica Acta – Molecular Cell Research*, **1542**, 149–159.

82 Roccaro, A.S., Blanco, A.R., Giuliano, F., Rusciano, D. and Enea, V. (2004) *Antimicrobial Agents and Chemotherapy*, **48**, 1968–1973.

83 Stermitz, F.R., Tawara-Masuda, J., Lorenz, P., Mueller, P., Zenewicz, L. and Lewis, K. (2000) *Journal of Natural Products*, **63**, 1146–1149.

84 Stermitz, F.R., Lorenz, P., Tawara, J.N., Zenewicz, L.A. and Lewis, K. (2000) *Proceedings of the National Academy of Sciences of the United States of America*, **97**, 1433–1437.

85 Chen, X., Schauder, S., Potier, N., Dorsselaer, A.V., Pelczer, I., Bassler, B.L. and Hughson, F.M. (2002) *Nature*, **415**, 545–549.

86 Winson, M.K., Camara, M., Latifi, A., Foglino, M., Chhabra, S.R., Daykin, M., Bally, M., Chapon, V., Salmond, G.P. and Bycroft, B.W. (1995) *Proceedings of the National Academy of Sciences of the United States of America*, **92**, 9427–9431.

87 Passador, L., Tucker, K.D., Guertin, K.R., Journet, M.P., Kende, A.S. and Iglewski, B.H. (1996) *Journal of Bacteriology*, **178**, 5995–6000.

88 Hentzer, M., Wu, H., Andersen, J.B., Riedel, K., Rasmussen, T.B., Bagge, N., Kumar, N., Schemberi, M.A., Song, Z., Kristoffersen, P., Manefield, M., Hoiby, N. and Givskov, M. (2003) *EMBO Journal*, **22**, 3803–3815.

89 Ahmad, I., Aqil, F., Ahmad, F., Zahin, M. and Musarrat, J. (2008), Quorum Sensing in Bacteria: Potential in Plant Health Protection in *Plant–Bacteria Interactions* (eds I. Ahmad, J. Pitchel and S. Hayat), Wiley-VCH Verlag GmbH, Weinheim, pp. 129–153.

90 Rasmussen, T.B. and Givskov, M. (2006) *International Journal of Medical Microbiology*, **296**, 149–161.

91 Nealson, K.H., Platt, T. and Hastings, J.W. (1970) *Journal of Bacteriology*, **104**, 313–322.

92 Lyczak, J.B., Cannon, C.L. and Pier, G.B. (2002) *Clinical Microbiology Reviews*, **15**, 1059–1067.

93 Hentzer, M. and Givskov, M. (2003) *Journal of Clinical Investigation*, **122**, 1300–1307.

94 McLean, R.J.C., Pierson, L.S. III and Fuqua, C. (2004) *Journal of Microbiological Methods*, **58**, 351–360.

95 Rasmussen, T.B., Bjarnsholt, T., Skindersoe, M.E., Hentzer, M., Kristoffersen, P., Kote, M., Nielsen, J.,

Eberl, L. and Givskov, M. (2005) *Journal of Bacteriology*, **187**, 1799–1814.

96 Steindler, L. and Venturi, V. (2007) *FEMS Microbiology Letters*, **266**, 1–9.

97 Adonizio, A.I., Downum, K., Bennett, B.C. and Mathee, K. (2006) *Journal of Ethnopharmacology*, **105**, 427–435.

98 Givskov, M., de Nys, R., Manefield, M., Gram, L., Maximilien, R., Eberl, L., Molin, S., Steinberg, P.D. and Kjelleberg, S. (1996) *Journal of Bacteriology*, **178**, 6618–6622.

99 Manefield, M., Rasmussen, T.B., Henzter, M., Andersen, J.B., Steinberg, P., Kjelleberg, S. and Givskov, M. (2002) *Microbiology*, **148**, 1119–1127.

100 Dong, Y.H., Wang, L.H., Xu, J.L., Zhang, H.B., Zhang, X.F. and Zhang, L.H. (2001) *Nature*, **411**, 813–817.

101 Hentzer, M., Riedel, K., Rasmussen, T.B., Heydorn, A., Anderson, J.B., Parsek, M.R., Rice, S.A., Eberl, L., Molin, S. and Holby, N. (2002) *Microbiology*, **148**, 87–102.

102 Carlier, A., Uroz, S., Smadja, B., Fray, R., Latour, X., Dessaux, Y. and Faure, D. (2003) *Applied and Environmental Microbiology*, **69**, 4989–4993.

103 Park, S.Y., Kang, H.O., Jang, H.S., Lee, J.K., Koo, B.T. and Yum, D.Y. (2005) *Applied and Environmental Microbiology*, **71**, 2632–2641.

104 Ren, D., Zuo, R. and Wood, T.K. (2005) *Applied Microbiology and Biotechnology*, **66**, 689–695.

105 Teplitski, M., Robinson, J.B. and Bauer, W.D. (2000) *Molecular Plant–Microbe Interactions*, **13**, 637–648.

106 Daniels, R., Vanderleyden, J. and Michiels, J. (2004) *FEMS Microbiology Reviews*, **28**, 261–289.

107 Fagerlind, M.G., Nilsson, P., Harlen, M., Karlsson, S., Rice, S.A. and Kjelberg, S. (2005) *BioSystems*, **80**, 201–213.

108 Qazi, S., Middleton, B., Muharram, S.H., Cockayne, A., Hill, P., O'Shea, P., Chhabra, S.R., Camara, M. and Williams, P. (2006) *Infection and Immunity*, **74**, 910–919.

109 Sio, C.F., Otten, L.G., Cool, R.H., Diggle, S.P., Braun, P.G., Bos, R., Daykin, M., Camara, M., Williams, P. and

Quax, W.J. (2006) *Infection and Immunity*, **74**, 1673–1682.

110 Choo, J.H., Rukayadi, Y. and Hwang, J.K. (2006) *Journal of Applied Microbiology*, **42**, 637–641.

111 Vattem, D.A., Mihalik, S.H., Crixell, S.H. and McLean, R.J.C. (2007) *Fitoterapia*, **78**, 302–310.

112 Ahmad, I., Hasan, S. and Zahin, M. (2007) Paper presented at the 9th Indian Agriculture Scientists and Farmers' Congress, Bioved Research and Communication Center, Allahabad, 29–30 January.

113 Qaseem, F., Zahin, M., Khan, M.S.A. and Ahmad, I. (2008) Paper presented at the 3rd J & K State Science Congress, Jammu, 26–28 February, abstract AGR-54.

114 Schoolnik, G.K., Summers, W.C. and Watson, J.D. (2004) *Nature Biotechnology*, **22**, 505–506.

115 Theil, K. (2004) *Nature Biotechnology*, **22**, 31–36.

116 Fischetti, V.A., Nelson, D. and Scuch, R. (2006) *Nature Biotechnology*, **24**, 1508–1511.

117 Alisky, J., Iczkowski, K., Rapoport, A. and Troitsky, N. (1998) *Journal of Infection*, **36**, 5–15.

118 Gould, I.M. (2005) *Medicine et Maladies Infectieuses*, **35**, S123–S124.

119 Felmingham, D., Feldman, C., Hryniewicz, W., Klugman, K., Kohno, S., Low, D.E., Mendes, C. and Rodloff, A.C. (2002) *Clinical Microbiology and Infection*, **8** (Suppl. 2), 12–42.

120 Victor, K.E. (2005) *International Journal of Antimicrobial Agents*, **26**, S1–S63.

121 Cizman, M., Beovic, B., Ktrcmery, V., Barsic, B., Tamm, E., Ludwig, E., Pelemis, M., Karovski, K., Grzesiowski, P., Gadovska, D., Volokha, A., Keuleyan, E., Stratchounski, L., Dumitru, C., Titov, L.P., Usonis, V. and Dvorak, P. (2004) *International Journal of Antimicrobial Agents*, **24**, 1–6.

122 Stratchounski, L.S., Taylor, E.W., Dellinger, E.P. and Pechere, J.C. (2005) *International Journal of Antimicrobial Agents*, **26**, 312–322.

3
Promising Current Drug Candidates in Clinical Trials and Natural Products Against Multidrug-Resistant Tuberculosis

Marcus Vinícius and Nora de Souza

Abstract

An important factor in the resurgence of tuberculosis (TB) worldwide is the advent of multidrug-resistant (MDR) strains, which are defined as being resistant to at least two current first-line drugs, that can be transmitted from an infectious patient to other people. Several elements are responsible for the MDR-TB problem, such as poor social conditions, incorrect prescriptions, abandoning treatment, immigration from countries having high rates of MDR-TB and no support of the public health system. Another serious problem nowadays is extensively drug-resistant (XDR)-TB, which are strains resistant to both first- and second-line anti-TB drugs. Due to the high impact of MDR and, recently, XDR in TB treatment, we urgently need new drugs and strategies to continue the fight against this disease. Considering that, this chapter highlights promising drug candidates in clinical trials and natural products that are being studied against MDR-TB.

3.1
Introduction

Tuberculosis (TB) has once again, since the mid-1980s, become an important infectious disease worldwide. Several factors are responsible for this, such as the acquired immunodeficiency syndrome (AIDS) epidemic, the advent of multidrug-resistant (MDR) strains, poor socioeconomic conditions, immigration and the lack of new drugs in the market. In the case of patients with AIDS, TB is the most common opportunistic infection and results in the death of one in every three patients. According to an estimate, between 10 and 12 million people are coinfected with *Mycobacterium tuberculosis* and human immunodeficiency virus (HIV). In addition, patients living with AIDS and contaminated with latent TB are 30 times more likely to develop active TB than noninfected patients [1].

Another important factor in the resurgence of TB worldwide is the advent of MDR strains, which are defined as being resistant to at least two current first-line drugs, that can be transmitted from an infectious patient to other people. Several elements are responsible for the MDR-TB problem, such as poor social conditions, incorrect prescriptions, abandonment of treatment, immigration from countries having high rates of MDR-TB and no public health support system. Due to the importance of MDR in TB treatment, in 1994 the World Health Organization (WHO) and the International Union against Tuberculosis and Lung Disease joined forces to establish a global survey of resistance to first-line anti-TB drugs. At present it is difficult to quantify, but it is estimated that at least 4% of all worldwide TB patients are resistant to at least one of the current first-line drugs. According to the WHO, there are between 300 000 and 450 000 new cases per year of MDR-TB worldwide, and nowadays 79% of MDR-TB is resistant to at least three out of the four main drugs used to treat TB [2].

In the context of MDR-TB, another serious problem is extensively drug-resistant (XDR) strains, which are strains resistant to both first- and second-line anti-TB drugs. XDR-TB is commonly defined as strains resistant to all current first-line drugs, as well as any fluoroquinolone and at least to one of three injectable second-line drugs (capreomycin, kanamycin or amikacin). A study by the United States Centers for Disease Control and Prevention (CDC) and WHO of 17 690 isolates from 49 countries during 2000–2004 indicated that 20% of the strains were MDR-TB and 2% were XDR-TB [3, 4]. Due to the high impact of MDR and, recently, XDR in TB treatment, we urgently need new drugs and strategies to continue the fight against this disease. Considering that, this chapter highlights promising drug candidates, new synthetic classes and natural products that are in development against MDR-TB.

3.2
Current MDR-TB Treatment

The treatment of MDR-TB is complex (Tables 3.1 and 3.2) and requires careful diagnosis using second-line drugs, which possess more side-effects and require much longer treatment times – up to 24 months as opposed to the 6–9 months in current TB treatment. The cure rate of patients with MDR-TB is very low, curing no more than 50% of patients, compared to more than 95% in current TB treatment. Another

Table 3.1 Treatment plans for resistant TB (www.projectinform.org).

Isoniazid-resistant TB	Rifampin-resistant TB
(1) Rifampin/rifabutin, pyrazinamide + ethambutol daily for at least 2 weeks	(1) isoniazid, streptomycin, pyrazinamide + ethambutol (daily for 8 weeks or daily for 2 weeks + twice a week for 6 weeks).
(2) The same three drugs twice a week for 6–9 months	(2) isoniazid, streptomycin + pyrazinamide 2–3 times a week for 7 months

Table 3.2 Suggested treatment regimens.

Susceptibility to essential drugs	Initial phase		Continuation phase	
	Drug	**Duration**	**Drug**	**Duration**
Resistance to isoniazid + rifampin	streptomycin^a + ethambutol + fluoroquinolone^b + pyrazinamide ± ethionamide	at least 6 months	ethambutol + fluoroquinolone^b + pyrazinamide ± ethionamide	12–18 months
Resistance to all essential drugs	one injectable + one fluoroquinolone + two of these three oral drugs: *p*-aminosalicylic acid, ethionamide, cycloserine	at least 6 months	the same drug except injectable	18 months

WHO/CDS/TB/2003.313: *Treatment of TB: Guidelines for National Programmes*, third edition revision approved by STAG, June 2004.
^{a}If resistance to streptomycin is confirmed, replace this drug with kanamycin, amikacin or capreomycin.
^{b}Fluoroquinolone (ciprofloxacin or ofloxacin).

important problem in MDR-TB treatment is its cost, estimated to be between US$19 000 and 50 000 per person. Furthermore, depending on the complexity of the treatment, it can cost US$1 million dollars per person, compared with only US$10–20 for normal TB treatment. Additionally, this kind of treatment needs extensive laboratory requirements necessary for following drug-susceptibility and hospitalization, at least in the beginning of the treatment [2, 5]. Due to the MDR-TB problem, the WHO and coworkers have developed guidelines for the treatment, management and surveying of MDR-TB, which are important measures in the fight against drug resistance. Other actions which also help in the combat against MDR-TB include The Green Light Committee based at the WHO Headquarters in Geneva, which focuses on finding alternatives as second-line TB drugs at lower costs [2, 5].

The drugs normally used to treat MDR-TB are amikacin, capreomycin, ciprofloxacin, cycloserine, ethionamide, anamycin, ofloxacin, *p*-aminosalicylic acid and protionamide (Table 3.2). According to the WHO, the retreatment regimen should include at least four drugs never used previously by the patient, including an injectable amikacin, capreomycin or kanamycin and a fluoroquinolone. Pyrazinamide and ethambutol can be added to the treatment because of the lower probability of resistance than with other essential drugs.

3.3
Current Drugs in Development Against MDR-TB

The importance of chemotherapy in TB treatment and in other diseases has saved countless lives and has brought hope to humankind [6]. However, despite the importance of current first-line TB treatment, people often interrupt it due to significant side-effects and the long time of the treatment, normally over a period of 6–9 months. With the abandonment of TB treatment or the presence of MDR-TB, it is necessary to use second-line compounds, which are the last line of defense in the fight against TB. However, these drugs possess more side-effects, longer treatment times and higher costs, in comparison with first-line drugs. Due to the disadvantages of first and second line drugs, and the increase of MDR strains worldwide, we urgently need new drugs to improve TB treatment. Unfortunately, since the introduction of rifampin in the market (1966), no new class of anti-TB drugs has been developed in the last four decades except for rifabutin (1992) and rifapentine (1998) – both very close analogs of rifampin. The lack of interest in the development of new anti-TB drugs is basically due to two incorrect opinions: (i) the current drugs used in TB treatment are adequate and sufficient for the control of TB, and (ii) the little interest of the pharmaceutical industry due to the high investment required to introduce a drug into the market compared with the projected low commercial return for TB treatment. Considering this problem, a number of actions have been taken and in 2000 the Global Alliance for TB Drug Development (GATB; www.tballiance.org) was established, with the main priority to develop TB drugs. In this context, the GATB has worked in partnership with several organizations, including academic institutions, government research laboratories, nongovernmental organizations, the pharmaceutical industry and contract research houses. Due to its important work, the GATB has changed the perspective of TB drug discovery.

There are a number of basic factors involved in the development of new anti-TB drugs, such as more effective treatment of latent TB infection, prevention and treatment of MDR-TB, and potent sterilizing activity, which is defined as the ability of a drug such as pyrazinamide and rifampin to destroy the bacteria known as persisters. The sterilizing activity is closely related to the time of the treatment. Other important factors can be mentioned, such as cost, new mechanisms of action, good pharmacokinetic distribution and permeation into cells and lung tissue, potent bactericidal activity, and selective activity against mycobacterial species [7]. At present, promising new drugs candidates are in clinical trials against TB [8–11] (Tables 3.3 and 3.4).

Table 3.3 New promising drugs candidates in clinical trials against MDR-TB.

Compound	Class of compound	Developer	Clinical phase
Gatifloxacin	fluoroquinolone	BMS	phase III
Moxifloxacin	fluoroquinolone	Bayer	phase II/II
OPC 67683	nitroimidazo-oxazole	Otsuka Pharmaceutical	phase II
TMC207	diarylquinoline	Johnson and Johnson	phase II
PA-824	nitroimidazole	PathoGenesis	phase I
SQ-109	diamine	Sequella	phase I
Pyrrole LL-3858	pyrrole	Lupin	phase I

Table 3.4 New promising drugs candidates in clinical trials against MDR-TB.

Compound	MIC (μg/ml)	Mechanism of action	Sponsor/coordinator
Gatifloxacin	0.03–0.12[a]	inhibition DNA replication transcription	European Commission; Institut de Recherche pour le Developpement; WHO/Special Programme for Research and Training in Tropical Diseases; Lupin
Moxifloxacin	0.06–0.5[a]	inhibition DNA replication transcription	Bayer; GATB; CDC; University College London; Johns Hopkins University
OPC 67683	0.006–0.012	inhibition of cell wall biosynthesis	Otsuka Pharmaceutical
TMC207	0.03–0.12	ATP depletion and imbalance biosynthesis in pH homeostasis	Johnson and Johnson (Tibotec)
PA-824	0.015–0.25	inhibition of protein synthesis Inhibition of cell wall lipids synthesis	GATB
SQ-109	0.16–0.32	inhibition of cell wall biosynthesis	Sequella
Pyrrole LL-3858	0.125–0.25[b]	no data available	Lupin

Effect on bacterial cell: bactericidal.
[a]MIC_{50}.
[b]MIC_{90}.

3.3.1
Fluoroquinolones (Gatifloxacin and Moxifloxacin)

Fluoroquinolones (7-fluoro-4-oxo-1,4-dihydro-quinoline-3-carboxylic acid) (Figure 3.1) are an important class of compound in the fight against different infectious diseases. This class possesses potent antibacterial activity with a broad spectrum of activity against Gram-positive, Gram-negative and mycobacterial organisms as well as anaerobes with great therapeutic potential, particularly against those organism resistant to other classes of antibacterial drugs.

3.3.1.1 Fluoroquinolones against MDR-TB
The advent of fluoroquinolones brought important advances in TB treatment and promising perspectives against MDR-TB; nowadays, much attention is paid to

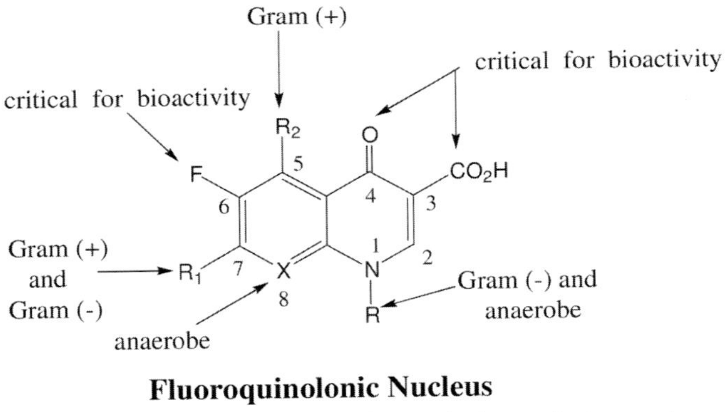

Fluoroquinolonic Nucleus

Figure 3.1 Structure–activity relationships in the fluoroquinolone nucleus.

newer fluoroquinolones due to their several advantages as antibiotics [12–14]. Their benefits include fewer toxic effects, and extensive and potent activity against Gram-positive and Gram-negative bacteria, including resistant strains, when compared with the earlier fluoroquinolones, and they can be administrated orally, with favorable pharmacokinetic profiles and good absorption, including proficient penetration into host macrophages. Their advantages can be explained by their mechanism of action. The efficacy of the new fluoroquinolones against resistant strains is due to their dual activity – inhibiting both DNA gyrase bacterial type II and topoisomerase IV that limit the emergence of fluoroquinolone resistance. Due to the promising perspectives of the newer fluoroquinolones against TB, several studies have suggested their utility as second-line drugs for MDR-TB. In this context, there are a number of fluoroquinolones and quinolones under study, including PD-161 148, CS-940, T-3811ME, gemifloxacin, sitafloxacin, sparfloxacin, moxifloxacin and gatifloxacin (Figure 3.2), of which the last two compounds are the most promising [12–14].

3.3.1.2 Gatifloxacin

Gatifloxacin is classified as a third-generation fluoroquinolone and was marketed in the United States in 1999 by Bristol-Myers Squibb as Tequin (Figure 3.2). This fluoroquinolone possesses broad-spectrum activity against bacterial infections, and it is available in tablets of 200 and 400 mg for oral administration and injection (gatifloxacin in 5% dextrose) for intravenous administration. The *in vivo* comparison of gatifloxacin with moxifloxacin alone and in combination with anti-TB drugs was evaluated in mice infected with *M. tuberculosis*, and gatifloxacin appears to have sufficient activity alone and in combination to deserve evaluation for treatment of TB. Different studies also have demonstrated that gatifloxacin could be useful against MDR-TB and TB/HIV coinfections (e.g. the combination of gatifloxacin with ethambutol and pyrazinamide could be useful against isoniazid and rifampin resistant TB) [12].

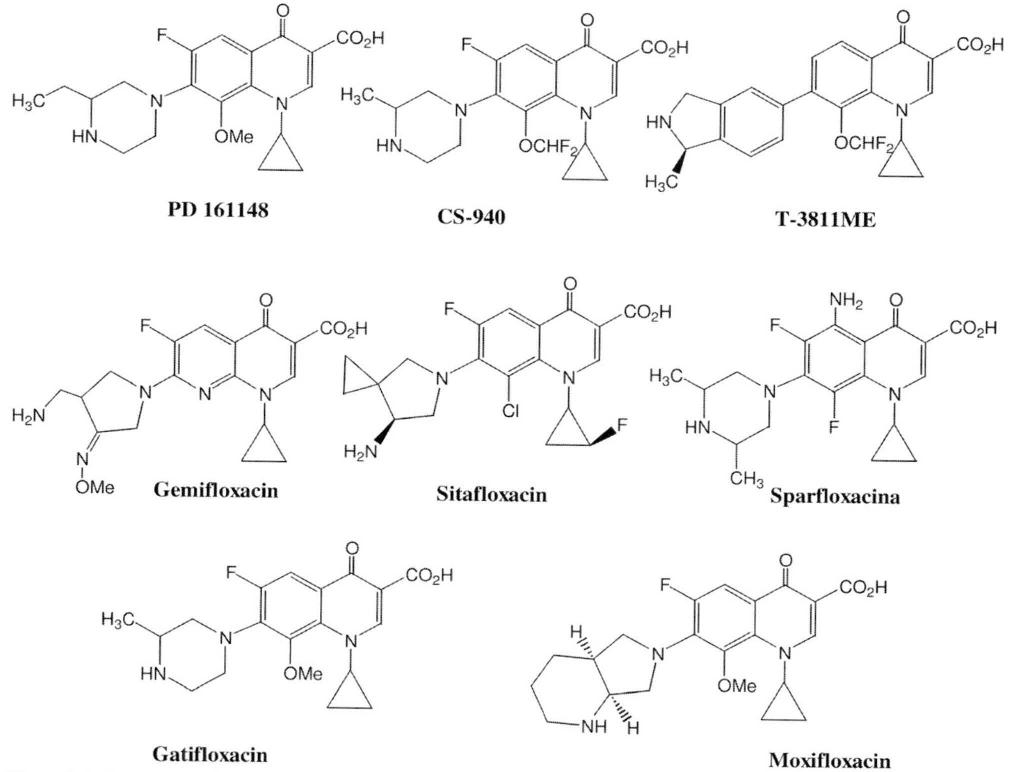

PD 161148   CS-940   T-3811ME

Gemifloxacin   Sitafloxacin   Sparfloxacina

Gatifloxacin   Moxifloxacin

Figure 3.2 Structures of promising fluoroquinolones against MDR-TB.

3.3.1.3 Moxifloxacin

Moxifloxacin (BAY12-8039) is a fourth-generation fluoroquinolone developed by Bayer and marketed as Avelox (moxifloxacin hydrochloride). It possesses excellent activities against a variety of different types of Gram-negative, Gram-positive and anaerobic bacteria, such as pneumonia, bronchitis and sinusitis (Figure 3.2). The excellent activities against *M. tuberculosis in vivo* and *in vitro* studies can be compared with those of other known fluoroquinolones (Table 3.5). For example, against

Table 3.5 Moxifloxacin compared with other known fluoroquinolones.

Drug	Tablet (mg)	MIC_{90} (µg/mL)	Area under curve	$t_{1/2}$ (hours-h)	C_{max}
Moxifloxacin	400	0.25	45–51	12	5
Gatifloxacin	400	0.25	29–40	08	5
Sparfloxacin	400	0.5	41–54	20	1
Levofloxacin	500	1.0	34	07	6

rifampicin-tolerant bacteria, moxifloxacin is much more effective in comparison with levofloxacin and ofloxacin, with better bactericidal and sterilizing activities (Figure 3.2). Moxifloxacin also has excellent oral bioavailability, a long $t_{1/2}$, and the elimination half-life of the drug in humans is about 12 h in comparison with 1–2 h for isoniazid. Moxifloxacin is under clinical studies for TB treatment with promising perspectives against MDR-TB and TB/HIV coinfections, because this fluoroquinolone has no cross-resistance to other TB drugs. Another promising feature is the reduction of the TB treatment period on using a combination of moxifloxacin and other TB drugs due to its great sterilizing effect [12].

3.3.2
PA-824

The nitroimidazole PA-824 was developed by PathoGenesis in 1995 for cancer treatment. However, the significance of this compound increased in 2000 with the discovery of its potent TB activity [15], based on the compound CGI-17341, which possesses promising *in vitro* and *in vivo* TB activity. However, the development of this compound was aborted due to its mutagenicity proprieties (Figure 3.3). Despite the potent TB activity of PA-824, Chiron, which bought PathoGenesis in 2000, decided not to continue the development of this compound and signed a license agreement that gave the GATB rights to develop nitroimidazole PA-824. This compound showed potent *in vitro* activity against *M. tuberculosis* with minimum inhibitory concentrations (MICs) of 0.015–0.25 μg/ml and also possesses high activity in mice with no toxicity in rodent models, as well as excellent sterilizing activity compared with isoniazid and rifampin. The mechanism of action of this prodrug involves the flavinoid, known as F-420 cofactor, that activates PA-824, which subsequently inhibits the synthesis of protein and cell wall lipids. While PA-824 has not yet been evaluated in humans, this compound has several advantages in TB treatment, such as effectiveness against MDR-TB strains, promising activity in latent-state TB and potent sterilization effects, compared to those of rifampicin and isoniazid. Furthermore, the reduction of the TB treatment period when administered with other classes of TB drugs is a great advantage. Another important perspective of PA-824 is that no significant inhibition of cytochrome P450 isozymes has been observed, suggesting that it could be used with HIV medications and thus be useful in TB/HIV treatment.

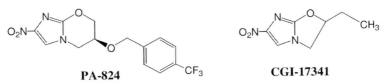

PA-824   **CGI-17341**

Figure 3.3 Structures of nitroimidazole PA-824 and CGI-17341.

3.3.3
Diarylquinoline TMC207

The diarylquinoline TMC207 (ex. R207910) (Figure 3.4) was developed by the researcher Koen Andries and coworkers at Johnson and Johnson Pharmaceutical R & D (Beerse, Belgium). They reported, in collaboration with other institutions and countries, for the first time in 2004 at the Interscience Conference on Antimicrobial Agents and Chemotherapy (ICAAC), the phase 1 human studies. This compound was the first of a new promising class of compounds against TB with a new mechanism of action [16] based on the attack on the enzyme ATP synthase, which is the energy source for the bacterium. TMC207 possesses a good pharmacokinetic/pharmacodynamic profile with potent activity against *M. tuberculosis*, as well as other *Mycobacterium* species, such as *M. marinum, M. fortuitum, M. abscessus, M. avium* and *M. smegmatis*. In the case of *M. tuberculosis*, the substitution of rifampicin, isoniazid or pyrazinamide with TCM207 in some combinations cleared the lungs in murine mouse models after about 2 months of treatment. It was also tested in combination with second-line drugs, such as amikacin, moxifloxacin and ethionamide. Due to the promising perspectives against MDR-TB, TCM207 is currently in phase 2 clinical trials.

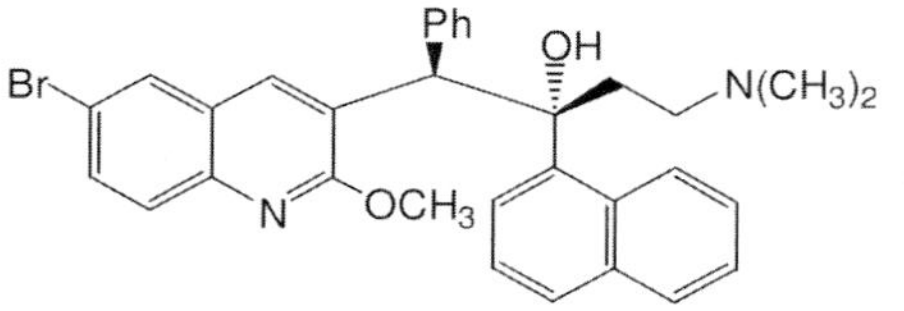

Figure 3.4 Structure of the diarylquinoline TMC207 (ex. R207910).

3.3.4
Nitroimidazo-Oxazole OPC-67683

The nitroimidazole OPC-67683 (Figure 3.5) was developed by Otsuka Pharmaceuticals as the lead of a library of 6-nitro-2,3-dihydroimidazo[2,1-*b*]oxazole compounds, based on PA-824, in the 5-nitroimidazole pharmacophore of the compound CGI-17341 [10, 17]. This library was evaluated with the goal of improving the

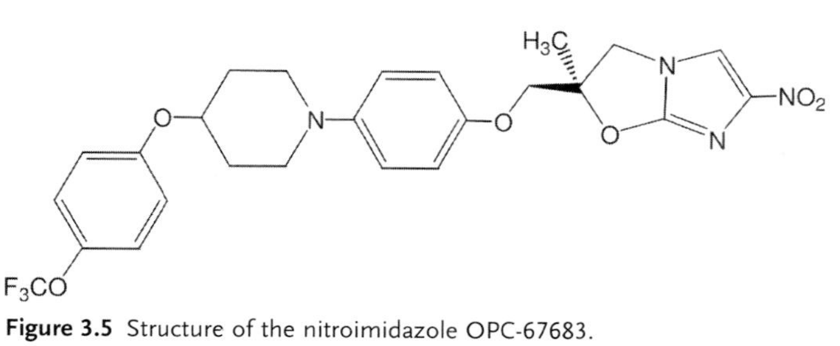

Figure 3.5 Structure of the nitroimidazole OPC-67683.

anti-TB activity and eliminating the mutagenic proprieties of this class of compound. In this context, the compound OPC-67683 was selected for preclinical development and the results were first described in 2005 at the ICAAC meeting. Like PA-824, this compound is a prodrug; however, its mechanism was not well defined. When compared with all known anti-TB drugs in the market or in development against different clinical strains, including MDR-TB, the OPC-67683 MIC (0.006–0.012 µg/ml) proved it to be the most potent compound discovered and, like PA-824, it has not demonstrated evidence of mutagenic proprieties. The combination of this compound with standard TB drugs has not indicated any cross-resistance and due to its promising results it could be used in shorter TB therapies.

3.3.5
SQ-109

An important work published in 2005 in the field of drug discovery was performed at Sequella, in collaboration with the National Institutes of Health/National Institute of Allergy and Infectious Diseases, which built a library of 63 238 compounds based on the pharmacophore of ethambutol, 1,2-ethylenediamine [18]. These compounds were evaluated against *M. tuberculosis*, and 26 demonstrated *in vitro* activity equal to or greater (up to 14-fold) than ethambutol and were also active against MDR-TB, including ethambutol-resistant TB. The compound SQ-109 (Figure 3.6) was selected for further development due to its potent activity with a MIC of 0.16–0.32 µg/ml, an SI (selectivity index) of 16.7 by inhibition of cell wall biosynthesis and 99% inhibition activity against intracellular bacteria. Additionally, it has potent activity *in vivo*, and limited toxicity *in vitro* and *in vivo* with synergistic interactions with rifampin and isoniazid and additive activity against streptomycin. The combinations of first-line TB drugs with SQ-109 were better at preventing TB-induced weight loss as well as providing a very effective treatment in comparison with the combination of first-line TB drugs.

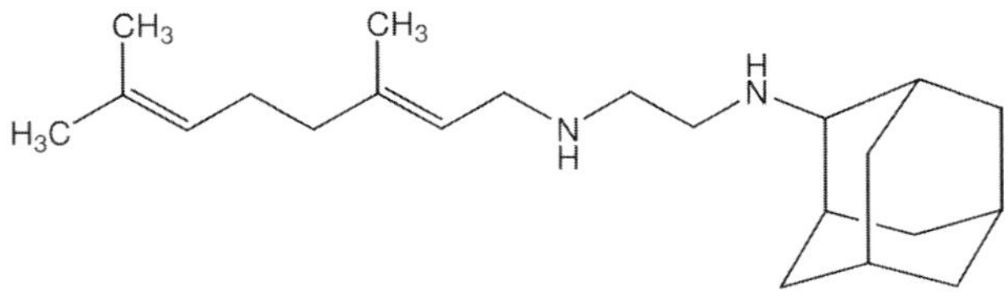

Figure 3.6 Structure of SQ-109.

3.3.6
Pyrrole LL-3858

The story of pyrrole LL-3858 started with pyrrole BM-212 described in the 1900s by Deidda *et al.* [19] and Biava *et al.* [20], who evaluated a series of pyrroles for antimicrobial properties (Figure 3.7). Lupin, in 2004, synthesized a new series of

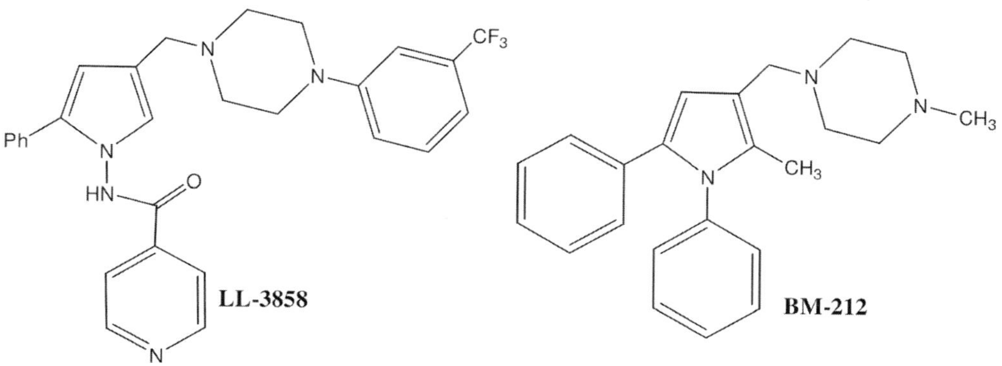

Figure 3.7 Structures of pyrroles LL-3858 and BM-212.

this class with improved anti-TB activity, the lead compound being LL-3858 [10], which is in phase 1 clinical trials. This compound is active against different strains including MDR-TB with a MIC range of 0.125–0.25 µg/ml and also possesses synergy with rifampin *in vitro*. However, the mechanism of this class has not yet been established. Another contribution made by Lupin in the TB field is the development of fixed-dose drug combinations of LL-3858 with first-line TB drugs.

3.4
Natural Products Against MDR-TB

Nature is an incredible source of both simple and complex compounds that play important roles in the treatment of human diseases. A vast number of drugs on the market today have been derived from living organisms (including plants, animals and insects). Nowadays, much attention is paid to the discovery of new molecules from nature with unexploited modes of action in the antibiotic field [21]. In this context, several studies have been reported on the development of new anti-TB agents derived from natural sources [22, 23, 25]. However, despite the great advances in this field, in general few studies have been reported on treating MDR-TB with natural products. Considering that, the aim of this section is to highlight new and promising anti-MDR-TB agents derived from plants, fungi and marine sources.

3.4.1
Fungi

3.4.1.1 Thiolactomycin
A promising natural product against MDR-TB is the antibiotic thiolactomycin (TLM) (Figure 3.8) obtained from the fermentation broth of a strain of actinomycetes belonging to the genre *Nocardia*, found in a Japanese soil sample and described for the first time in 1982 by Oishi *et al.* [26]. This natural product

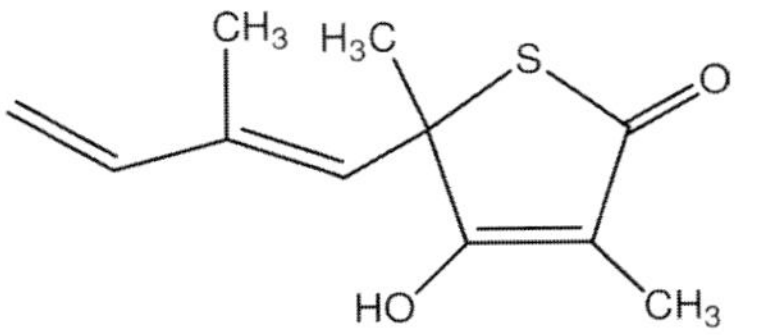

Figure 3.8 Structure of TLM. MIC = 62.5 μg/ml.

exhibits broad-spectrum activity *in vivo* against many pathogenic microorganisms such as Gram-positive, Gram-negative and anaerobic bacteria. Despite this broad spectrum of activity, TLM shows modest activity against laboratory strains of *M. tuberculosis* (MIC = 62.5 μg/ml). However, this natural product possesses relevant activities against several strains of *M. tuberculosis* that are resistant to other drugs. Other characteristics make TLM an attractive lead molecule for TB, such as low molecular weight (210 g/mol), high water solubility, appropriate lipophilicity (log P = 3), good oral absorption and a low toxicity profile in mice. The mechanism of action of this natural product is based on inhibiting KasA and KasB, two KAS enzymes that are components of the specialized FASII system involved in the synthesis of the very long-chain meromycolic acids, constituent building blocks for bacterial cell walls. It is the only thiolactone to possess anti-TB activity by inhibiting fatty acid and mycolic acid biosynthesis.

The success of this natural product has increased synthetic efforts by different groups to produce even more effective agents [6].

3.4.1.2 Pacidamycin Family

Pacidamycins are a class of antibiotics isolated from *Streptomyces coeruleorubidus*, which together with mureidomycins and napsamycins are known as uridyl peptide antibiotics (Figure 3.9). The importance of these classes in the fight against MDR

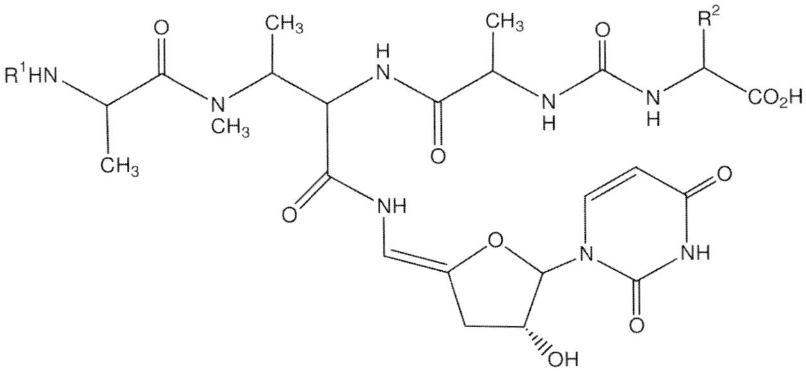

R^1 = H; alanyl or glycyl; R^2 = 3-indoyl; benzyl or 3-hydroxybenzyl

Figure 3.9 General structure of the pacidamycin family.

strains is due to several of their advantages, including favorable pharmacokinetic, pharmacodynamic and toxicological proprieties and their mechanism of action, which is based on translocase I inhibitors of the cell wall biosynthetic enzyme MraY (transferase, translocase I) acting on polymer peptidoglycan biosynthesis – an essential component in the cell wall of bacteria. Considering that, several synthetic modifications have been reported with promising activity against MDR-TB, possessing a MIC range of 4–8 μg/ml [27].

3.4.1.3 Capuramycin

The antibiotic capuramycin was isolated in 1986 by Yamaguchi *et al.* from the culture of *Streptomyces griseus* 446-S3, possessing in this structure a uracil nucleoside and a caprolactam subunit [28] (Figure 3.10). This natural product and its derivatives are very active against several diseases caused by mycobacteria including MDR-TB and its mechanism of action is based on translocase I inhibition acting on polymer peptidoglycan biosynthesis [29].

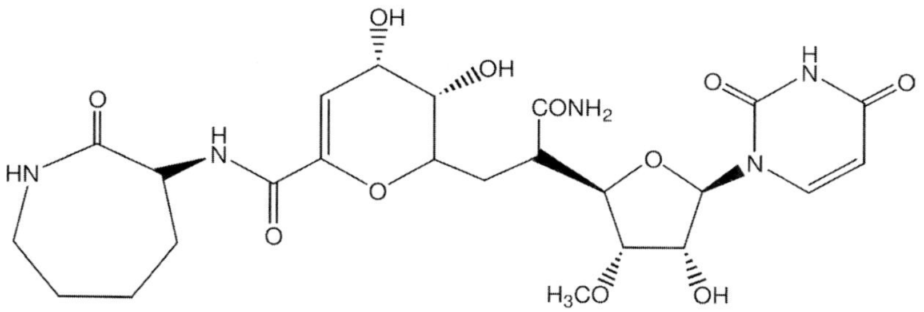

Figure 3.10 Structure of capuramycin.

3.4.1.4 Caprazamycin

Caprazamycin is a novel class of lipo-nucleoside antibiotic agent acting against acid-fast bacteria by inhibiting the biosynthesis of the cell wall (Figure 3.11). This class was isolated from the culture broth of *Streptomyces* ssp. MK730-62F2 by Igarashi *et al.*, who also contributed to the elucidation of the structures of caprazamycin A–G [30]. Caprazamycin B, which is the major component of this species, possesses bactericidal activity only against mycobacterial species, MIC range 3.13–50 μg/ml, including *M. tuberculosis*, MDR-TB and *Mycobacterium avium* complex, being a specific natural product against mycobacterial species [30].

3.4.2
Plants

3.4.2.1 Calanolide

Calanolides are coumarins isolated from *Calophyllum lanigerum* var. *austrocoriaceum* (Guttiferae). These are the first class of reported natural compounds to possess important antimycobacterial and antiretroviral activities. Xu *et al.* found important

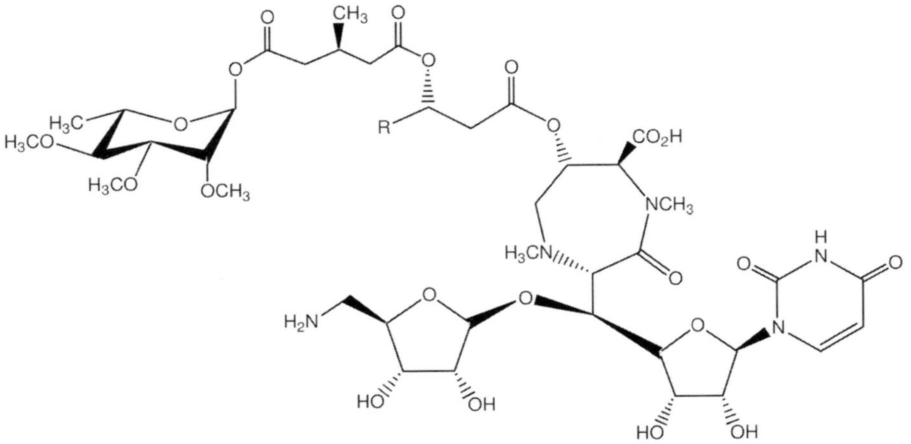

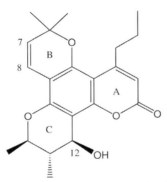

A CH$_3$(CH$_2$)$_{11}$CH$_2$ **B** CH$_3$CH(CH$_3$)(CH$_2$)$_9$CH$_2$ **C** CH$_3$(CH$_2$)$_{10}$CH$_2$ **D** CH$_3$CH(CH$_3$)(CH$_2$)$_8$CH$_2$

E CH$_3$(CH$_2$)$_9$CH$_2$ **F** CH$_3$CH(CH$_3$)(CH$_2$)$_7$CH$_2$ **G** CH$_3$CH$_2$CH(CH$_3$)(CH$_2$)$_7$CH$_2$

Figure 3.11 Structures of caprazamycin A–G.

Figure 3.12 Structure of (+)-calanolide A. Inhibition activity 96%. MIC = 3.13 μg/ml.

anti-TB activity in the natural product (+)-calanolide A [31] (Figure 3.12). This coumarin was active *in vitro* against different strains of *M. tuberculosis*, including some resistant to standard anti-TB drugs. Due to this important result, Xu and his research group screened other related pyranocoumarins [31] and demonstrated that this class of compound contains a new pharmacophore group in the search for new lead compounds against MDR-TB.

3.4.2.2 Diospyrin
Diospyrin is a bisnaphthoquinonoid compound isolated from the stem bark of *Diospyros montana* Roxb. with a wide range of activities, such as anticancer,

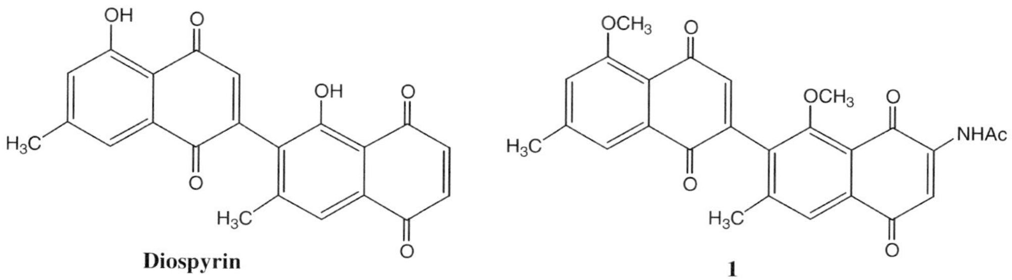

Figure 3.13 Structure of diospyrin and its aminoacetate derivative **1**.

Diospyrin

(MIC = 100 μg/ml)

H37Rv

1

(MIC = 50 μg/ml)

MDR-TB

antimalaria, anti-*Leishmania* and anti-*Trypanosoma* (Figure 3.13). Considering that, Meyer *et al.* have evaluated the biological activity of diospyrin and some synthetic derivatives against drug-susceptible strain H37Rv and MDR strains of *M. tuberculosis* [32]. In this context, the derivative **1** (Figure 3.13) was found to be active with an MIC of 50 μg/ml against MDR-TB.

3.4.3
Marine Natural Products

3.4.3.1 Aerothionin
Marine natural products are today an outstanding source of compounds and inspiration in drug discovery, which play an important role in the treatment of human diseases. However, despite this importance and the studies performed in the TB field using marine sources, few studies with MDR-TB have been undertaken. A good example of the importance of marine sources as agents against MDR-TB is the marine sponge *Aplysina gerardogreeni*, which furnished two bromotyrosine derivatives, aerothionin and calafianin, isolated by Encarnación-Dimayuga (Figure 3.14). Aerothionin was tested against MDR-TB with a MIC ranging from 6.5 to 25 μg/ml; however, its close derivative calafianin was inactive, indicating the necessity of certain pharmacophore groups [33].

3.5
Future Perspectives Against MDR-TB

The advent of MDR-TB and, more recently, XDR-TB poses serious problems for the treatment and effective control of this disease. Considering that, strategies such as 'DOTS' (directly observed treatment short-course), fixed-dose combinations, diagnostics, molecular understanding, international scientific collaboration and financial support are being created. Another important point in the fight against the rapid spread of MDR-TB strains against all major anti-TB drugs present in the market is the

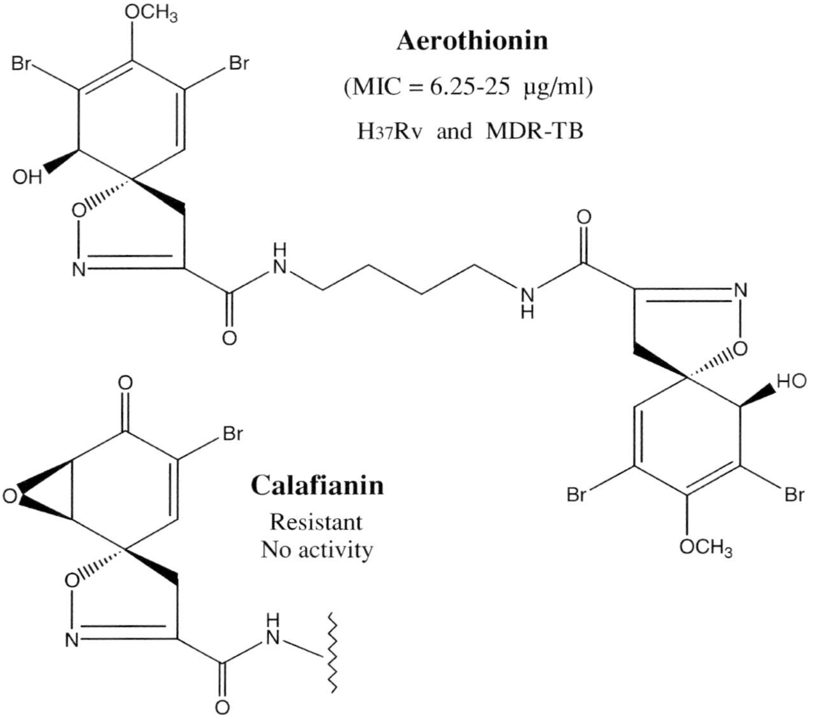

Figure 3.14 Structures of aerothionin and calafianin.

development of news drugs able to reduce the total length of treatment, with fewer toxic side-effects, improved pharmacokinetic properties, and extensive and potent activity against Gram-positive and Gram-negative bacteria.

References

1 http://www.who.int/tb/en.

2 http://www.who.int/tb/dots/dotsplus/en.

3 http://www.who.int/mediacentre/news/notes/2006/np23/en/index.html.

4 Raviglione, M.C. and Smith, I.M. (2007) *New England Journal of Medicine*, **356**, 656–659.

5 Mukherjee, J.S., Rich, M.L., Socci, A.R., Joseph, J.K., Virú, F.A., Shin, S.S., Furin, J.J., Becerra, M.C., Barry, D.J., Kim, J.Y., Bayona, J., Farmer, P., Smith Fawzi, M.C. and Seung, K.J. (2004) *Lancet*, **363**, 474–481.

6 Janin, Y.L. (2007) *Bioorganic and Medicinal Chemistry*, **15**, 2479–2513.

7 Spigelman, M. and Ginsbrg, A.M. (2007) *Nature Medicine*, **13**, 290–294.

8 De Souza, M.V.N. (2006) *Current Opinion in Pulmonary Medicine*, **12**, 167–171.

9 WHO (2007) *WHO Drug Information*, **20**, 239–245 (http://www.who.int/medicinedocs/index/assoc/s14182e/s14182e.pdf).

10 Protopopova, M., Bogatcheva, E., Nikonenko, B., Hundert, S., Einck, L. and Nacy, C.A. (2007) *Medicinal Chemistry*, **3**, 301–316.

11 De Souza, M.V.N. (2006) *Recent Patents on Anti-Infective Drug Discovery*, **1**, 33–45.

12 De Souza, M.V.N., Vasconcelos, T.A., Cardoso, S.H. and Almeida, M.V. (2006) *Current Medicinal Chemistry*, **13**, 455–463.

13 De Souza, M.V.N. (2005) *Mini-Reviews in Medicinal Chemistry*, **5**, 1009–1018.

14 De Souza, M.V.N., Almeida, M.V., Couri, M.R.C. and Silva, A.D. (2003) *Current Medicinal Chemistry*, **10**, 21–40.

15 Stover, C.K., Warrener, P., VanDevanter, D.R., Sherman, D.R., Arain, T.M., Langhorne, M.H., Anderson, S.W., Towell, J.A., Yuan, Y., McMurray, D.N., Kreiswirth, B.N., Barry, C.E. and Baker, W.R. (2000) *Nature*, **405**, 962–966.

16 Cole, S.T. and Alzari, P.M. (2005) *Science*, **307**, 214–215.

17 Matsumoto, M., Hashizume,1 H., Tomishige,1 T., Kawasaki, M., Tsubouchi, H., Sasaki, H., Shimokawa, Y. and Komatsu, M. (2006) *PLoS Medicine*, **3**, 2131–2144.

18 Protopopova, M., Hanrahan, C., Nikonenko, B., Samala, R., Chen, P., Gearhart, J., Einck, L. and Nacy, C.A. (2005) *Journal of Antimicrobial Chemotherapy*, **56**, 968–974.

19 Deidda, D., Lampis, G., Fioravanti, R., Biava, M., Porretta, G.C., Zanetti, S. and Pompei, R. (1998) *Antimicrobial Agents and Chemotherapy*, **42**, 3035–3037.

20 Biava, M., Porretta, G.C. and Manetti, F. (2007) *Mini-Reviews in Medicinal Chemistry*, **7**, 65–78.

21 Butler, M.S. and Buss, A.D. (2006) *Biochemical Pharmacology*, **71**, 919–929.

22 Copp, B.R. and Pearce, A.N. (2007) *Natural Product Reports*, **24**, 278–297.

23 De Souza, M.V.N. (2006) *Scientific World Journal*, **6**, 847–861.

24 De Souza, M.V.N. (2005) *Scientific World Journal*, **5**, 609–628.

25 Guido, F., Pauli, G.F., Case, R.J., Inui, T., Wang, Y., Cho, S., Fischer, N.H. and Franzblau, S.G. (2005) *Life Sciences*, **78**, 485–494.

26 Oishi, H., Noto, T., Sasaki, H., Suzuki, K., Hayashi, T., Okazaki, H., Ando, K. and Sawada, M. (1982) *Journal of antibiotics*, **35**, 391–395.

27 Boojamra, C.G., Lemoine, R.C., Blais, J., Vernier, N.G., Stein, K.A., Magon, A., Chamberland, S., Hecker, S.J. and Lee, V.J. (2003) *Bioorganic and Medicinal Chemistry Letters*, **13**, 3305–3309.

28 Yamaguchi, H., Sato, S., Yoshida, S., Takada, K., Itoh, M., Seto, H. and Otake, N. (1986) *Journal of Antibiotics*, **39**, 1047–1053.

29 Koga, T., Fukuoka, T., Doi, N., Harasaki, T., Inoue, H., Hotoda, H., Kakuta, M., Muramatsu, Y., Yamamura, N., Hoshi, M. and Hirota, T. (2004) *Journal of Antimicrobial Chemotherapy*, **54**, 755–760.

30 Igarashi, M., Takahashi, Y., Shitara, T., Nakamura, H., Naganawa, H., Miyabe, T. and Akamatsu, Y. (2005) *Journal of Antibiotics*, **58**, 327–337.

31 Xu, Z.-Q., Barrow, W.W., Suling, W.J., Westbrook, L., Barrow, E., Lin, Y.-M. and Flavin, M.T. (2004) *Bioorganic and Medicinal Chemistry*, **12**, 1199–1207.

32 Lall, N., Das Sarma, M., Hazra, B. and Meyer, J.J.M. (2003) *Journal of Antimicrobial Chemotherapy*, **51**, 435–438.

33 Encarnacion-Dimayuga, R., Ramirez, M.R. and Luna-Herrera, J. (2003) *Pharmaceutical Biology*, **41**, 384–387.

4

Non-antibiotics – An Alternative for Microbial Resistance: Scope and Hope

Debprasad Chattopadhyay, Soumen Kumar Das, Arup Ranjan Patra, and Sujit Kumar Bhattacharya

Abstract

The antimicrobial activity of non-chemotherapeutic compounds, such as methylene blue, phenothiazine, thioxanthene and related agents, has been known since the time of Paul Ehrlich (1854–1915). In this context the term 'nonantibiotics' was introduced to include a variety of compounds used in the management of pathological conditions of noninfectious etiology. Despite the availability of hundreds of anti-infective drugs the emergence of 'antibiotic resistance' and new infectious agent's create the therapeutic challenge to the medical community. Hence, the search for newer agents to tackle the global problem is continuing. It has been noted that many of the phenothiazines, thioxanthenes, other neurotropics, antihistamines, anesthetics, analgesics, antihypertensives, muscle relaxants, some cardiovascular agents and so on can inhibit diverse classes of microbes, as well as the drug-resistant strains at different dose levels, by modifying the architecture of the microbial membrane and its permeability. A review of the literature suggests that some of these membrane-active compounds can enhance the activity of conventional antibiotics, eliminate natural resistance to specific antibiotics (reversal of resistance) and exhibit strong activity against multidrug-resistant forms of *Staphylococcus aureus, Escherichia coli, Salmonella* spp., *Mycobacterium tuberculosis, Plasmodium falciparum* and so on. This chapter covers the antimicrobial activity of some nonantibiotics, especially against drug-resistant microbes, that cause therapeutic challenge with an emphasis on the group of drugs used as antihistamines, sedatives, hypnotics and so on, and their stereoisomers.

4.1
Introduction

Antibiotics are substances produced by microorganisms that can destroy or inhibit the growth of other microorganisms and can even act against some cancer cells. On the

New Strategies Combating Bacterial Infection. Edited by Iqbal Ahmad and Farrukh Aqil
Copyright © 2009 WILEY-VCH Verlag GmbH & Co. KGaA, Weinheim
ISBN: 978-3-527-32206-0

other hand, substances of nonmicrobial origin that have similar effects on microbes or malignant cells are described as chemotherapeutic agents [1]. Drugs that are neither antibiotics nor antimicrobial chemotherapeutic agents, but which possess antimicrobial properties, are termed 'nonantibiotics' [2]. Pharmacologically, drugs have been classified on the basis of their apparently major and predominant pharmacological action, such as antihistamines, analgesics, antihypertensives, neuroleptics, local anesthetics, anti-inflammatory agents and so on. Nevertheless many of the drugs are basically multifunctional and show different activity at different dose levels [3]. This multiplicity of function, other than their first described one, may be quite significant. Hence, redescriptions of such drugs are possible (e.g. Prontosil, an azo dye containing a sulfonamide group, was later developed as an antibacterial agent) [4, 5]. Similarly, the synthetic nitrofurantoin is a selective antibacterial against urinary tract bacteria, but can also damage mammalian DNA [6]. Likewise, the phenazine dye clofazamine possesses antihistaminic, anti-inflammatory as well as powerful antibacterial action against leprosy bacilli by inhibiting DNA template formation [7]. Similarly, the traditional analgesic aspirin (acetyl salicylic acid) is now often prescribed as an anticoagulant for cardiovascular diseases [8]. Metronidazole, a nitroimidazole, is a powerful broad-spectrum amebicide as well as a DNA degrader in obligate anaerobic bacteria [9], while cyproheptidine (periactin) is an antihistamine as well as an anabolic stimulant [10]. Hence, the term 'nonantibiotics' is considered to include a variety of compounds like antihistamines, anesthetics, hypnotics, sedatives, antpsychotics, analgesics, diuretics, antihypertensives, muscle relaxants, cardiovascular agents and many more which are used in the management of pathological conditions of a noninfectious etiology, but can modify the cellular permeability of microbes and exhibit broad-spectrum antimicrobial activity. The antimicrobial properties of several nonantibiotic compounds have been investigated sporadically and their application for the management of microbial infections has not been systematically evaluated. It has been reported that a variety of 'nonantibiotic' compounds used in the management of noninfectious etiology have broad-spectrum antimicrobial activity against viruses [11–16], *Mycoplasma* [17, 18], bacteria [19–74], *Mycobacteria* [7, 75–92], yeast [93–95], protozoa such as *Amoeba* [96–97], *Plasmodium* [98–101], *Leishmania* [102–107], *Trypanosoma* [108–110] and helminths [111] both *in vitro* and *in vivo* [2, 3, 19, 23, 26, 41, 49, 51, 53, 56, 61, 67–69, 71–74, 77, 82–89, 91, 105, 108, 109]. Many of these nonantibiotics can modify the cellular permeability of the microbes [2, 3, 18–20, 23, 25, 27, 30, 34, 38–40, 46, 54, 59, 60, 79, 81, 95, 101–103, 107, 112]. This chapter will present the antimicrobial activity of some potential 'nonantibiotics' especially against drug-resistant microbes like viruses, bacteria, fungi and protozoa, with an emphasis on the psychotherapeutic, antihistaminic, sedative and hypnotic agents, and their stereoisomers.

4.2
Historical Development: Non-antibiotics with Antimicrobial Potential

The antimicrobial activity of synthetic, nonchemotherapeutic compounds, such as methylene blue, phenothiazines and so on, has been known since the time of Paul

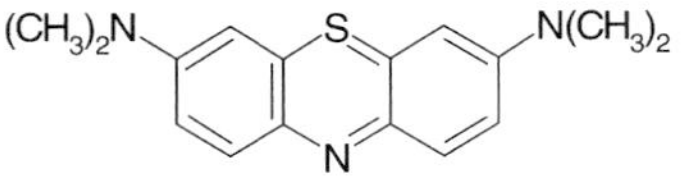

Reported by: H. Caro (1834–1911).
Stains yarn, fabric and bacteria blue.
Antimicrobial *in vitro* and *in vivo*.
Analgesic   [116].
Neuroleptic   [118].

Figure 4.1 Methylene blue.

Ehrlich (1854–1915). During the second half of the nineteenth century aniline dye was extracted from coal tar in the dyeing industry [113], but in 1876 the German dye-stuff expert Heinrich Caro extracted methylene blue from aniline, while in 1883 methylene blue was synthesized as a derivative of phenothiazine [114]. Methylene blue (Figure 4.1) was first used by Paul Ehrlich as a bacteriological stain, and he discovered that methylene blue and its analogs had pronounced affinity for nerve and brain tissue [115]. Later, Ehrlich applied methylene blue in the treatment of neuritic and rheumatic pain [116], and in 1891 he used this dye in the treatment of malaria [117].

Simultaneously, the Italian physician Pietro Bodoni discovered that methylene blue had a positive effect on psychotic patients [118]. In 1939 it was found that sterile urine from rabbits fed with phenothiazine (Figure 4.2) remained sterile for several weeks even when exposed to air, which led to the therapeutic trial of phenothiazine in urinary tract infections [21] with positive results. Interestingly, it was observed that suture material prepared from the gut of phenothiazine-treated sheep was more stable and stronger than the known material. Hence, phenothiazines were used by veterinarians in the treatment of intestinal parasites, especially as antihelminthics [111]. However, the interest in phenothiazines as antimicrobial candidates declined with the discovery of penicillin by Alexander Fleming in 1928 [119] and, later, a number of 'classic' antibiotics like streptomycin [120]. However, after World War II the French military surgeon Henri Marie Laborit discovered that phenothiazines could be used in the therapy of shock and pain after surgery as analgesics [121], thus confirming the initial observation made by Ehrlich and Leppmann in 1890 with methylene blue [116].

Laborit [121] also found that phenothiazines had a calming effect on patients and proposed that these agents might be used as sedatives in psychiatric treatment. However, in the context of emerging microbial diseases like severe acute respiratory syndrome, human immunodeficiency virus (HIV)/acquired immunodeficiency syndrome (AIDS), and re-emerging diseases like tuberculosis and malaria as well

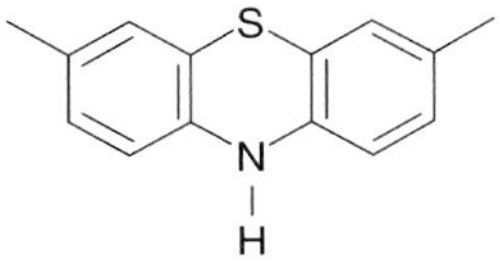

Reported by: A.H. Bernthsen (1855–1931).
Antimicrobial *in vitro* and *in vivo* [21].

Figure 4.2 Phenothiazines.

as drug-resistant microbes create a therapeutic challenge to the clinicians which needs more new and improved drugs.

To counter the challenge of microbial resistance, drug companies and scientific community are expanding work on new antibiotics, as well as modifications of the naturally occurring antibiotics that exhibit little or no toxicity to humans [120]. The alarming worldwide increase in the frequency of antibiotic resistance [120, 122] suggests that more antibiotics are needed for the initial or adjuvant management of patients with infections caused by high-frequency multidrug-resistant (MDR) pathogens. There is evidence that certain nonantibiotic compounds, alone or in combination with conventional antibiotics, may play a useful role in the management of specific bacterial infections associated with a high risk of resistance to conventional antibiotics [2, 3, 18, 48, 53, 56, 64, 83, 84, 86–88, 92, 123–128]. A review of the literature, coupled with some recent investigations, suggests that some of these and other membrane-active compounds enhance the activity of conventional antibiotics or chemotherapeutic agents [2, 18, 25, 56, 64, 65, 74, 77, 84, 86, 89–92, 123, 125, 126, 128–140], eliminate natural resistance to specific antibiotics (reversal of resistance) and exhibit strong activity against multi-drug resistant strains of *Staphylococcus aureus, Staphylococcus epidermidis, Streptococcus pyogenes, Streptococcus pneumoniae, Escherichia coli, Enterococcus faecalis, Burkholderia pseudomallei, Mycobacterium tuberculosis* and *Plasmodium falciparum* [141–156]. Thus nonantibiotics may have a significant role in the management of certain bacterial infections and a list of some well-studied nonantibiotics with their antimicrobial activity against microbes is presented in Table 4.1.

4.3
Psychotherapeutics as Non-antibiotics

Out of different categories of drugs, psychotherapeutics have long-standing antimicrobial potential. The major group of psychotherapeutic drugs includes phenothiazines, thioxanthenes and phenylpiperidines. 'Phenothiazines' [10H-phenothiazine, $C_{12}H_9NS$; (Figure 4.2)], developed as synthetic dyes in 1883, are the largest of the five main classes of neuroleptic antipsychotics. Chemically, phenothiazines are dibenzothiazine or thiodiphenylamine with a three-ring structure in which two benzene rings are joined by sulfur and nitrogen atoms at nonadjacent positions, obtained by fusing diphenylamine with sulfur. Phenothiazine is a semivolatile organic compound used as an intermediate of various antipsychotic neuroleptic drugs [157], introduced as an insecticide by DuPont in 1935, and as an antihelminthic in livestock and for the manufacture of rubber additives [158]. Phenothiazine pesticides work by affecting the nervous system of insects, inhibiting the acetylcholine breakdown by blocking the enzyme acetylcholinesterase and is a potent α-adrenergic blocker. Hence, many of their side-effects are due to their anticholinergic blocking effects [159].

Chlorpromazine (Figure 4.3) was the first neuroleptics used in psychiatry [160]. The structure–activity relationship study revealed that stereoisomerism is responsible for the antimicrobial activity of optically inactive chlorpromazine (Figure 4.3) and

Table 4.1 Antimicrobial activity of some important nonantibiotics.

Class	Compounds	Activity against	Reference
Dyes	methylene blue (aniline)	*Plasmodium falciparum*	[1, 101]
		Plasmodium vinckei petteri,	[152]
		Plasmodium yoeli nigeriensis	[54]
		Helicobacter pylori	[171]
	dimethyl methylene blue	DNA and RNA virus	[13]
		Mycoplasma spp.	[17]
Antipsychotic (Aliphatic)	phenothiazine	Mycobacteria (slow growing)	[79]
		Mycobacterium tuberculosis	[76, 78, 86]
		Mycobacteria (Isoniazid resistant)	[156, 181]
		Mycobacteria spp.	[86]
		Mycobacteria (MDR)	[87]
		Yeast	[95]
		Amoeba proteus	[97, 96]
		Helminthes	[111]
		Trypanosoma, Leishmania spp.	[108, 109]
		Escherichia coli (UTI)	[21]
		intestinal anaerobes	[43]
		Bacteroides spp., *Prevotella* spp., *Fusobacterium* spp., *Vibrio cholerae, Plasmodium falciparum, Leishmania* spp., *Candida* spp., *Amoeba* spp.	[50, 59]
		plasmid replication	[43]
	chlorpromazine	*Mycobacterium tuberculosis* (4–12 μg/ml)[a]	[25, 77–79, 83, 87, 173]
		influenza virus	[11]
		bacteria (5–10 μg/ml)[a]	[22, 31]

(Continued)

Table 4.1 (Continued)

Class	Compounds	Activity against	Reference
		Escherichia coli, Vibrio cholerae, Salmonella typhimurium	[47]
		Bordetella spp.	[40]
		Candida spp.	[93, 94]
		Leishmania donovani	[102, 103, 106]
		Staphylococcus aureus, Enterococcus faecalis,	[105]
		R factor elimination	[31]
		Plasmodium falciparum	[98]
	promazine	Mycobacterium tuberculosis (10 µg/ml) antimicrobial	[78, 70]
	triflupromazine	antibacterial	[70]
	levomepromazine	antibacterial	[47]
(Piperazines)	prochlorpromazine	Plasmodium falciparum	[100]
(Piperidines)	thioridazine	Mycobacterium tuberculosis (10)[a]	[88, 156]
		Mycobacterium tuberculosis (6–32)[a]	
		Mycobacterium avium	[90]
	tricyclic antipsychotics	Proteus vulgaris	[62]
Antihistaminic	alimemazine	bacteria (37)[a]	[50]
	methdilazine	Staphylococcus aureus, Escherichia coli,	[51, 61, 133, 134]
		Shigella spp., Salmonella spp., Klebsiella spp., Proteus spp.,	
		Pseudomonas aeruginosa	
	promethazine	Mycobacterium tuberculosis (5–15)[a]	[82]
	phenergan	Trypanosoma spp., Escherichia coli adhesion	[41, 110]
(Ethanolamine)	bromodiphenhydramine	Mycobacterium tuberculosis	[75]
		Staphylococcus aureus, Streptococcus pneumoniae,	[32]
		Escherichia coli, Klebsiella spp., Pseudomonas spp.,	
		Mycobacterium spp.	
	diphenhydramine		
(Alkyl amine)	clofizamine	bacteria	[33, 80]

(Phenyl piperadines)	tripolidine	dye, steroid	[120]
	femoxetine	bacteria	[63]
	paroxetine	enterobacteria (<400)[a]	[47, 79]
		Enterococcus faecalis, Streptococcus progenes, Streptococcus pneumoniae	
	cyproheptadine	*Plasmodium berghei, Plasmodium yoelii, Plasmodium falciparum*	[154]
		virus	
Neuroleptics (Piperazines)	prochlorperazine	bacteria (25–200 µg/ml)[a]	[73]
		Burkholderia pseudomalleus	[72]
(Thioxanthene)	trifluoperazine	*Mycobacterium tuberculosis* (5–32)[a]	[180]
		Chlamydomonas spp.	[79, 81, 84, 85]
		Moloney leukemia virus	[178]
			[14]
	clopenthixol:		
	trans (*E*)	*Vibrio cholera*	[39, 47, 154]
		Plasmodium berghei	
		P. aeruginosa	
		Plasmodium aureus, Staphylococcus epidermidis	
	cis (*Z*)	*Plasmodium falciparum* (758 ng/ml)[a]	[98]
	flupenthixol	bacteria	[73]
Antidepressant	trimipramine	*Campylobacter pylori*	[55]
Benzo[*a*]phenothiazine	aminoperazine	HSV-2	[15, 16]
		HIV-1	
Trifluromethyl ketone	*t*-phenyl-butanedione	*Bacillus megaterium*	[65]
	t-dimethyl-propanone	*Corynebacterium* spp.	
	t-benzoxazdyl-propanone	*Escherichia coli*, yeast	

(Continued)

Table 4.1 (Continued)

Class	Compounds	Activity against	Reference
Antihypertensive	methyl-L-DOPA	*Mycobacterium* spp. *S. aureus, Escherichia coli; Shigella* spp., *Vibrio Cholerae, P. aeruginosa; Bacillus subtilis; C. albicans*	[48, 91]
Cardiovascular	propanolol	Bacteria (5–200)[a], *Escherichia coli, S. aureus*	[44]
	amlodipine	antibacterial	[68, 69, 139]
	dobutamine		
Ca^{2+} channel blocker	oxyfedrine	antibacterial	[67, 140]
Antispasmodic	dicyclomine	bacteria	[71]
Anti-inflammatory	diclofenac sodium	*Escherichia coli*	[74, 89, 92]
		Mycobacterium tuberculosis	
Anesthetics	lignocaine, procaine, lidocaine	*Staphylococcus aureus, Escherichia coli*	[195]
		Shigella spp., *Vibrio cholerae, Proteus mirabilis, P. aeruginosa,*	[28, 37, 38, 52, 195]
		Enterococcus faecalis, Bacillus cereus	
Diuretics	amiloride	*Pseudomonas* spp.	[128]
		Streptococcus spp.	[46]
	probenecid	*Neisseria* spp.	[45]
	triamiterene	*Streptococcus* spp.	[46]
	β-adrenergic blocker	bacteria	[42]
Others	acetyl salicylic acid	*Staphylococcus aureus*	[65, 128]
Proton pump inhibitors	omeprazole	*Staphylococcus aureus, Escherichia coli; Corynebacterium diphtheriae, Serratia marcescens, Proteus vulgaris, Pseudomonas aeruginosa*	[180, 181]
	lansoprazole	*Helicobacter pylori*	

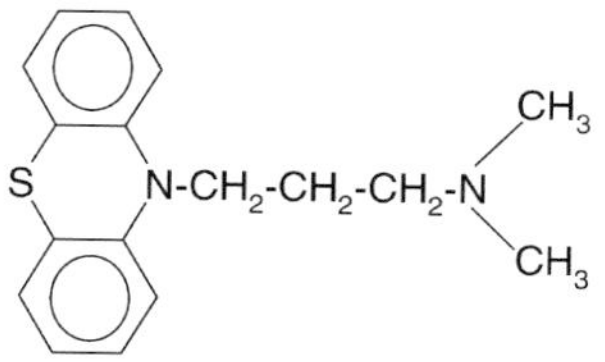

Antihistaminic (IC_{50} 23 nM).
Antihypersecretory (ED_{50} 6.2 µM/kg).
Antimicrobial effect *in vitro*.

Figure 4.3 Chlorpromazine.

active levomepromazine (Figure 4.4) [161]. Thioxanthenes differ from phenothiazines by having a carbon atom instead of nitrogen in the central ring, and exist in two geometric stereo-isomers, *cis* (Z) and *trans* (E) form, but only *cis* (Z) compounds have neuroleptic activity [162]. The third group, 4-phenylpiperidines (Figure 4.5), have two chiral centers and four stereoisomeric forms, (+) *trans*, (−) *trans*, (+) *cis* and (−) *cis*, having antidepressant activity [163].

These antipsychotics often have antiemetic properties, although they may cause akathisia, tardive dyskinesia, extrapyramidal symptoms and rarely fatal neuroleptic malignant syndrome with substantial weight gain. Pharmacologically, phenothiazine neuroleptic antipsychotics are closely related to the thioxanthenes. There are three

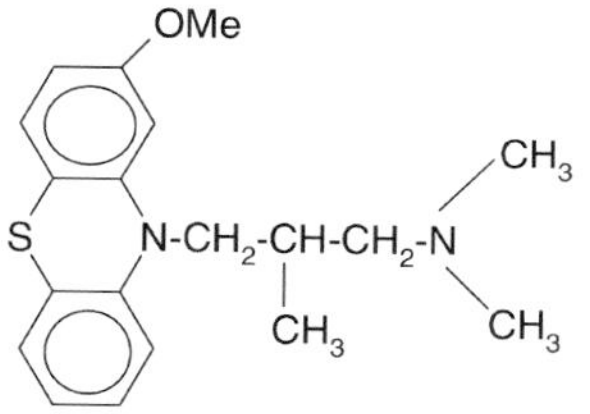

Antihistaminic (IC_{50} 20 nM).
Antihypersecretory (ED_{50} 5.5 µM/kg).
Neuroleptic (IC_{50} 23 nM).
Antimicrobial effect *in vitro*.

Figure 4.4 Levomepromazine.

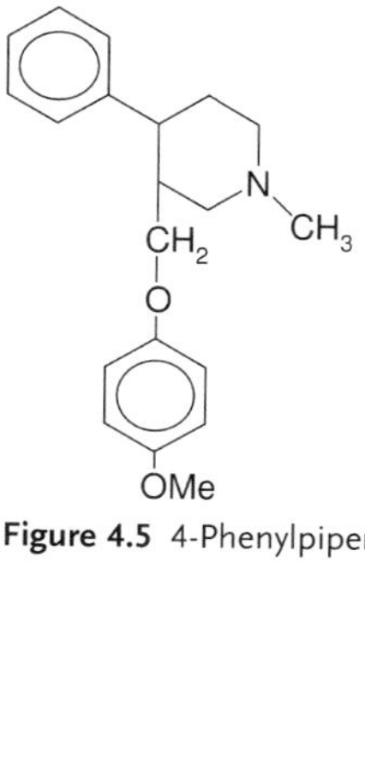

Figure 4.5 4-Phenylpiperidines.

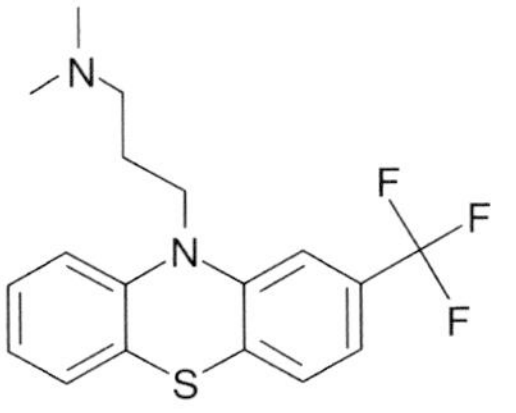

Figure 4.6 Triflupromazine.

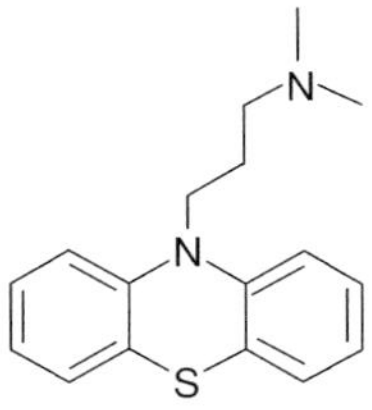

Figure 4.7 Promazine.

groups of phenothiazine antipsychotics (aliphatic, piperidines and piperazines), differing by their chemical structure and pharmacological effects. An aliphatic, piperidine or piperazine moiety is added to the phenothiazine molecule for the purpose of enhancing absorption and bioavailability of the phenothiazine. The common aliphatic phenothiazines includes strong sedatives like chlorpromazine (Figure 4.3), triflupromazine (Figure 4.6), levomepromazine (Figure 4.4) and methotrimeprazine, and the moderate sedative promazine (Figure 4.7). The piperidines include strong sedatives like mesoridazine (Figure 4.8) and thioridazine (Figure 4.9),

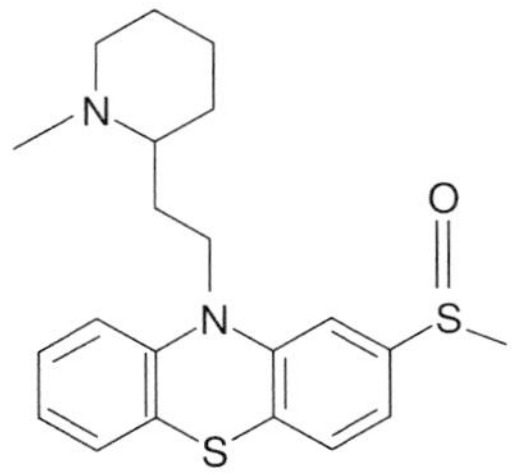

Figure 4.8 Mesoridazine.

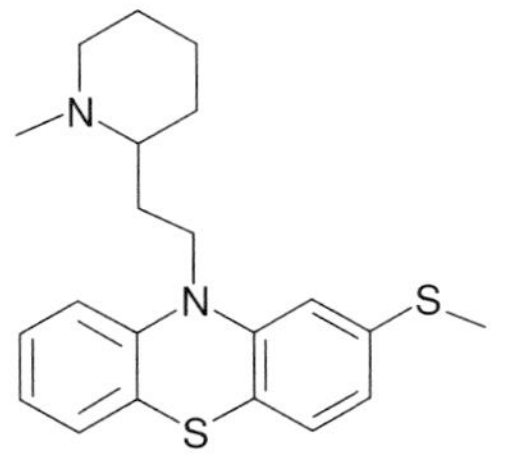

Figure 4.9 Thioridazine.

(a)

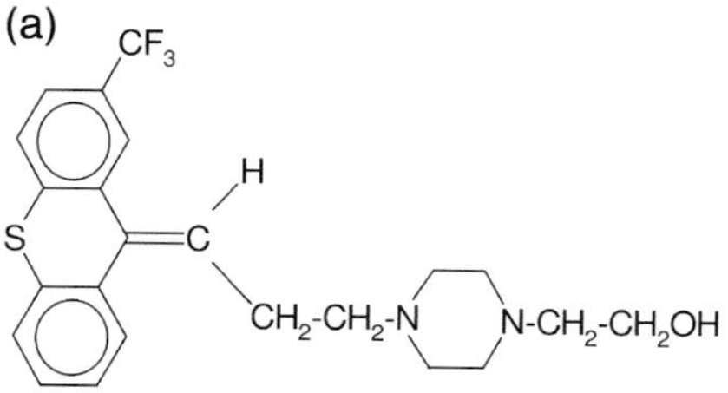

Antihistaminic (IC_{50} 260 nM).
Antihypersecretory (ED_{50} >9.8 µM/kg).
Neuroleptic (IC_{50} 500 nM).
Antimicrobial effect *in vitro.*

(b)

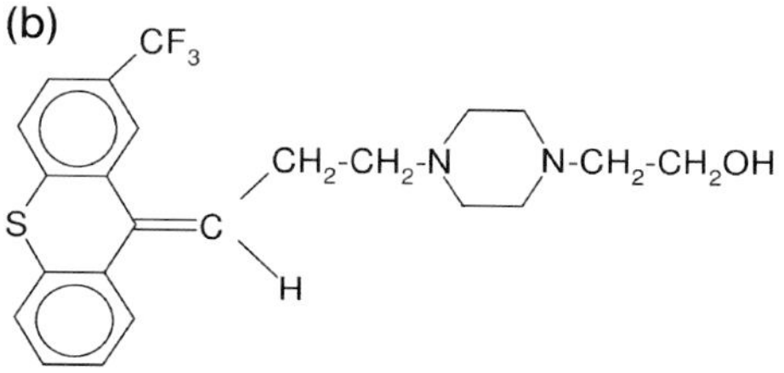

Antihistaminic (IC_{50} 290 nM).
Antihypersecretory (ED_{50} 3.0 µM/kg).
Neuroleptic (IC_{50} 3.2 nM).
Antimicrobial effect *in vitro.*

Figure 4.10 (a) *Trans (E)*-flupentixol. (b) *Cis (Z)*-flupentixol.

while piperazines includes moderate sedatives flupentixol (Figure 4.10a and b) and trifluoperazine (TFP) (Figure 4.11) and weak sedatives like fluphenazine (Figure 4.12) and perphenazine (Figure 4.13). Chlorpromazine is a phenothiazine that contains an aliphatic side-chain, typical for low to middle potency neuroleptics, having an oral bioavailability of 30–50% due to extensive first-pass metabolism in the liver, with an elimination half-life of 16–30 h, and has many active metabolites with greatly varying half-lives and pharmacological profiles [162]. The cytochrome P450 isoenzymes 1A2 and 2D6 are needed for metabolization of chlorpromazine, and the 2D6 subtype is inhibited by chlorpromazine [161–163].

The antischizophrenic activity of phenothiazine drugs and their tendency to elicit extrapyramidal symptoms are due to blockade of synaptic dopamine receptors in the brain [18, 161]. Space filling molecular models shows that van der Waals interactions

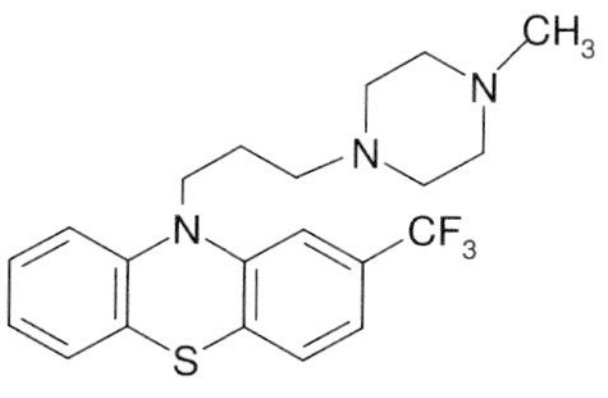

Figure 4.11 Trifluoperazine.

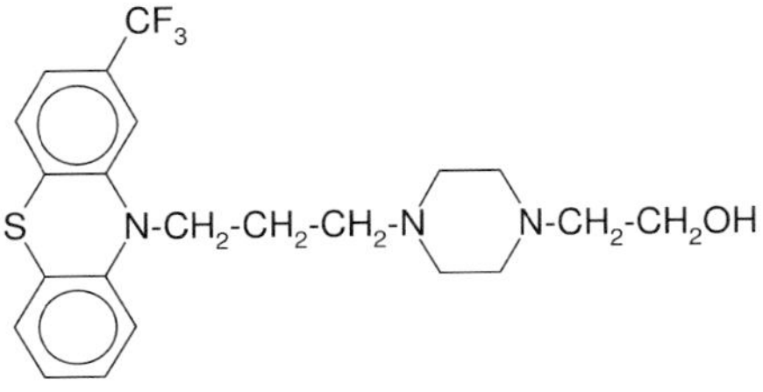

Antihistaminic (IC$_{50}$ 73 nM).
Antihypersecretory (ED$_{50}$ 7.3 µM/kg).
Neuroleptic (IC$_{50}$ 4.4 nM).
Antimicrobial effect *in vitro*.
Figure 4.12 Fluphenazine.

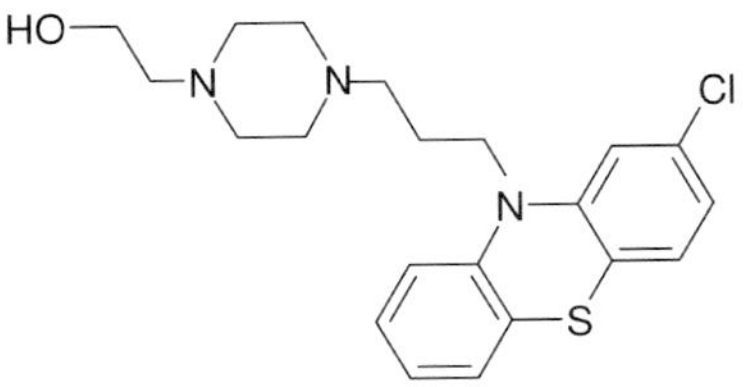

Figure 4.13 Perphenazine.

between the amino side chain of phenothiazines and the 2-substituent on ring A can promote a conformation mimicking dopamine. These van der Waals attractive forces can explain (i) the greater potency of drugs with trifluoromethyl rather than chlorine as a 2-substituent, (ii) the enhanced activity of phenothiazines with piperazine instead of alkylamino side-chains, (iii) the increased potency associated with hydroxyethylpiperazines as contrasted to piperazine side-chains, (iv) the greater potency of *cis* rather than *trans* thioxanthenes and (v) the crucial location of the ring A substituent at carbon 2, and molecular models suggest an active conformation for the phenothiazines [18, 54, 164].

4.4
Sedatives and Tranquilizers as Non-antibiotics

Barbiturates and benzodiazepines are drugs of the sedative-hypnotic group that depresses the central nervous system (CNS), thereby causing calmness, relaxation, reduction of anxiety, sleepiness, slurred speech and poor judgment. Barbiturates were developed in the 1860s and the first barbiturate drug, barbital, was used in 1903. The addictiveness of barbiturates and the high incidence of their misuse for addictiveness forced scientists to look for safer drugs [161]. The well-known barbiturates are pentobarbital, mephobarbital, secobarbital, amobarbital and phenobarbital; while diazepam, chlordiazepoxide, alprazolam, lorazepam and clorazepate are benzodiazepines. Benzodiazepines, formulated in 1957 and used in the 1960s, are safer than barbiturates and have now replaced barbiturates in many cases. Barbiturates, benzodiazepines and opioids are CNS depressants (i.e. slow down the CNS and then the rest of the body), and some also reveal antimicrobial activity [2, 11, 14–16, 18, 21–23, 25, 39, 54, 55, 66, 73, 77, 79, 84, 88, 93, 94, 98, 100, 103, 106, 107, 123, 126, 156, 165].

4.5
Antihistaminics as Non-antibiotics

Antihistamines are histamine antagonist that reduce or eliminate effects mediated by histamine (an endogenous chemical mediator released during allergic reactions) through action at the H_1-receptor. The main therapeutic effect is mediated by negative modulation of histamine receptors, thus termed as antihistamines. Hence, 'antihistamine' refers only to H_1-receptor antagonists, also called H_1-antihistamines, as these compounds are actually inverse agonists at the histamine H_1-receptor, rather than antagonists per se [166]. The H_1-antihistamines are used in the treatment of histamine-mediated allergic conditions like allergic rhinitis, allergic conjunctivitis, contact dermatitis, urticaria, angioedema, diarrhea, atopic dermatitis, insect bites, anaphylactic reactions, nausea and vomiting [167]. First-generation or classical antihistamines can relieve allergic symptoms, but are typically potent anticholinergic agents, acting on α-adrenergic receptors and/or 5-hydroxytryptamine receptors. This lack of receptor selectivity of some first-generation agents such as diphenhydramine, carbinoxamine, clemastine, chlorpheniramine and brompheniramine is the basis of the poor tolerability profile. Several antihistamines, including some first-generation agents like diphenhydramine, bromodiphenhydramine, chlorpheniramine and so on, showed *in vitro* and *in vivo* antimicrobial activity, and some can enhance the activity of antibiotics or chemotherapeutic agents in combination by altering the membrane fluidity or permeability of sensitive and resistant strains [2, 3, 21, 24, 32, 33, 35, 51, 53, 54, 63, 70, 75, 85, 99, 129, 133, 135, 136, 146, 147, 165, 168–170].

4.6
In vitro and *In vivo* Antimicrobial Potential of Non-antibiotics

Many of the nonantibiotic compounds including dyes like methylene blue [1, 54, 101, 117, 152, 171] and acridine [24, 36, 142]; atabrine [141]; chlorproma-zine [11, 17, 22–27, 29, 31, 40, 54, 59, 61, 66, 76, 77, 93, 94, 98, 100, 102, 105, 106, 123, 124, 137, 165, 172–176], promazine [34, 35], femoxetin [49], triflupromazine [70], promethazine [36, 41, 110, 129, 136, 168], imipramine [177], trimiprime [55], prochlorperazine [100], thioridazine [83, 88, 90, 156], thioxanthene clopenthixol [73, 79, 132] and flupenthixol [39, 73]; TFP [14, 81, 84, 85, 178], aminoperazine [15], benzo-phenothiazine [16] and quinacrine [143]; antihistamines like diphenhydra-mine, bromodiphenhydramine [33, 135], methdilazine [3, 51, 53, 82, 85, 133, 134, 168–170] and triprolidine [63, 150]; dobutamine [68] and dicyclomine [71]; anti-inflam-matory diclofenac sodium [74, 89, 92, 155]; antihypertensives methyl-DOPA [48, 91] and propranolol [44]; cardiovascular agents amlodipine [69, 139] and oxyfedrine [67, 140]; anticoagulant analgesic like aspirin [57, 64, 130]; calcium channel blockers verapamil [144] and desipramine [145, 146, 148, 149]; β-adrenergic blockers [42]; general and local anesthetics lidocaine, lignocaine and procaine [28, 37, 38, 52, 171]; *N*-acetylcysteine [125, 179]; and diuretics like amiloride [46, 128] and probenecid [45] are reported to have antimicrobial activities as these agents can inhibit influenza virus [11, 12], herpes simplex virus [16] and HIV [15], *Mycoplasma* [17, 18], both Gram-positive and Gram-negative bacteria [19–74], *Mycobacterium* spp. [7,75–92], *Candida* spp. [93–95], *Amoeba proteus* [96, 97], *P. falciparum* [98–101], *Leishmania donovani* and *Leishmania major* [102–107], *Trypanosoma* [108–110] and helminths [111] *in vitro* and *in vivo* [2, 3, 19, 23, 26, 41, 49, 51, 53, 56, 61, 67–69, 71–74, 77, 82–89, 91, 105, 108, 109] by altering the permeability of the microbial cell membrane [2, 3, 18–20, 23, 25, 27, 30, 34, 38–40, 46, 54, 59, 60, 79, 81, 95, 101–103, 107, 112] and enhance the *in vitro* and *in vivo* activity of certain antibiotics against specific bacteria [2, 18, 39, 128, 130] to make antibiotic-resistant bacteria susceptible to the previously used antibiotics [86, 112, 141, 152, 154, 156, 180] and exhibit strong *in vitro* antimycobacterial activity against clinical strains resistant to one or more conventional antibiotics [83–88, 131, 143, 152, 154, 156, 181]. These compounds of diverse classes, primarily belonging to the phenothiazines, thioxanthenes and related groups with affinities for cellular transport systems, are characterized by their effects on the plasma membrane of eukaryotic cells (e.g. local anesthetic activity) and have been termed membrane stabilizers, which can alter the stability of inner membrane of bacteria. The *in vitro* and *in vivo* activities of some well-studied nonantibiotics with their mode of action are depicted in Table 4.2.

Table 4.2 Mode of action of some important nonantibiotics.

Activity on	Compounds	References
Adhesion of *Escherichia coli*	promethazine, imipramine	[41]
Escherichia coli	tranquilizer	[197]
Salmonella enteritidis	chlorpromazine	[23]
Antimicrobial	antipsychotics, dyes	[54]
Bacterial plasmid replication	phenothiazine	[43]
Plasmid elimination	phenothiazine, imipramine	[177]
Reversal of resistance		
chloroquine resistant *Plasmodium falciparum*	verapamil	[144]
antibiotic-resistant *Escherichia coli*	diclofenac sodium	[155]
Antibiotic-resistant bacteria	nonantibiotics	[154]
Mechanism of action		
bactericidal	phenothiazine	[19]
bactericidal	chlorpromazine	[27]
potassium efflux of *Staphylococcus aureus*	chlorpromazine	[194]
cell wall permeability	chlorpromazine	[174]
membrane permeability	methdilazine	[170]
lipid layer of membrane	chlorpromazine	[196]

4.6.1
Antiviral Activity of Non-antibiotics

It has been reported that some nonantibiotic compounds inhibit both DNA and RNA viruses [11–16]. When testing the phenothiazine dye methylene blue and its derivatives for their ability to photoinactivate viruses in red blood cell suspensions, the dye 1,9-dimethyl-3-dimethylamino-7-dimethylaminophenothiazine (1,9-dimethylmethylene blue) exhibited good intracellular and extracellular virucidal activity against several RNA and DNA viruses in the presence of light and oxygen. In addition, the hydrophobicity/hydophilicity of the compounds varied with the partition coefficients (2-octanol : water). It was found that compound 4-140 had a high affinity for protein but low affinity for DNA, while the high affinity of 1,9-dimethylmethylene blue for DNA and its efficient singlet oxygen yield suggest viral nucleic acid as a potential target [13].

The enzyme reverse transcriptase plays an essential role in the early steps of the replicative cycle of retroviruses and because of resistance against nucleoside analog inhibitors the importance of non-nucleoside analog inhibitors is increasing. It was reported that phenothiazine TFP and its metal complexes (TFP-VO, TFP-Cu, TFP-Ni, TFP-Pd, TFP-Sn) had much higher inhibitory activity on Moloney murine leukemia virus reverse transcriptase than that of TFP. Therefore, TFP and its metal complexes could be a new non-nucleoside analogs against the retrovirus replication [14].

T cell anergy, apoptosis and chronic activation of T lymphocytes are prevailing features in HIV infection. In infected patients an efficient natural antiviral activity cannot develop probably due to the failure of the antigen presentation by dendritic cells in chronically activated lymphoid tissues. It was reported that a phenothiazine derivative aminoperazine [2-amino-10-[3'-(1-methyl-4-piperazinyl)propylphenothiazine, $C_{20}H_{26}N_4S$; MW 354.51] can increase (effective dose 0.1–100 nM) the antigen-specific dendritic cell-driven proliferation and differentiation of *in vitro* HIV-infected and uninfected T cells as well as T cells from HIV-1 patients. This immunomodulatory effect of aminoperazine-sensitized dendritic cells is found to be due to soluble factors derived from dendritic cells. Aminoperazine can also increase HIV gag-p24-specific proliferation and anti-HIV cytotoxic activity of patient's CD8$^+$ T cells against autologous B lymphoblastoid cell lines expressing a HIV *gag* gene, resulting in the suppression of both proviral DNA and supernatant viral RNA in the HIV-1-infected patient's T cell culture. Thus, aminoperazine might be used for boosting the immune response of vaccinated individuals and for restoring the immunity of immunocompromised patients [15].

The combined antiviral effects of some benzo[*a*]phenothiazines and acyclovir on the multiplication of HSV-2 in Vero cells showed that the antiviral effect of acyclovir on a wild strain of HSV-2 was enhanced in the presence of 5-oxo-5H-benzo[*a*] phenothiazine and 6-methyl-5-oxo-5H-benzo[*a*]phenothiazine in a yield reduction test. The combinations of 5-oxo-5H-benzo[*a*]phenothiazine or 6-methyl-5-oxo-5H-benzo[*a*]phenothiazine or their derivatives with acyclovir reduced the infective virus population. When the above two most effective derivatives were simultaneously used with acyclovir against a wild HSV-2 strain during consecutive passages, the infective virus titers were moderately decreased, suggesting that a combination of these benzo [*a*]phenothiazines with acyclovir might enhance their antiviral activity, probably by reduction of the mutagenic rate in the virus populations [16].

4.6.2
Antibacterial Activity of Non-antibiotics

The nonantibiotics are reported to possess a variety of activities that range from modulation of function of the nervous system to antimicrobial and antiplasmid activity. Several reports indicated that both Gram-positive and Gram-negative bacteria are inhibited by nonantibiotics [19–74]. It was evident that Gram-positive bacteria are more susceptible to psychotherapeutic drugs [21, 25, 26, 31, 35, 182] at the range of 3–250 μg/ml [174]; while Gram-negative bacteria are inhibited at higher doses [21, 25, 26, 31, 35, 40, 182].

When 188 clinical isolates were tested against clopenthixol it was found that *trans* (*E*)-clopenthixol (Figure 4.14a) is the most active antibacterial, especially against *Pseudomonas aeruginosa*. On the other hand, when two neuroleptic phenothiazines [fluphenazine (Figure 4.12) and chlorpromazine (Figure 4.3)] and two antihistaminic phenothiazines [alimemazine (Figure 4.15) and promethazine (Figure 4.16)] were tested against 61 bacterial isolates *in vitro*, all four had antibacterial activity, but these

(a)

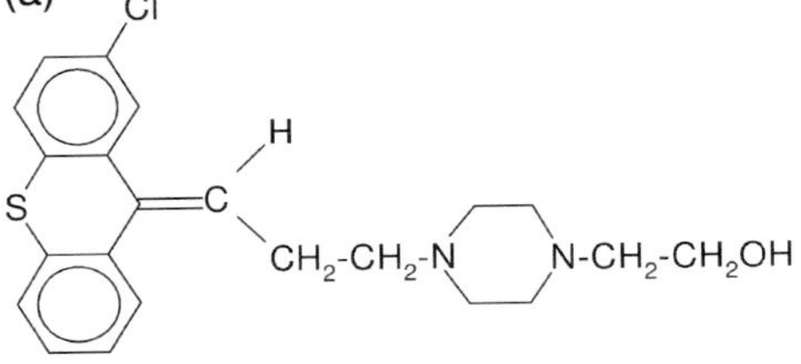

Antihistaminic (ID$_{50}$ 73 nM).
Antihypersecretory (ED$_{50}$ >11 μM/kg).
Neuroleptic (IC$_{50}$ 110 nM).
Antimicrobial effect *in vitro*.

(b)

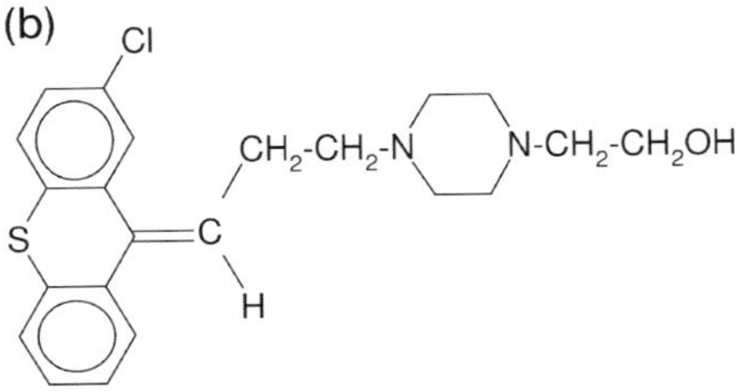

Antihistaminic (IC$_{50}$ 86 nM).
Antihypersecretory (ED$_{50}$ 1.6 μM/kg).
Neuroleptic (IC$_{50}$ 6.4 nM).
Antimicrobial effect *in vitro*.

Figure 4.14 (a) *Trans* (*E*)-clopenthixol. (b) *Cis* (*Z*)-clopenthixol.

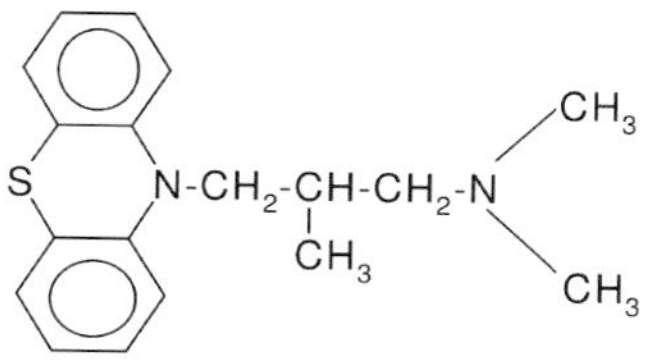

Antihistaminic (IC$_{50}$ NT).
Antihypersecretory (ED$_{50}$ NT).
Neuroleptic (IC$_{50}$ NT).
Antimicrobial activity *in vitro*.

Figure 4.15 Alimemazine.

agents were more potent against Gram-positive bacteria. The range of antibacterial potency (IC$_{50}$) was fluphenazine 29 μM (15 μg/ml), alimemazine 49 μM (37 μg/ml), promethazine 88 μM (28 μg/ml) and chlorpromazine 92 μM (29 μg/ml), and their antibacterial potency was not linked to their neuroleptic or antihistaminic potency. Thus, these phenothiazines may represent a pool of potentially new antimicrobial drugs [50]. A therapeutic application of these results, however, requires additional *in vitro* and *in vivo* testing, using s bacterial model as a model system to study the interaction of neuropharmacological and membrane-active compounds on biomembranes.

The neuroleptically inactive stereoisomeric analogs of the thioxanthene *trans* (*E*)-clopenthixol showed *in vitro* antimicrobial activity in other microbes also. In a murine *Pneumococcus peritonitis* model *trans* (*E*)-clopenthixol showed antibacterial activity at 0.3–0.9 mg per mouse, while at higher doses it can enhance the bacterial virulence. When combined with subtherapeutic doses of penicillin, a significantly higher survival rate was obtained compared with either drug given alone. *In vitro* studies demonstrated a similar combined effect [132]. These results indicate that the non-neuroleptic thioxanthene stereoisomer has antibiotic potential *in vitro* and *in vivo*. It has been reported that the *trans* (*E*)-clopenthixol (Figure 4.14a) has no neuroleptic effect, but has different antibacterial patterns from classical antibiotics [39], and clopenthixol (median IC$_{50}$ 18 μg/ml) is more active than chlorpromazine (median IC$_{50}$ 30 μg/ml) [18]. A general evaluation of the bacterial susceptibility in respect to their IC$_{50}$ values indicated that there is a positive correlation between the

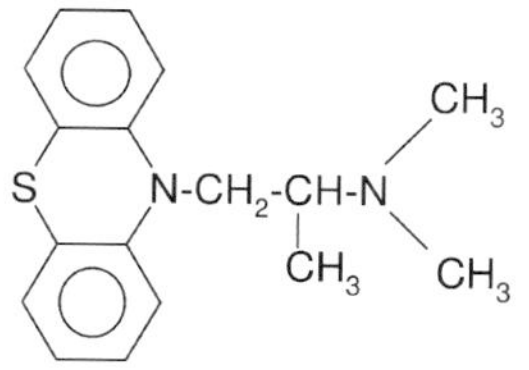

Antihistaminic (IC$_{50}$ 17nM).
Antihypersecretory (ED$_{50}$ > 16 μM/kg).
Neuroleptic (IC$_{50}$ 1600 nM).
Antimicrobial effect *in vitro*.

Figure 4.16 Promethazine.

growth-inhibiting effects of these drugs related to the individual bacterial strains. If a bacterial strain had low IC_{50} with one drug, it shows similar activity with other drugs [18, 39].

Dicyclomine hydrochloride, an antispasmodic agent, can rapidly kill *S. aureus* NCTC 6571, 8530 and several other strains, while the drug was bacteriostatic against *Shigella boydii* 8 NCTC 254/66 and *Salmonella typhimurium* NCTC 74, and high rate of killing was achieved for Gram-positive bacteria within 2 h [71]. In animal an model, the drug at 30 and 60 µg/g body weight per mouse could significantly protect the animals challenged with 50% mouse lethal dose (MLD_{50}) of *S. typhimurium* NCTC 74 [71]. Another study with the antipsychotic flupenthixol revealed that the drug can significantly protect ($P < 0.001$) mice challenged with *S. typhimurium*, and remarkably reduced the number of viable bacteria in organ homogenates and blood [73]. Diclofenac sodium, an anti-inflammatory agent, also showed remarkable inhibitory action against sensitive and resistant clinical isolates of various Gram-positive and Gram-negative bacteria. At 1.5 and 3.0 mg/g body weight per mouse diclofenac significantly protected the animals from the lethality of *S. typhimurium* infection; and in combination with streptomycin *in vitro* and *in vivo* it was more effective, indicating that diclofenac has potential for the management of problematic antibiotic-resistant bacterial infections [74, 89, 92, 155]. Interestingly, a variety of 'nonantibiotic' pharmaceutical preparations used in the management of noninfectious diseases have shown *in vitro* antimicrobial activity. Kruszewska *et al.* [20] studied over 160 different pharmaceutical preparations on standard American Type Culture Collection strains of *S. aureus, E. coil, P. aeruginosa* and *C. albicans*. It was found that many of these drugs (acyclovir, alendronate, alverine, butorphanole, clodronate, diclofenac, emadastine, etodolac, fluvastatine, ketamine, levocabastine, losartan, matipranolol, mesalazine, naproxen, oxaprosine, oxymethazoline, proxymetacaine, ribavirin, rutoside with ascorbate, sulodexide, tegaserole, telmisartan, temosolomide, ticlopidine, tolfenamic acid, tramadole and tropicamide) inhibited growth of at least one of the examined strains. *S. aureus* was susceptible to most of the above drugs, while ticlopidine was active against *S. aureus, E. coli* and *C. albicans* [minimum inhibitory concentrations (MICs) of 0.45, 0.45 and 0.65 mg/ml, respectively]. Oxymetazoline showed activity against *S. aureus* and *E. coli* (MICs of 0.005 and 0.025 mg/ml, respectively), while *P. aeruginosa* was sensitive to alendronate, clondronate, oxaprozine, ribavirin and tramadole at an MICs of 10, 63, 60, 3 and 43 mg/ml, respectively [20].

4.6.3
Activity Against Diarrheagenic Bacteria

When several nonantibiotics are exposed to diarrhea-causing bacteria it was found that a number of *Vibrio cholerae* are inhibited by phenothiazines (Figure 4.2), chlorpromazine (Figure 4.3) and thioxanthene [2, 23, 31–33, 35, 47, 49, 51, 57, 112, 133, 135, 140, 182]. Dash *et al.* [35] reported that diarrheagenic bacteria *Shigella* spp. were inhibited by the phenothiazine promazine hydrochloride; while Kristiansen *et al.* [49] found that *Shigella* and *Yersinia* are most sensitive to femoxetine

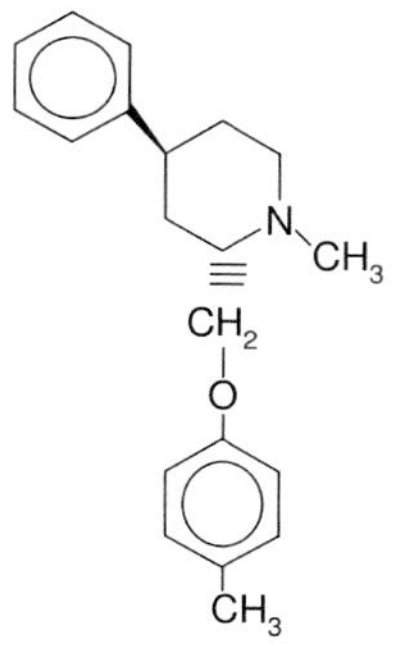

Antihistaminic (IC_{50} 17 μM).
Antihypersecretory (ED_{50} NT).
Neuroleptic (IC_{50} 12 μM).
Serotonin reuptake inhibitory (IC_{50} 80 nM).
Antimicrobial effect *in vitro*.

Figure 4.17 Femoxetine.

(Figure 4.17), while *E. coli* 10 407, *Salmonella typhi* and *Salmonella typhi paratyphi* are moderately sensitive. It was found that clopenthixol has the strongest antibacterial effect against *V. cholerae*, but chlorpromazine has both antibacterial and antisecretory activity. On the other hand, both clopenthixol *cis* (Z) and *trans* (E) isomers (Figure 4.14a and b) have antibacterial effect, but the *cis* form has antisecretory and neuroleptic effects [47]. When 20 diarrhea-producing bacteria of enterobacteriaceae were tested against the antidepressant femoxetine and its stereoisomeric analogs it was observed that the *trans* forms of femoxetine are more than twice as active as the *cis* forms and inhibited all the strains below 400 μg/ml (1.2 mM). The two *cis* compounds inhibited 11 and nine strains of the 20 strains, respectively, between 100 and 800 μg/ml (0.3–2.4 mM) [49]. Flupenthixol, an antipsychotic thioxanthene, containing a trifluoromethyl substituent at position 2, exhibited distinct antibacterial activity against 352 strains of bacteria from three Gram-positive and 13 Gram-negative genera with an MIC from 10 to 100 μg/ml; and is found to be bacteriostatic against *S. aureus* and *V. cholerae* [73]. These results pointed out that the bacterial cell is a target for many of the psychopharmacologically active antipsychotic and their stereoisomers, and these agents may represent a pool of potentially new antimicrobial drugs.

4.6.4
Activity Against Mycobacteria

Earlier reports indicated that the nonantibiotics used in the management of psychosis can also inhibit the *in vitro* growth of *M. tuberculosis* at significantly greater concentrations in patients [7, 75–92]. It was found that chlorpromazine is concentrated by human macrophages at 10–100 times its concentration in plasma and can inhibit mycobacteria phagocytosed by these cells [156]. Phenothiazines have significant *in vitro* activity against susceptible, polydrug-resistant and MDR strains of *M. tuberculosis*, and can enhance the activity of some first-line antitubercular agents [156]. Thioridazine, a phenothiazine with very mild antipsychotic effect and drowsiness, showed equal antituberculosis activity *in vitro* like chlorpromazine, thus thioridazine was studied as an adjuvant to the four- or five-drug regimens employed for the management of a freshly diagnosed tuberculosis infection of unknown antibiotic susceptibility, at least for first 2–3 months when the antibiotic susceptibility was made. As thioridazine also enhances the activity of antimycotic antibiotics rifampicin and streptomycin, that frequently have adverse effects, additional studies evaluating its use as an adjuvant may allow a reduction in the dosages of these antibiotics as well as a decreased frequency of adverse effects [156]. Although chlorpromazine and thioridazine have equal *in vitro* activities against antibiotic-sensitive and -resistant *M. tuberculosis*, they are not used as antituberculosis agents as their *in vitro* activities require higher concentrations which cannot be clinically achievable and chronic use of chlorpromazine produces severe side-effects. Thus, thioridazine, a phenothiazine with less side-effects than chlorpromazine, which can kills antibiotic-sensitive and -resistant *M. tuberculosis* within human macrophages at concentrations well below those present in the plasma of patients (nontoxic), and does not affect *in vitro* cellular immune processes, can be a serious candidate for the management of a freshly diagnosed pulmonary tuberculosis or as an adjunct to conventional antituberculosis therapy if the patient originates from an area known to have a high prevalence of MDR *M. tuberculosis* isolates [83, 88, 156]. Nevertheless, further studies and clinical trials are required to determine whether thioridazine itself may be safely and effective. It is well known that patients presenting with AIDS are predisposed to coinfection with *Mycobacterium avium*. The management of such patients is problematic due to underlying immuno-incompetence and the high resistance of *M. avium* to most nontoxic compounds. Therefore, the need for effective agents is obvious. When several phenothiazines and related compounds, like chlor-promazine (Figure 4.3), thioridazine (Figure 4.9), promazine (Figure 4.7) prometha-zine (Figure 4.16) and desipramine, were investigated *in vitro* against reference and clinical strains of *M. avium*, it was found that all these phenothiazines could inhibit *M. avium* (MIC $10 \geq 50$ mg/l), but relatively mild thioridazine (Figure 4.9) was the most active against *M. avium* [90]. Similarly, the nonsteroidal anti-inflammatory drug diclofenac sodium exhibited remarkable inhibitory action against both drug-sensitive and -resistant clinical isolates of *M. tuberculosis*, as well as other mycobacteria. When diclofenac was tested *in vitro* against 45 different strains of mycobacteria, it inhibited most strains at 10–25 µg/ml concentration. At 10 µg/g body weight of mice, it could significantly ($P < 0.01$) protect the animals challenged with MLD_{50} of *M. tuberculosis* H_{37} Rv102. Further study revealed that diclofenac can produce statistically significant ($P < 0.05$) synergism with streptomycin against *Mycobacterium smegmatis* 798 having a fractional inhibitory concentration (FIC) index of 0.37 [92], and was bactericidal at 40 µg/ml ($4 \times$ MIC) against *M. tuberculosis* H37Rv. Albino male mice intravenously infected with 2.3×10^7 *M. tuberculosis* H37Rv and then treated with diclofenac (10 mg/g/day) or streptomycin (150 mg/g/day), alone or in combination for 4 weeks, showed significantly ($P < 0.05$) reduced bacterial counts, mean spleen weight, and counts in the lungs and spleen compared with control(s) [89]. Again, the antihypertensive agent methyl-L-DOPA showed significant *in vitro* activity against atypical species including *M. avium* complex, *Mycobacterium scrofulaceum*, *Mycobacterium xenopi* and

Mycobacterium marinum, and the rare pathogen *Mycobacterium fortuitum.* The *in vitro* and *in vitro* activity against 19 wild mycobacteria and 34 clinical isolates of both drug-sensitive and -resistant *M. tuberculosis* showed that these strains were inhibited at 10–25 µg/ml. When methyl-L-DOPA was injected into male mice challenged with MLD_{50} of *M. tuberculosis* H37Rv102 at 10 µg/g body weight (20 g each), it could significantly ($P < 0.01$) protect the challenged mice [91].

4.6.5
Activity Against Protozoa

It has been reported that induced pinocytosis in *A. proteus* can be inhibited by membrane-stabilizing drugs [96] and some phenothiazines inhibit the *in vitro* growth of pathogenic free-living amebae [97]. To tackle the increasing resistance of *P. falciparum* to chloroquine, membrane stabilizers like chlorpromazine and its stereoisomers *cis* (*Z*)- and *trans* (*E*)-clopenthixol (Figure 4.14a and b) showed *in vitro* antimalarial activity, using a modified Desjardins' [^{3}H]hypoxanthine assay [98]. The IC_{50} of chlorpromazine was found to be 1.028 ng/ml (3.2 µM), while for *trans* (*E*)-clopenthixol it was 758 ng/ml (1.6 µM) and 436 ng/ml (0.9 µM) for *cis* (*Z*)-clopenthixol [98]. The inhibitory effect of *trans* (*E*)-clopenthixol in low concentrations on *P. falciparum in vitro* seems particularly promising, as *trans* (*E*)-clopenthixol has no neuroleptic effect [98–101]. Verapamil (a calcium channel blocker) [144], desipramine (antidepressant) and cyproheptidine (an anabolic stimulant and antihistamine) can reverse the chloroquine resistance in *P. falciparum* [145–149], while a combination of chlorpromazine or prochloroperazine with chloroquine can alter the resistance pattern of *P. falciparum in vitro* [100]. Interestingly methylene blue and some of its analogs exert an antimalarial effect on *P. falciparum* in culture, and inhibit *Plasmodium vinckei petteri* and *P. yoeli nigeriensis in vivo* [101, 117]. One of the problematic protozoan parasites of tropical countries is *Leishmania.* It was observed that chlorpromazine and its derivatives have a lethal effect of on *L. donovani* [102, 103] and *Leishmania tropica* in human macrophages *in vitro* [104], while diffuse cutaneous leishmaniasis can be treated with chlorpromazine ointment [105]. The antileishmanial activity of chlorpromazine and other tricyclic drugs [106] may be due to reduced proton motive force in *L. donovani* promastigotes [107]. Similarly, phenothiazines and related polycyclics had both antimalarial and trypanocidal activities [108], as phenothiazines can inhibit trypanothione reductase of trypanosoma [109], as was observed with trypanocidal effects of promethazine [110]. Studies with some phenothiazines indicated that they also have antihelminthic activity [111].

4.7
Single and Combined Activity Against Bacteria

The nonantibiotic agent dobutamine hydrochloride is reported to have powerful *in vitro* bacteriostatic activity at 5–200 mg/ml and *in vivo* studies revealed that the drug offered significant protection ($P < 0.001$) to mice challenged with virulent

S. typhimurium [68]. The cardiovascular drug amlodipine exhibited remarkable *in vitro* antibacterial activity against diverse bacterial strains, and a mouse-virulent *S. typhimurium*. Twelve bacterial strains, sensitive to amlodipine and antibiotics like benzyl penicillin, streptomycin, chloramphenicol, tetracycline, erythromycin and ciprofloxacin, showed that the combination of amlodipine and streptomycin revealed marked synergism (FIC index of 0.24) and can significantly ($P < 0.001$) protect the challenged mouse, indicating that this combination can enhance the scope of prolonged antibiotic therapy in bacterial infections, especially for drug-resistant bacteria [69, 139]. Another cardiovascular drug, oxyfedrine, showed *in vitro* antibacterial activity against 501 strains of both Gram-positive and Gram-negative bacteria with significant protection to mice challenged with *S. typhimurium*. Distinct and statistically significant ($P < 0.01$) synergism was observed between oxyfedrine and tetracycline with 10 bacterial strains sensitive to oxyfedrine as well as benzyl penicillin, chloramphenicol, ciprofloxacin, erythromycin, streptomycin and tetracycline, with an FIC index of 0.37 [67, 140]. Again, the antipsychotic drug prochlorperazine (Figure 4.18) the MIC was found to have an MIC from 25 to 200 µg/ml when screened against bacteria. A further study revealed that the efficacy of this drug could be enhanced in the presence of antihistaminic methdilazine and the combination has remarkable antimicrobial action [72].

Four bacterial strains sensitive to prochlorperazine, methdilazine, fluphenazine (Figure 4.12) and thioridazine (Figure 4.9) tested with the prochlorperazine-methdilazine combination revealed marked synergism (FIC index 0.37), indicating that this combination might be a new therapeutic approach against drug resistance in bacterial infections [72]. *E. coli*, the causative agent of uncomplicated urinary tract infections (UTIs), accounts 85% of recurrent cystitis and 35% recurrent pyelonephritis in man. A recent study with 136 urine samples from UTI patients showed that 33% of *E. coli* isolated from 85 samples were resistant to ampicillin, tobramycin, augmentin, nalidixic acid, cefuroxime, nitrofurantoin, kanamycin, pipemidic acid, chloramphenicol, cefotaxime, cefamendol, ofloxacin, ceftizoxime, norfloxacin and amikacin. The anti-inflammatory drug diclofenac sodium exhibited significant antibacterial activity against these *E. coli* strains, both *in vitro* and *in vivo*, with an MIC of 5–50 µg/ml ($MIC_{90} = 25$ µg/ml). Thus, diclofenac alone or in combination with other antibiotics can be effective to treat UTIs caused by *E. coli* [74]. A list of some well studied synergism is presented in Table 4.3.

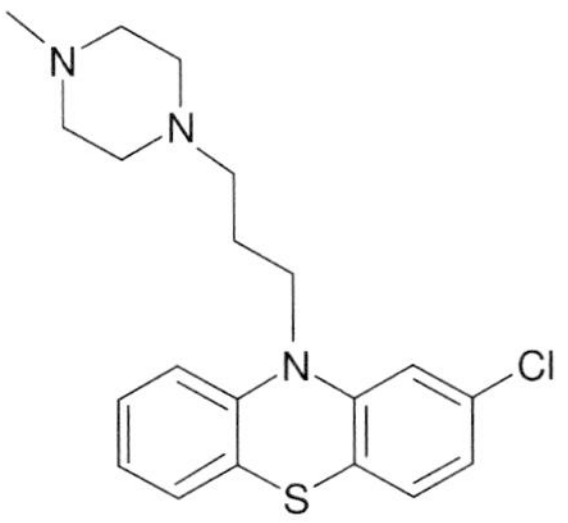

Figure 4.18 Prochlorperazine.

Table 4.3 Synergism between nonantibiotics and antibiotics or chemotherapeutic agents.

Nonantibiotic	Antibiotic/nonantibiotic	Organism	Reference
Amlodipine	streptomycin	*Staphylococcus aureus, Escherichia coli*	[139]
Amiloride	tobramycin	*Pseudomonas* spp.	[128]
Bromodiphenhydramine	penicillin		[135]
Chlorpromazine	antibiotic		[137]
Clopenthixol	penicillin	*Streptococcus pneumoniae*	[132]
Diclofenac sodium	streptomycin	*Mycobacterium tuberculosis*	[89, 92, 155]
Methdilazine	streptomycin		[133]
	aminoglycosides		[134]
	prochlorperazine		[72]
Oxyfedrine HCl	tetracyclines		[140]
Promethazine	gentamicin		[136]
Phenothiazine	aminoglycosides isoniazid, streptomycin		[131, 134, 181]
Phenothiazine	antibiotic	*Burkholderia pseudomallei*	[126, 180]
Phenothiazine	quinacrine		[143]
Piperazine	chemotherapeutic agent		[124]
Thioxanthene	flupenthixole		[73]
Trimethoprim	sulfamethoxazole		[185]

4.8
Synergism and 'Reversal of Resistance': A Special Synergy

Synergism is said to occur when two antimicrobial drugs in combination produce a greater inhibitory effect than the sum of the individual activities of each drug of that combination, keeping the total drug concentration constant (Figures 4.19 and 4.20). Synergy does not always require that the drugs used in combination against a given microbe be active at the concentrations used [183, 184]. Some of the important synergistic combinations between nonantibiotics and antibiotics are presented in Table 4.2. In a few cases, the *in vitro* synergism is also observed *in vivo*, which resulted in the development of drug combinations for the management of difficult-to-treat infections with one antibiotic alone [185, 186].

Clinically, synergistic combinations of antibiotics offer certain theoretical advantages, as when one of the drugs of that combination has potential toxicity [137]. Another advantage is the use of a drug that has no useful antimicrobial activity at the concentration used, but its presence increases the activity of a second antibiotic to which the organism was previously resistant [123, 136]. Unfortunately, these two potential advantages of *in vitro* synergic combinations have rarely been studied *in vivo* [18, 64]. Synergism between conventional

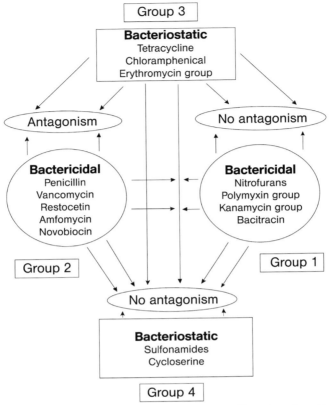

Figure 4.19 Antagonism between different groups of bacteriostatic and bactericidal agents.

antibiotics and 'nonantibiotics' has also been suggested and supported by a number of reports [1, 18, 136, 137, 143].

Several nonantibiotic compounds produced such synergism against a wide range of bacteria, mycobacteria and protozoa. Some of these combinations are: chlorpromazine with chloroquine [100], chlorpromazine with several antibiotics [137], chlorpromazine with perphenazine [124], promethazine with gentamicin [136], thioxanthene with penicillin [132], prochlorperazine with methdilazine [72], methdilazine with aminoglycosides [51, 133, 134, 169], amlodipine with streptomycin [69, 139], oxyfedrine with tetracycline [67, 140], diclofenac with streptomycin [89, 92, 155], trifluromethylketone with promethazine [65], phenothiazine with antibiotics [127, 134, 180], *N*-acetylcysteine with carbenicillin [125], amiloride with tobramycin [128], bromodiphenhydramine with penicillin [135]; between tricyclic antidepressants [18, 51, 136, 137, 141–143]; and other membrane stabilizing compounds used in psychoses, pain and so on [1, 12, 57, 125, 128, 179]. The synergy produced by chlorpromazine in combination with aminoglycosides, β-lactams, quinolones and so on has been limited to bacteria that are sensitive either to the antibiotic or to chlorpromazine [137].

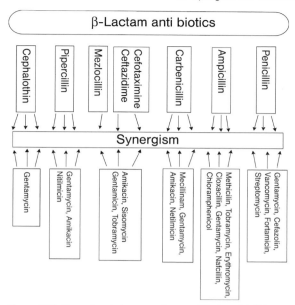

Figure 4.20 Synergism between various antibiotics with β-lactam agents.

4.8.1
Clinical Synergism

In children with recurrent pyelonephritis, aminoglycoside and chlorpromazine has shown a significant synergic effect compared with the aminoglycoside alone [136]. It was reported that the disruption of BpeAB–OprB function could attenuate *B. pseudomallei* virulence in cell invasion and cytotoxicity [187], and the use of a subinhibitory concentration ($0.06 \times$ MIC) of erythromycin together with a phenothiazine inhibited *B. pseudomallei* and protected the mammalian cells from infection and cytotoxicity. It is known that erythromycin entered eukaryotic cells and is concentrated within polymorphonuclear leukocytes to kill intracellular pathogens. Chlorpromazine is also concentrated by human macrophages and exerted bactericidal action against *S. aureus* phagocytosed by human monocyte-derived macrophages [66]. The combination of a phenothiazine with a subinhibitory concentration of erythromycin might accumulate within the mammalian cells and enhance their bacteriostatic/bactericidal action or inhibit the efflux pump, and consequently reduce its virulence by inhibiting the efflux of a key metabolite(s) required for virulence and pathogenicity of *B. pseudomallei* [180, 188].

4.8.2
Reversal of Resistance

A specific type of synergy, termed 'reversal of resistance' [154], is noticed when a β-lactam antibiotic is used with a phenothiazine against methicillin-resistant

S. aureus (MRSA; MIC 100 mg/l). Interestingly, this MRSA become sensitive to methicillin (at 6.2 mg/l) when cultured in a medium containing chlorpromazine [123]. The other chlorpromazine-related compounds can also reduce the MIC to 1.6 mg/l, and the same effect is observed when penicillin-resistant *Corynebacterium* are cultured with low concentrations of chlorpromazine and penicillin [123, 151]. This type of 'reversal of resistance' has also been produced by combinations of nonantibiotics against *P. falciparum*, *L. donovani* and other parasites [2, 165]. Phenothiazines have been known to exhibit antimycobacterial activities *in vitro* and *in vivo* [18, 83, 189], and the well-studied phenothiazine chlorpromazine have significant antimycobacterial activity [76–78, 86, 87, 90, 131, 137, 154, 156, 181]. However, chlorpromazine has several undesirable side-effects and had not been considered as a potential antimycobacterial drug before the development of multidrug resistance mycobacteria. Since phenothiazines are concentrated almost 100-fold by macrophages [83, 172, 189] and pulmonary cells [173] the concentration required in tissue fluid for intracellular antimycobacterial activity would be similar to that reached in the treatment of psychosis (0.5 mg/l). Another phenothiazine, thioridazine, having less side-effects than chlorpromazine, is also as active as chlorpromazine against various wild and resistant strains of *M. tuberculosis*, including those that are resistant to one or more conventional antimycobacterial agents [18, 83]. These qualities of some phenothiazine derivatives, having lower activities in the nervous system, mean that they can be considered as potential candidates for the initial management of fresh cases of tuberculosis. As phenothiazines enhance the activity of antimycobacterial agents [83, 189] they may have an additional adjuvant function. However, clinical studies are needed to evaluate the potential role of thioridazine or other phenothiazines in the management of mycobacteria resistant to conventional antimycobacterial agents.

Mechanisms of antibiotic resistance of bacteria include efflux pumps which extrude the antibiotic prior to reaching its target. Many phenothiazine compounds inhibit the activity of efflux pumps and thereby alter the susceptibility of bacteria. It was observed that chlorpromazine (Figure 4.3) and thioridazine (Figure 4.9) reduce the susceptibility of MRSA, but not of methicillin-susceptible *S. aureus* (MSSA) to oxacillin (MIC of oxacillin reduced from more than 500 to 10 mg/l). Reserpine, an inhibitor of antibiotic efflux pumps, also reduced the resistance of MRSA to oxacillin, suggesting the presence of an efflux pump that contributes to the antibiotic resistance of MRSA [153]. In MDR organisms and cancer cells the intracellular efflux pump is one of the major targets of many drugs, although drug efflux mechanisms other than pumps are reported to be affected by nonantibiotics (e.g. depressant phenothiazines, thixenes and antidepressant phenylpiperidines), alone or in combination with antimicrobials on MDR *S. aureus*, *S. epidermidis*, *E. faecalis*, *S. pyogenes* and *Streptococcus pneumoniae* [153, 154]. Of the nonantibiotics, L-thioridazine, *trans*-clopenthixol and phenylpiperidine isomers femoxetine and paroxetine were found to be potent antimicrobials, pointing to a possible general isomeric structure–activity relationship. Hence, these compounds can be regarded as new efflux inhibitors. Moreover, these isomers had less neurotropism with reduced toxicity that rules out concerns about the adverse effects and therapeutic safety for infected patients in life-threatening situations. The combination of appropriate isomeric nonantibiotics having low neurotropism and toxicity with classical antimicrobials promises early, new therapeutic strategies salutary against microbial resistance, resistance development, pathogenicity and virulence [154].

4.9
Mechanism of Action of Non-antibiotics

The precise mechanism by which nonantibiotics exert their effects on susceptible bacteria is not yet fully known. However, the effects noted from ultrastructural and biochemical studies strongly support that these agents primarily affect the cell wall of both susceptible and resistant bacteria. *S. aureus* susceptible to chlorpromazine (MIC 10–50 mg/l) exhibits a wide array of changes in its cell wall very similar to changes produced by a β-lactam antibiotic [59]. The ultrastructure of *S. aureus* exposed to a sub-MIC concentration of chlorpromazine with or without oxacillin showed that chlorpromazine can mimic the action of a β-lactam on the *S. aureus* cell wall [2, 18, 172, 176, 190]. The β-lactam antibiotic action on the bacterial cell wall involves inhibition of penicillin-binding proteins, which helps in the synthesis and construction of cell wall. Hence, a similar action may be produced by the phenothiazines, either directly or indirectly [43, 151, 190–192]. The effects of chlorpromazine on the morphology of *E. coli* also mimic the effects of ampicillin, as both produce significant elongation of the cell [43, 190, 192]. However, salmonellae, which are very resistant to chlorpromazine (MIC 110–200 mg/l), are not inhibited by chlorpromazine below the MIC, but lose the 'rough' texture of their cell walls due to loss of a 50-kDa protein and this interaction is direct because the organisms are not agglutinated by antiserum to O antigen [193]. These studies indicated that chlorpromazine exerts its effect on bacterial cell walls by affecting the efflux of potassium via the plasma membrane [18, 194]. The mechanism by which synergy may result from the combination of nonantibiotics and conventional antibiotics is also unknown.

However, Chattopadhyay *et al.* [170] demonstrated that the growth inhibition of *E. coli* and *S. aureus* was accompanied by significant release of K^+ and ultraviolet-absorbing small molecules upon exposure to methdilazine, an antihistaminic phenothiazine. A severe decrease in [^{14}C]glucose uptake and a rapid efflux of hexose from sugar-preloaded bacteria were observed without visible cellular lysis. Hence, considerable damage to membrane permeability by methdilazine was proposed to explain the rapid loss in colony-forming units per milliliter of the bacteria [170]. Again, the effect of anesthetics, pentobarbital, chlorpromazine and other tranquilizers on the physiology and membrane lipid composition of *E. coli* [195–197] indicated that many nonantibiotics probably interfere with the activity of the microbial membrane. Since the activity of many nonantibiotics are manifested at the bacterial plasma membrane, like in sensitive eukaryotic cells, it would be reasonable to expect that the action of the nonantibiotic is to influence the permeability of the cell. *In vitro* studies with *S. aureus* showed that chlorpromazine had bacteriostatic and bactericidal activity along with depigmentation, and chlorpromazine at a bacteriostatic concentration affects the transport of potassium across membrane in the same manner as described for mammalian muscle tissue [198, 199]. One of the targets for synergistic interaction between phenothiazines and the susceptible antibiotics seems to be the proton gradient, and chlorpromazine is shown to affect ion flux across the membrane in *S. aureus* and *Saccharomyces cerevisiae*, and alter the transmembrane potential in *L. donovani* [2, 18, 107, 172]. Interestingly, the antimicrobial effect of trifluoromethyl ketones was observed against *Bacillus megaterium* and *Corynebacterium michiganese*, but not against *P. aeruginosa* and *Serratia marcescens*. It was found that trifluoro-dimethyloxazol-propanone and benzoxazolyl-trifluoro-propanone inhibited *E. coli*, while benzoxazolyl-trifluoro-propanone was also effective against yeasts [65]. The combination of promethazine with benzoxazolyl-trifluoro-propanone was significantly synergistic against *E. coli*, especially the proton pump-deficient mutant, which suggest that membrane transporters are the target of trifluoromethyl ketones, as the inhibition was more marked in the proton pump-deficient *E. coli* mutant [65]. As the rates of MDR *M. tuberculosis* continue to escalate globally and no new effective drug has been made available, compounds that enhance the killing activity of monocytes against those bacteria is of interest. Human neutrophils are known to be effective and efficient killers of bacteria, but macrophages derived from monocytes are almost devoid of killing activity. Interestingly, these monocytes can be transformed into effective killers of mycobacteria or staphylococci when exposed to clinical concentrations of a phenothiazine or efflux pump inhibitor (reserpine and verapamil), or K^+ transport inhibitor (ouabain).

A review correlates the mechanisms by which these agents manifest their effects, and a model describing the mechanisms by which efflux pumps of the phagosome–lysosome complex are inhibited by K^+ flux inhibitors and predicting the existence of a K^+-activated exchange (pump) probably located in the membrane that delineates the lysosomes is of great interest [181]. This putative pump is immune to inhibitors of K^+ flux and can cause the acidification of the lysosome, thereby activating its hydrolytic enzymes. As the nonkiller macrophage can be transformed into an effective killer by a variety of compounds that inhibit K^+ transport, perhaps it would be wise to develop drugs that enhance the killing of these cells as such drugs would not be subject to any resistance, like conventional antibiotics. Another study showed that chlorpromazine and thioridazine can reduce the susceptibility of MRSA to oxacillin, and prochlorperazine, chlorpromazine or promazine have a synergistic interaction with many antimicrobial agents, thereby enhancing their antimicrobial potency against *B. pseudomallei*. The antibiotics that interacted synergistically with phenothiazines include streptomycin, erythromycin, oleandomycin, spectinomycin, levofloxacin, azithromycin and amoxicillin-clavulanic acid, where the MICs of these antibiotics were reduced as much as 8000-fold in the presence of these phenothiazines [138, 180, 187, 188]. Omeprazole, a proton pump inhibitor, augmented the antimicrobial activities similar to phenothiazines, suggesting that the phenothiazines might have interfered with the proton gradient at the inner membrane, as *B. pseudomallei* accumulated more erythromycin in the presence of phenothiazines, an effect similar to that of carbonyl cyanide *m*-chlorophenylhydrazone, a proton gradient uncoupler [187]. Hence, in the presence of phenothiazines, a much reduced concentration of erythromycin (0.06 × IC) can protect human lung epithelial cells and macrophages from *B. pseudomallei* infection and its cytotoxicity [187, 188]. In Gram-negative bacteria multidrug resistance is primarily caused by overexpression of efflux pumps that extrude unrelated antibiotics from the periplasm or cytoplasm of the bacterium prior to reaching their intended target. A variety of efflux pump inhibitors can be used as 'helper compounds' in combination with antibiotics to which the organism is initially resistant or some inhibitors with higher toxicity may serve as lead compounds [200].

4.10
Conclusions

Many of the compounds that are used for the therapy of noninfectious pathology have antimicrobial properties [20, 201] and these so called 'nonantibiotics' showed potential for the therapy of problematic infections [152, 202]. Functionally, the 'nonantibiotics' consists of two groups. The first group is made up of antimicrobial nonantibiotics that have direct antimicrobial activity [152]. The second group contains two subclasses – some 'help' to alter the permeability of the microorganism to a given antibiotic [59, 131, 155, 170] and are known as 'helper compounds', while others that enhance the killing activity of macrophages that have phagocytosed the microorganism [181] are known as 'macrophage modulators'. Although the antimicrobial activity of various 'nonantibiotic' compounds has been known since the 1880s, the enhancement of the activity of conventional antibiotics by these compounds has only studied been recently [116, 145]. It is now clear that the mechanism by which phenothiazines, thioxanthenes and related agents inhibit the function of MDR efflux pumps is multifactorial, including perturbation of membrane energetics and possibly a direct interaction with the transporters themselves. A single amino acid change in human P-glycoprotein affects the inhibitory activity of both *cis* (*Z*)- and *trans* (*E*)-flupentixol, suggesting that these stereoisomers directly interact with that pump [180, 187]. However, further work is necessary to establish direct evidence from bacterial efflux pumps. It would also be of great interest to determine whether these compounds can inhibit the activity of other bacterial drug transporters, such as the MexAB–OprM system of *P. aeruginosa* or ATP-dependent macrolide efflux pump MsrA of *S. epidermidis* and *S. aureus*. In that case structure–activity relationships of phenothiazines and thioxanthenes, especially *trans* isomers, would be a logical step toward finding potent and minimally toxic, broad-spectrum bacterial MDR efflux pump inhibitor. It is reasonable to evaluate other clinically used drugs, like selective serotonin reuptake inhibitors that possess antimicrobial activity, as potential efflux pump inhibitors.

The antimicrobial activities of nonantibiotics, their ability to enhance the activity of antibiotics, reverse resistance to conventional antibiotics and their activity against MDR bacteria (e.g. *M. tuberculosis*, *P. falciparum*, *L. donovani*, *Trypanosoma*, etc.) may be of considerable significance, as many of these agents are either developing multidrug resistance to the existing antimicrobials or are beyond the control of modern therapeutic regimen. However, with the exception of one study [173], the *in vitro* antimicrobial activity of phenothiazines required concentrations that exceed those employed for the management of other conditions. In this chapter we have discussed the antimicrobial potential of diverse classes of nonantibiotics, the mechanisms by which some nonantibiotics have direct antibacterial activity and assist the penetration of antibiotics by their ability to inhibit efflux of the antibiotic prior to reaching its intended target, while some nonantibiotics directly transform nonkiller cells into effective killers of pathogenic bacteria that are problematic to manage, as well as the ability of some compounds to cure bacterial plasmids. Hence, the significance of 'nonantibiotics' may be as complementary or helper compounds in combating microbial resistance in both human diseases and animal husbandry. Nevertheless, the potential offered by nonantibiotics is sufficiently great to merit further studies by investigators in the field of infectious diseases.

Acknowledgments

The authors are dedicating this compilation to Late Professor A.N. Chakrabarty, who started the work in Indian Subcontinent in the late 1960s, realizing the future importance of these compounds in infectious disease control. Our acknowledgement is to Dr (Mrs) Jette Elisabeth Kristiansen (Denmark) and Professor Joseph Molnar (Hungary), whose clinical eye, legendary observation and wisdom was the initiator of scientific endeavor of 'nonantibiotics' in the international arena. We pay deep homage to Late Professor J.D. Williams (United Kingdom), who helped in directing the outcome of these studies in clinical management. We also acknowledge the consistent and persistent scientific effort of Professor Sujata Ghosh Dastidar to uphold the research on nonantibiotics. The authors also acknowledge the help of Sonali Das, a PhD student and the Officer In-Charge of ICMR Virus Unit, Kolkata, during the preparation of this manuscript.

References

1 Ehrlich, P. (1913) *Lancet*, **ii**, 445–451.

2 Kristiansen, J.E. (1992) *Acta Pathologica, Microbiologica et Immunologica Scandinavica Supplementum*, **30**, 7–14.

3 Chattopadhyay, D. (1989) PhD Thesis, Jadavpur University, Kolkata.

4 Domagk, G. (1935) *Deutsche Medizinische Wochenschrift*, **61**, 250–253.

5 Colebrook, L. and Kenny, M. (1936) *Lancet*, **i**, 1279–1286.

6 Andride, V.T. (1990) Laboratory diagnosis and therapy of infectious diseases in *Principles and Practice of Infectious Diseases*, 3rd edn (eds G.L. Mandell, R.G. Douglas Jr and J.E. Bennett), Churchill Livingstone, New York, pp. 345–349.

7 Singh, M., Kaur, S., Kumar, B., Kaur, I. and Sharma, V.K. (1987) The associated diseases with leprosy. *Indian Journal of Leprosy*, **59**, 315–321.

8 Hennekens, C.H., Dyken, M.L. and Fuster, V. (1997) *Circulation*, **96**, 2751–2753.

9 Oldenburg, B. and Speck, W.T. (1983) *Pediatric Clinics of North America*, **30**, 71–75.

10 Feldman, J.M., Plonk, J.W. and Bivens, C.H. (1976) *Clinical Endocrinology*, **5**, 71–78.

11 Krizanová, O., Ciampor, F. and Veber, P. (1982) *Acta Virologica*, **26**, 209–216.

12 Kristiansen, J.E., Andersen, L.P., Vestergaard, B.F. and Hvidberg, E.F. (1991) *Pharmacology and Toxicology*, **68**, 399–403.

13 Wagner, S.J., Skripchenko, A., Robinette, D., Foley, J.W. and Cincotta, L. (1998) *Photochemistry and Photobiology*, **67**, 343–349.

14 Nacsa, J., Nagy, L. and Molnar, J. (1998) *Anticancer Research*, **18** (3A), 1373–1376.

15 Lu, W., Achour, A., Arlie, M., Cao, L. and Andrieu, J.M. (2001) *Journal of Immunology (Baltimore)*, **167**, 2929–2935.

16 Mucsi, I., Molnár, J. and Motohashi, N. (2001) *International Journal of Antimicrobial Agents*, **18**, 67–72.

17 Lind, K. and Kristiansen, J.E.H. (1982) Proceedings of the 4th IOM Congress, Tokyo, Abstract.

18 Kristiansen, J.E. (1990) *Danish Medical Bulletin*, **37**, 165–182.

19 Thomas, J.O., DeEds, F. and Eddy, C.W. (1936) *Journal of Pharmacology and Experimental Therapeutics*, **64**, 280–297.

20 Kruszewska, H., Zareba, T. and Tyski, S. (2002) *Acta Poloniae Pharmaceutica*, **59**, 436–439.

21 DeEds, F., Stockton, A.B. and Thomas, J.O. (1939) *Journal of Pharmacology and Experimental Therapeutics*, **65**, 353–371.

22 Cutinelli, C. and Messore, G. (eds) (1958) On the antibacterial action of chlorpromazine. II. The Origin, strength and duration of actively and adoptively acquired immunity. Congress for Microbiology, Stockholm (ed. G. Tunevall), 332.

23 Grosy, H.J. and Norton, J. (1959) *Science*, **129**, 784–785.

24 Hirota, Y. (1960) *Proceedings of the National Academy of Sciences of the United States of America*, **46**, 57–64.

25 Bourdon, J.L. (1961) *Annals de l' Institute Pasteur*, **101**, 876–886.

26 Brunelli, A. and Giorgi, M. (1965) *Atti Acadica Medicine Lombard*, **20**, 437–442.

27 Agarwal, R.P. and Guha, A. (1965) *British Journal of Pharmacology*, **24**, 466–469.

28 Schmidt, R.M. and Rosenkranz, H.S. (1970) *Journal of Infectious Diseases*, **121**, 596–607.

29 Orlowski, M. and Goldman, N. (1974) *Canadian Journal of Microbiology*, **20**, 1689–1693.

30 Brown, N.W. (1975) *Journal of Pharmaceutical Sciences*, **64**, 700–701.

31 Molnar, J., Manadi, Y. and Kiraly, J. (1976) *Acta Microbiologica Academiae Scientiarum Hungaricae*, **23**, 45–54.

32 Saha, P.K. and Dastidar, S.G. (1976) *Indian Journal of Medical Research*, **64**, 1677–1679.

33 Dastidar, S.G., Saha, P.K., Sanyamat, B. and Chakrabarty, A.N. (1976) *Journal of Applied Bacteriology*, **41**, 209–214.

34 Sohair, A.M. (1976) *Zentralblatt für Bakteriologie, Abt II*, **131**, 501–505.

35 Dash, S.K., Dastidar, S.G. and Chakrabarty, A.N. (1977) *Indian Journal of Experimental Biology*, **15**, 324–326.

36 William, F., Burke, J.P. and Spizizen, S. (1977) *Journal of Bacteriology*, **129**, 1215–1221.

37 Silva, M.T., Sousa, J.C.F., Polónia, J.J. and Macedo, P.M. (1979) *Journal of Bacteriology*, **137**, 461–468.

38 Johson, B.H. and Eger, E.I. (1979) *Anesthesia and Analgesia*, **58**, 136–138.

39 Kristiansen, J.E. and Mortensen, I. (1981) *Acta Pathologica et Microbiologica Scandinavica [B]*, **89**, 437–438.

40 Hewlett, E.L., Myers, G.A. and Pearson, R.D. (1983) *Antimicrobial Agents and Chemotherapy*, **23**, 201–206.

41 Molnàr, J., Mucsi, I. and Kása, P. (1983) *Zentralblatt für Bakteriologie Hygiene, I Abt Orig A*, **254**, 388–396.

42 Takahashi, N., Murota, H. and Sutoh, I. (1983) *Ophthalmic Research*, **15**, 277–279.

43 Molnàr, J. (1984) DSc Thesis, Hungarian Academy of Sciences, Budapest.

44 Manna K.K., Dastidar S.G. and Chakrabarty F A.N. (1984) Proceedings of the National Congress of IAMM, Calcutta, pp. 137–141.

45 Catlin, B.W. (1984) *Antimicrobial Agents and Chemotherapy*, **25**, 676–682.

46 Giunta, S., Galeazzi, L., Turchetti, G., Sampaoli, G. and Groppa, G. (1985) *Antimicrobial Agents and Chemotherapy*, **28**, 419–420.

47 Kristiansen, J.E. and Gaarslev, K. (1985) *Acta Pathologica et Microbiologica Scandinavica Section B Microbiology*, **93**, 49–51.

48 Dastidar, S.G., Mondal, U., Niyogi, S. and Chakrabarty, A.N. (1986) Indian Journal of Medical Research, **84**, 142–147.

49 Kristiansen, J.E., Mortensen, I. and Gaarslev, K. (1986) *Acta Pathologica et Microbiologica Scandinavica* Section B Microbiology, **94**, 103–106.

50 Kristiansen, J.E. and Mortensen, I. (1987) *Pharmacology and Toxicology*, **60**, 100–103.

51 Chattopadhyay, D., Dastidar, S.G. and Chakrabarty, A.N. (1987) *Indian Journal of Medical Microbiology*, **5**, 171–177.

52 Dastidar, S.G., Das, S., Chattopadhyay, D., Mookerjee, M., Ray, S. and Chakrabarty, A.N. (1988) *Indian Journal of Medical Research*, **87**, 506–508.

53 Dastidar, S.G., Acharya, D.P., Bhattacharya, S. and Chakrabarty, A.N. (1988) *Bangladesh Journal of Microbiology*, **5**, 1–6.

54 Kristiansen, J.E. (1989) *Danish Medical Bulletin*, **36**, 178–185.

55 Kristiansen, J.E., Justesen, T., Hvidberg, E.F. and Andersen, L.P. (1989) *Pharmacology and Toxicology*, **64**, 386–388.

56 El-Sokkary, M.A., Ibrahim, R.H. and El-Khouly, A.S. (1989) *Journal of Chemotherapy, Infectious Diseases and Malignancy*, **1** (Suppl.), 158.

57 Domenico, P., Schwartz, S. and Cunha, B.A. (1989) *Infection and Immunity*, **57**, 3778–3782.

58 Powel, M., Majcherczik, K.P.A. and Williams, J.D. (1990) 1st International Conference on Antimicrobial Activity of Non-Antibiotics, Copenhagen.

59 Amaral, L. and Lorian, V. (1991) *Antimicrobial Agents and Chemotherapy*, **30**, 556–558.

60 Iwahi, T., Satoh, H., Nakao, M., Iwasaki, T., Yamazaki, T., Kubo, K., Tamura, T. and Imada, A. (1991) *Antimicrobial Agents and Chemotherapy*, **353**, 490–496.

61 Dastidar, S.G., Chattopadhyay, D., Chakrabarty, P., Roy, S. and Chakrabarty, A.N. (1991) Assessment of antibacterial activity spectra of chlorpromazine and some related phenothiazines in *Thiazines and Related Compounds* (ed. G.M. Echert *et al.*), Krieger, Los Angeles, CA, pp. 203–212.

62 Molnàr, J., Ren, J., Kristiansen, J.E. and Nakamura, M.J. (1992) *Antonie van Leeuwenhoek*, **62**, 315–320.

63 Roy, K. and Chakrabarty, A.N. (1994) *Indian Journal of Medical Microbiology*, **12**, 9–18.

64 Nicolau, D.P., Marangos, M.N., Nightingale, C.H. and Quintiliani, R. (1995) *Antimicrobial Agents and Chemotherapy*, **39**, 1748–1751.

65 Kawase, M., Motohashi, N., Sakagami, H., Kanamoto, T., Nakashima, H., Ferenczy, L., Wolfard, K., Miskolci, C. and Molnár, J. (2001) *International Journal of Antimicrobial Agents*, **18**, 161–165.

66 Ordway, D., Viveiros, M., Leandro, C., Jorge Arroz, M., Molnar, J., Kristiansen, J.E. and Amaral, L. (2002) *Journal of Infection and Chemotherapy*, **8**, 227–231.

67 Mazumdar, K., Ganguly, K., Asok Kumar, K., Dutta, N.K., Chakrabarty, A.N. and Dastidar, S.G. (2003) *Microbiological Research*, **158**, 259–264.

68 Sarkar, A., Kumar, K.A., Dutta, N.K., Chakraborty, P. and Dastidar, S.G. (2003) *Indian Journal of Medical Microbiology*, **21**, 172–178.

69 Asok Kumar, K., Ganguly, K., Mazumdar, K., Dutta, N.K., Dastidar, S.G. and Chakrabarty, A.N. (2003) *Acta*

Microbiologica Polonica, **52**, 285–292, PMID: 14743981.

70 Dastidar, S.G., Debnath, S., Mazumdar, K., Ganguly, K. and Chakrabarty, A.N. (2004) *Acta Microbiologica et Immunologica Hungarica*, **51**, 75–83.

71 Karaka, P., Asok Kumar, K., Basu, L.R., Dasgupta, A., Ray, R. and Dastidar, S.G. (2004) *Biological and Pharmaceutical Bulletin*, **27**, 2010–2013.

72 Basu, L.R., Mazumdar, K., Dutta, N.K., Karak, P. and Dastidar, S.G. (2005) *Microbiological Research*, **160**, 95–100.

73 Jeyaseeli, L., Das Gupta, A., Asok Kumar, K., Mazumdar, K., Dutta, N.K. and Dastidar, S.G. (2006) *International Journal of Antimicrobial Agents*, **27**, 58–62.

74 Mazumdar, K., Dutta, N.K., Dastidar, S.G., Motohashi, N. and Shirataki, Y. (2006) *In Vivo*, **20**, 613–619.

75 Levaditi, C., Chaigneau-Erhard, H. and Henry-Eveno, J. (1951) *Comptes Rendues de l' Societe Biologique*, **145**, 1454–1456.

76 Alcozar, G. and Lingiardi, G. (1957) *Archivio e Maragliano di Patologia e Clinica*, **13**, 1247–1264.

77 Kaminska, M. (1967) *Folia Medica Cracoviensia*, **9**, 115–143.

78 Molnar, J., Beladi, I. and Foldes, I. (1977) *Zentralblatt fur Backteriologie A*, **239**, 521–526.

79 Kristiansen, J.E. and Vergmann, B. (1986) *Acta Pathologica, Microbiologica et Immunologica Scandinavica*, **94**, 393–398.

80 Hastings, R.C., Gillis, T.P., Krahenbuhl, J.L. and Franzblau, S.G. (1988) *Clinical Microbiology Reviews*, **1**, 330–348.

81 Ratnakar, P. and Murthy, P.S. (1992) *FEMS Microbiology Letters*, **76**, 73–76.

82 Chakrabarty, A.N., Bhattacharya, C.P. and Dasdidar, S.G. (1993) *Acta Pathologica et Microbiologica Scandinavica*, **101**, 449–454.

83 Amaral, L., Kristiansen, J.E., Abebe, L.S. and Millet, W. (1996) *Journal of Antimicrobial Chemotherapy*, **38**, 1049–1053.

84 Gadre, D.V., Talwar, V., Gupta, H.C. and Murthy, P.S. (1998) *International Clin Psychopharmacol*, **13**, 129–131.

85 Gadre, D.V. and Talwar, V. (1999) *Journal of Chemotherapy*, **11**, 203–206.

86 Amaral, L. and Kristiansen, J.E. (2000) *International Journal of Antimicrobial Agents*, **14**, 173–176.

87 Viveiros, M.B., Bosne-David, S. and Amaral, L. (2000) *International Journal of Antimicrobial Agents*, **16**, 69–71.

88 Ordway, D., Viveiros, M., Leandro, C., Bettencourt, R., Almeida, J., Martins, M., Kristiansen, J.E., Molnar, J. and Amaral, L. (2003) *Antimicrobial Agents and Chemotherapy*, **47**, 917–922.

89 Dutta, N.K., Dastidar, S.G., Asok Kumar, K., Mazumdar, K., Ray, R. and Chakrabarty, A.N. (2004) *Brazilian Journal of Microbiology*, **35**, 316–323.

90 Viveiros, M., Martins, M., Couto, I., Kristiansen, J.E., Molnar, J. and Amaral, L. (2005) *In Vivo*, **19**, 733–736.

91 Dutta, N.K., Mazumdar, K., Dastidar, S.G., Chakrabarty, A.N., Shirataki, Y. and Motohashi, N. (2005) *In Vivo*, **19**, 539–545.

92 Dutta, N.K., Mazumdar, K., Dastidar, S.G. and Park, J.H. (2007) *International Journal of Antimicrobial Agents*, **30**, 336–340.

93 Romano, C. and Galdiero, F. (1975) *Nuovi Annali d' Igiene e Microbiologia*, **26**, 461–464.

94 Wood, N.C. and Nugent, K.M. (1985) *Antimicrobial Agents and Chemotherapy*, **27**, 692–694.

95 Eilam, Y. (1983) *Biochimica et Biophysica Acta*, **733**, 242–248.

96 Josefsson, J.O., Johansson, G. and Hansson, S.E. (1975) *Acta Physiologica Scandinavica*, **95**, 270–285.

97 Schuster, F.L. and Mandel, N. (1984) *Antimicrobial Agents and Chemotherapy*, **25**, 109–112.

98 Kristiansen, J.E. and Jepsen, S. (1985) *Acta Pathologica Microbiologica Immunologica Scandinavica B*, **93**, 249–251.

99 Xing-Qing, P., Min-Xian, Z. and Wen-Hou, H. (1990) 1st International

Conference on Antimicrobial Activity of Non-Antibiotics, Copenhagen.

100 Basco, L.K. and LeBras, J. (1992) *Antimicrobial Agents and Chemotherapy*, **36**, 209–213.

101 Atamna, H., Krugliak, M., Shalmiev, G., Deharo, E., Pescarmona, G. and Ginsburg, H. (1996) *Biochemical Pharmacology*, **51**, 693–700.

102 Pearson, R.D., Hall, D., Harcus, J.L., Manian, A.A. and Hewlett, E.L. (1981) *Clinical Research*, **29**, 837.

103 Pearson, R.D., Manian, A.A., Harcus, J.L., Hall, D. and Hewlett, E.L. (1982) *Science*, **217**, 369–371.

104 Berman, J.D. and Lee, L.S. (1983) *American Journal of Tropical Medicine and Hygiene*, **32**, 947–951.

105 Henriksen, T.H. and Lende, S. (1983) *Lancet*, **i**, 126.

106 Pearson, R.D., Manian, A.A., Hall, D., Harcus, J.L. and Hewlett, E.L. (1984) *Antimicrobial Agents and Chemotherapy*, **25**, 571–574.

107 Zilberstein, D., Liveanu, V. and Gepstein, A. (1990) *Biochemical Pharmacology*, **39**, 935–940.

108 Loiseau, P.M., Mettey, Y. and Vierfond, J.M. (1996) *International Journal for Parasitology*, **26**, 1115–1117.

109 Chan, C., Yin, H. and Garforth, J. (1998) *Journal of Medicinal Chemistry*, **41**, 148–156.

110 Fernandez, A.R., Fretes, R., Rubiales, S., Lacuara, J.L. and Paglini-Oliva, P. (1997) *Medicina*, **57**, 59–63.

111 Harwood, P.D. (1946) *Science Monthly*, **62**, 32–42.

112 Amaral, L., Kristiansen, J.E., Thomsen, V.F. and Markovich, B. (2000) *International Journal of Antimicrobial Agents*, **14**, 225–229.

113 Lund, H. (1928) *Hvad man får af stenkulstjaere* in Frem. Gyldendals Boghaudel-Nordisk Forlag, København, Bind I. (eds H.A. Syane and V. Madsen), pp. 596–661.

114 Bernthsen, A. (1883) *Berichte der Deutschen Chemischen Gesellschaft*, **16**, 2896–2904.

115 Ehrlich, P. (1956) Über die Methylenblaureaktion der lebenden Nervensubstanz in *The Collected Papers of Paul Ehrlich* (eds F. Himmelweit, M. Marquardt and H. Dale), Pergamon Press, London, vol. 1, pp. 500–508.

116 Ehrlich, P. and Leppmann, A. (1956) Über Schmerzstilleude wirkung des Methylenblau in *The Collected Papers of Paul Ehrlich* (eds F. Himmelweit, M. Marquardt and H. Dale), Pergamon Press, London, vol. 1, pp. 555–558.

117 Guttmann, P. and Ehrlich, P. (1891) *Berliner Klinische Wochenschrift*, **39**, 953–956.

118 Bodoni, P. (1899) *Clinica Medica Italia*, **21**, 217–222.

119 Ratcliff, J.D. (1945) *Eventyret om penicillin et moderne Mirakel*, NYT Nordisk Forlag/Arnold Busck, Copenhagen.

120 Garrod, L.P., Lambert, H.P., O'Grady, F. and Waterworth, P.M. (1981) *Antibiotic and Chemotherapy*, Churchill Livingstone, Edinburgh.

121 Laborit, H.A. (1950) *Presse Medicale*, **58**, 416.

122 Neu, H.C. (1994) *Antimicrobica and Infectious Diseases Newsletter*, **13**, 1–8.

123 Kristiansen, J.E.H. (1993) *Current Opinion in Investigational Drugs*, **2**, 587–591.

124 Yamabe, S. (1978) *Chemotherapy*, **24**, 81–86.

125 Roberts, D. and Cole, P. (1981) *Journal of Infection*, **3**, 353–359.

126 Shibl, A.M., Hammouda, Y. and Al-Sowaygh, I. (1984) *Journal of Pharmaceutical Science*, **73**, 841–843.

127 Ray, S., Dastidar, S.G. and Chakrabarty, A.N. (1986) *Indian Journal of Medical Microbiology*, **1**, 7–12.

128 Cohn, R.C., Jacobs, M. and Aronoff, S.C. (1988) *Antimicrobial Agents and Chemotherapy*, **32**, 395–396.

129 Chakrabarty, A.N., Acharya, D.P., Neogi, D. and Dastidar, S.G. (1989) *Indian Journal of Medical Research*, **89**, 233.

130 Aumercieris, M., Murray, D.M. and Rosner, J.L. (1990) *Antimicrobial Agents and Chemotherapy*, **34**, 786–791.

131 Viveiros, M. and Amaral, L. (2001) *International Journal of Antimicrobial Agents*, **17**, 225–228.

132 Kristiansen, J.E., Sebbesen, O., Frimodt-Møller, N., Aaes-Jørgensen, T. and Hvidberg, E.F. (1988) *Acta Pathologica, Microbiologica et Immunologica Scandinavica*, **96**, 1079–1084.

133 Chattopadhyay, D., Dastidar, S.G. and Chakrabarty, A.N. (1988) *Arzneimittel Forschung/Drug Research*, **38** (II), 869–872.

134 Chattopadhyay, D., Dastidar, S.G. and Chakrabarty, A.N. (1989) *Journal of Chemotherapy, Infectious Diseases and Malignancy*, **1** (Suppl), 154.

135 Roy, S., Chattopadhyay, D., Dastidar, S.G. and Chakrabarty, A.N. (1990) *Indian Journal of Experimental Biology*, **28**, 253–258.

136 Molnàr, J., Haszon, I., Bodrogi, T., Martonyi, E. and Turi, S. (1990) *International Journal of Urology and Nephrology*, **22**, 405–411.

137 Amaral, L., Kristiansen, J.E. and Lorian, V. (1992) *Journal of Antimicrobial Chemotherapy*, **30**, 556–558.

138 Chan, Y.Y., Ong, Y.M. and Chua, K.L. (1997) *Antimicrobial Agents and Chemotherapy*, **51**, 623–630.

139 Asok Kumar, K., Mazumdar, K., Dutta, N.K., Karak, P., Dastidar, S.G. and Ray, P. (2004) *Biological and Pharmaceutical Bulletin*, **27**, 1116–1120.

140 Mazumdar, K., Dutta, N.K., Asok Kumar, K. and Dastidar, S.G. (2005) *Biological and Pharmaceutical Bulletin*, **28**, 713–717.

141 Sevag, M.G. (1964) *Archives of Biochemistry and Biophysics*, **108**, 85–88.

142 Heller, C.S. and Sevag, M.G. (1966) *Applied Microbiology*, **14**, 879–885.

143 Manion, R.E., Bradley, S.G. and Hall, W.H. (1969) *Proceedings of the Society of Experimental Biology and Medicine*, **130**, 206–209.

144 Martin, S., Oduola, M. and Milhous, W. (1987) *Science*, **235**, 899–901.

145 Bitonti, A., Sjoerdsma, A. and McCann, P. (1988) *Science*, **242**, 1301–1303.

146 Bitonti, A. and McCann, P. (1989) *Lancet*, **ii**, 1282–1283.

147 Peters, W., Ekong, R., Robinson, B. and Warhurst, D. (1989) *Lancet*, **ii**, 334–335.

148 Basco, L. and Le Bras, J. (1990) *Lancet*, **335**, 422.

149 Salama, A. and Facer, C.A. (1990) *Lancet*, **8662**, 164–165.

150 Dastidar, S.G., Datta, A., Ganguly, M., Chattopadhyay, D. and Chakrabarty, A.N. (1990) 1st International Conference on Antimicrobial Activity of Non-Antibiotics, Copenhagen.

151 Kristiansen, J.E., Amaral, L. and Thomsen, V.F. (1993) 18th International Congress of Chemotherapy, Stockholm.

152 Kristiansen, J.E. and Amaral, L. (1997) *Journal of Antimicrobial Chemotherapy*, **40**, 319–327.

153 Kristiansen, M.M., Leandro, C., Ordway, D., Martins, M., Viveiros, M., Pacheco, T., Kristiansen, J.E. and Amaral, L. (2003) *International Journal of Antimicrobial Agents*, **22**, 250–253.

154 Kristiansen, J.E., Hendricks, O., Delvin, T., Butterworth, T.S., Aagaard, L., Christensen, J.B., Flores, V.C. and Keyzer, H. (2007) *Journal of Antimicrobial Chemotherapy*, **59**, 1271–1279.

155 Dutta, N.K., Subramanian, A., Mazumdar, K., Dastidar, S.G., Kristiansen, J.E., Molnar, J., Martins, M. and Amaral, L. (2007) *International Journal of Antimicrobial Agents*, **30**, 242–249.

156 Amaral, L., Kristiansen, J.E., Viveiros, M. and Atouguia, J. (2001) *Journal of Antimicrobial Chemotherapy*, **47**, 505–511.

157 Whitaker, R. (2004) *Medical Hypotheses*, **62**, 5–13.

158 *NPL Site Narrative for Forest Glen Mobile Home Subdivision, Niagara Falls*, New York Federal Register Notice, November 21, 1989.

159 *Facts and Comparisons III*, W. Port Plaza, St. Louis, MO, 1990, pp. 3146–3098.

160 Delay, J. and Deniker, P. (1952) *Compte Rendu du Congres Alién Neurol*, **50**, 497–503.

161 Smith, D.M. (1985) *CRC Handbook of Stereo-isomers: Drugs in Psychopharmacology*, CRC Press, Boca Raton, FL.

162 Petersen, P.V., Nielsen, I.M., Pedersen, V., Jørgensen, A. and Lassen, N. (1977) Thioxanthenes, in *Psychotherapuetic drugs* (eds E. Usdin and I. Forrest), Marcel Dekker, New York, pp. 827–867.

163 Jones, P.G. and Kennard, O. (1979) *Acta Crystallographica B Structural Science*, **35**, 1732–1735.

164 Feinberg, A.P. and Snyder, S.H. (1975) *Proceedings of the National Academy of Sciences of the United States of America*, **72**, 1899–1903.

165 Kristiansen, J.E. (1991) *ASM News*, **57**, 135–139.

166 Leurs, R., Church, M.K. and Taglialatela, M. (2002) *Clinical and Experimental Allergy*, **32**, 489–498.

167 Rossi, S. (ed.) (2004) *Australian Medicines Handbook*, Australian Medicines Handbook, Adelaide.

168 Chakrabarty, A.N., Sen, T., Banerjee, S., Chattopadhyay, D. and Dastidar, S.G. (1991) Effects of structural modifications on antimicrobial actions of promethazine, methdilazine and related phenothiazines in *Thiazines and Related Compounds* (ed. G.M. Echert *et al.*) Krieger, Los Angeles, CA, pp. 219–233.

169 Chattopadhyay, D., Dastidar, S.G. and Chakrabarty, A.N. (1990) Methdilazine, a new Antibacterial. Pharma Project, CAS No. 1982-37-2 Product Profile, England.

170 Chattopadhyay, D., Mukherjee, T., Pal, P., Saha, B. and Bhadra, R. (1998) *Journal of Antimicrobial Chemotherapy*, **42**, 83–86.

171 Hack, H.M., Parsonnet, J. and Triadalopoulos, G. (1994) *Gastrointestinal Endoscopy*, **40**, 397–398.

172 Forrest, F.M., Forrest, I.S. and Roizin, L.E. (1963) *Revue Agressologie*, **3**, 259–264.

173 Crowle, J.A., Douvas, G.S. and May, H.M. (1992) *Chemotherapy*, **38**, 410–419.

174 Kristiansen, J.E.H. (1979) *Acta Pathologica et Microbiologica Scandinavica Section B Microbiology*, **87**, 317–319.

175 Kristiansen, J.E.H. (1981) *Ugeskrift for Laeger*, **143**, 1900–1904.

176 Kristiansen, J.E., Amaral, L. and Thomson, V.F. (1993) Abstracts of the 18th International Congress of Chemotherapy, Stockholm.

177 Molnàr, J., Mándi, Y. and Földeák, S. (1982) *Acta Microbiologica Academiae Scientiarum Hungaricae*, **29**, 17–25.

178 Detmers, P.A. and Condeelis, J. (1986) *Experimental Cell Research*, **163**, 317–326.

179 Parry, M.F. and Neu, H.C. (1977) *Journal of Clinical Microbiology*, **5**, 58–61.

180 Chan, Y.Y., Ong, Y.M. and Chua, K.L. (2007) *Antimicrobial Agents and Chemotherapy*, **51**, 623–630.

181 Amaral, L., Martins, M. and Viveiros, M. (2007) *Journal of Antimicrobial Chemotherapy*, **59**, 1237–1246.

182 Decourt, P., Gastal, R. and Grenat, R. (1953) Etude de l'action narcobiotique sur les germes microbiens *Compte Rendus de l'Académie des Sciences*, **237**, 1109–1111.

183 Elipoulos, G.M. and Moellering, R.C. (1980) Antimicrobial combinations, in *Antibiotics in Laboratory Medicine* (ed. V. Lorian), Williams & Wilkins, Baltimore, MD, pp. 432–480.

184 Mayer, K.H., Opal, S.M. and Medeiros, A. (1990) Mechanisms of antibiotic resistance in *Principles and Practice of Infectious Diseases*, 3rd edn (eds G.L. Mandell, R.G. Douglas and J.E. Bennett), Churchill Livingstone, New York, pp. 218–228.

185 Acar, J.F., Goldstein, F. and Chabbart, Y.A. (1973) *Journal of Infectious Diseases*, **128** (Suppl), 470–477.

186 Gutschik, E. (1984) Thesis, University of Copenhagen, Copenhagen.

187 Chan, Y.Y. and Chua, K.L. (2005) *Journal of Bacteriology*, **187**, 707–719.

188 Chan, Y.Y., Tan, T.M., Ong, Y.M. and Chua, K.L. (2004) *Antimicrobial Agents and Chemotherapy*, **48**, 1128–1135.

189 Williams, J.D. (1995) *Journal of Antimicrobial Chemotherapy*, **35**, 721–737.

190 Gemmel, C.G. and Lorian, V. (1991) Effects of low antibiotic concentrations on bacteria: effect on ultrastructure, their virulence, and suspectibility to immunodefences: clinical significance in *Antibiotics in Laboratory Medicine*, 2nd edn (ed. V. Lorian), Williams & Wilkins, Baltimore, MD, pp. 397–452

191 Walter, A.M. and Heilmeyer, L. (1965) *Antibiotika-Fibel, Antibiotik and Chemotherapie*, Georg Thieme Verlag, Stuttgart, pp. 117–129.

192 Amaral, L., Lee, Y., Schwarz, U. and Lorian, V. (1986) *Journal of Bacteriology*, **167**, 492–495.

193 Amaral, L., Kristiansen, J.E. and Thomsen, V.F. (1993) Abstract of the 18th International Congress of Chemotherapy, Stockholm.

194 Kristiansen, J.E., Mortensen, I. and Nissen, B. (1982) *Biochimica et Biophysica Acta*, **685**, 379–382.

195 Kramer, K., Sorgatz, H. and Höppe, F.U. (1994) *Hygiene Medizin*, **19**, 527–534.

196 Ingram, L.O., Ley, K.D. and Hoffmann, E.M. (1978) *Life Sciences*, **22**, 489–494.

197 Mostafa, S.A. (1976) Effects of a tranquilizer (Phenothiazine Derivative) on Growth and Physiology of *Escheria coli* B., *Zentralblatt für Bakteriologie Abt. II.*, **131**, 101–109.

198 Karreman, G., Isenberg, I. and Szent-Gyorgyi, A. (1959) *Science*, **130**, 1191–1192.

199 Kwant, W.O. and Seeman, P. (1969) *Biochimica et Biophysica Acta*, **193**, 338–349.

200 Martins, M., Dastidar, S.G., Fanning, S., Kristiansen, J.E., Molnar, J., Pagès, J.M., Schelz, Z., Spengler, G., Viveiros, M. and Amaral, L. (2008) *International Journal of Antimicrobial Agents*, **31**, 198–208.

201 Haug, B.E., Strom, M.B. and Svendsen, J.S. (2007) *Current Medicinal Chemistry*, **14**, 1–18.

202 Amaral, L., Viveiros, M. and Kristiansen, J.E. (2006) *Current Drug Targets*, **7**, 887–891.

5
Use of Natural Products to Combat Multidrug-Resistant Bacteria

Christine M. Slover, Larry H. Danziger, Bolanle A. Adeniyi, and Gail B. Mahady

Abstract

Shortly after the introduction of penicillin, scientists discovered penicillin-resistant strains of *Staphylococcus aureus* – a common bacterium that makes up part of the normal human bacterial flora. From this initial case of resistant *Staphylococcus*, the problem of antimicrobial resistance has grown into a serious public health concern with economic, social and medical implications. These concerns are global in scope, and cross all environmental and ethnic boundaries. As much as 60% of hospital-acquired infections are caused by multidrug-resistant microbes and these infections, many of which are caused by vancomycin-resistant enterococci and methicillin-resistant *S. aureus* (MRSA), are now no longer confined to hospital wards, but are now being seen at an alarming rate at the community level as well. Although most antibiotics are still active, the rapid progression of resistance suggests that many of these drugs may not be effective in the future. Therefore, research and development of new strategies and alternative therapies for treating bacterial infections are urgently needed. Currently, the published global scientific data reveals over 2500 plant species that are reported to have activity against *S. aureus*, and there is growing body of evidence that specific plant species and their extracts may be of benefit for the treatment of MRSA. This chapter will focus on the recent advances in natural products research in the search for plant-based medicines active against MRSA by evaluating some of the abundantly available *in vitro* data, but focusing on clinical data, and further propose a direction for future research.

5.1
Introduction

According to the World Health Organization, infectious diseases are a significant cause of morbidity and mortality worldwide, accounting for approximately one-half of all deaths in tropical countries [1, 2]. Infectious and parasitic diseases remain the

New Strategies Combating Bacterial Infection. Edited by Iqbal Ahmad and Farrukh Aqil
Copyright © 2009 WILEY-VCH Verlag GmbH & Co. KGaA, Weinheim
ISBN: 978-3-527-32206-0

major killers of children in developing countries, partly as a result of the human immunodeficiency virus (HIV)/acquired immunodeficiency syndrome (AIDS) epidemic [2]. Approximately 10.5 million children less than 5 years of age died in 2002, of which 98% lived in developing countries [2]. While it may not be surprising to see such statistics in developing countries, it should also be noted that infectious disease mortality rates are also increasing in western nations, such as the United States. Death from infectious disease ranked fifth in 1981, but by 1992 it was the third leading cause of death – an increase of 58% [3]. This is alarming given that it was once believed that it would be possible to eliminate infectious disease by the end of the twentieth century [1]. Contributing factors to these increases include the HIV/AIDS pandemic and resultant secondary infections. Other contributing factors include an increase in antibiotic resistance in both nosocomial and community acquired pathogens [3]. Furthermore, despite the progress made in the understanding of microbiology and the control of microorganisms, incidents of epidemics due to drug-resistant microorganisms and the emergence of hitherto unknown disease-causing microbes pose an enormous threat to public health. Such negative health trends call for a new global initiative for the development of new strategies for the prevention and treatment of infectious disease [4]. Proposed solutions are outlined by the Centers for Disease Control and Prevention as a multipronged approach that includes prevention (such as vaccination), improved monitoring and the development of new treatments. It is this last solution that would encompass the development of new antimicrobials [5].

5.1.1
Penicillin-Resistant Microorganisms

Just a few years after the introduction of penicillin, scientists began noticing the emergence of a penicillin-resistant strain of *Staphylococcus aureus* – a common bacterium that makes up part of the normal human bacterial flora. From that first case report of resistant *Staphylococcus*, the problem of antimicrobial resistance has grown into a serious public health threat [6, 7]. Today, penicillin-resistant pneumococci and resistant malaria are on the rise, disabling and killing millions of children and adults each year [8].

5.1.2
Multidrug-Resistant Microorganisms

In 1990, almost all cholera isolates gathered around New Delhi (India) were sensitive to inexpensive, first-line drugs such as furazolidone, ampicillin, cotrimoxazole and nalidixic acid. However, by 2000, most of these formerly effective drugs were useless and unable to contain cholera epidemics [2]. In the developed world, as many as 60% of hospital-acquired infections are caused by multidrug-resistant (MDR) microbes. Many of these infections are caused by Gram-positive bacteria such as, vancomycin-resistant enterococci (VRE) and methicillin-resistant *S. aureus* (MRSA), and unfortunately these infections are now no longer confined to hospital wards, but have spread into communities at large [9]. Although many antibiotics are still active, the rapid progression of resistance to many of the Gram-positive bacteria suggests that many of these drugs may not be effective for much longer. Formerly, first-line antimicrobials were both effective and affordable; however, with the onset of resistance, newer treatments are proving too costly to the vast majority of those living in poor developing nations [1, 2].

5.2
Natural Product Development

Conventional medicine is increasingly receptive to the research and development of antimicrobial and other plant-based medicines, as microorganisms become resistant to traditional antibiotics [7]. Another driving factor for the renewed interest in plant antimicrobials over the past 20 years has been the rapid rate of (plant) species extinction. There is a sense of urgency among environmentalists, natural-products chemists and microbiologists alike that the multitude of potentially useful antimicrobial plants and novel phytochemicals are at risk of being irretrievably lost. In addition, over the past 20 years the number of new antibiotics in the research and development pipeline has dramatically declined. Thus, pharmaceutical companies need to become more receptive to novel research and development of antimicrobials and other drugs from 'nontraditional' sources (plants) [7]. Furthermore, the general public has become increasingly aware of the overuse and misuse of antibiotics, and is now very interested in alternative medicines such as 'medicinal plants'. It has been estimated that between 20 and 80% of the populations many countries worldwide use botanical products, and consider them to be safe and effective [10].

5.2.1
Natural Products as Medicines

Historically, man has used plants to supply many of his basic needs such as food, clothing, shelter and medicines [10, 11]. Plants have always played a central role in the prevention and treatment of disease since the beginning of time [10]. Synthetic chemistry dominated the twentieth century and allowed the pharmaceutical industry to become tremendously successful, as well as make significant advances in the prevention and treatment of disease. However, even in the 1970s, approximately 25% of all prescription drugs dispensed in the United States were plant-based and in 2000, 11% of 252 drugs on the World Health Organization's essential drugs list were exclusively obtained from flowering plants [12, 13].

While it is difficult to determine the exact number, it is estimated that there are between 250 000 and 500 000 plant species worldwide, of which approximately 10% (25 000–50 000 species) of these plants are used as foods and medicines by both humans and animal species [7, 14]. Interestingly, less than 10% of these known plants have ever been thoroughly scientifically investigated.

5.2.2
Use of Natural Medicines Worldwide

For over 100 years, natural products have been a source of inspiration for novel drug development [4]. Naturally derived compounds have made considerable contributions to human health and well-being. The role of natural products has been 2-fold in the development of new drugs: (i) they may become the base for the development of a medicine, a natural blueprint for the development of new drugs, or (ii) a phytomedicine to be used directly for the treatment of disease [4]. While western countries are more eager to develop new drugs for patenting purposes, herbal extracts and phytomedicines are used primarily in developing countries and Europe. The first generation of plant drugs were usually simple botanicals employed in more or less their crude form. Several effective medicines used in their natural state such as cinchona, opium, belladonna and aloe were selected as therapeutics agents based on empirical evidence of their clinical application by traditional societies from different parts of the world. However, following the industrial revolution, a second generation of natural drugs emerged based on the scientific practice of processing the plant extracts to isolate their active chemical constituents [4]. Second-generation natural products were purified chemical compounds, with an identified structure that have been isolated from plant extracts. Notable examples are quinine from *Cinchona* for the treatment of malaria, reserpine from *Rauvolfia serpentine* for the treatment of hypertension and more recently taxol from *Taxus brevifolia* for the treatment of ovarian cancer [4]. These compounds followed the same method of development and evaluation as other pharmaceutical drugs.

5.2.3
Natural Products as Antimicrobial Agents

There are vast amounts of published scientific information from around the globe describing the antimicrobial activities of plant extracts [15, 16]. A search of the NAPRALERT database (www.napralert.org), a natural products database housed within the University of Illinois at Chicago, has shown that of the 58 725 plant species listed in the database, 6350 species had experimental antimicrobial activity of which almost 4000 species had ethnomedical data supporting the use of these plants to treat infectious disease [15, 16]. The majority of the plants had activity against a range of bacteria, *Mycobacterium* or fungi. The NAPRALERT database cites more than 105 articles that describe the experimental antimicrobial activities of over 155 plant species from Thailand alone [15]. Many of the medicinal plant extracts purported to have antibacterial activity have only been subjected to *in vitro* screening and the vast majority of these extracts have never been tested rigorously in animal models or controlled clinical trials [17]. While thousands of plant species have been tested against many of different strains of bacteria, the most common bacteria used in susceptibility tests are: *Escherichia coli*, *Bacillus subtilis*, *Chlamydia pneumonia*, *S. aureus*, MRSA, *Streptococcus pneumoniae*, *Bacillus cereus*, *Enterococcus faecalis*, *Klebsiella pneumoniae*, VRE, *Pseudomonas aeruginosa* and *Helicobacter pylori* [15].

While there are volumes of *in vitro* data describing the *in vitro* antimicrobial activities of many plants, this chapter will focus on the MDR bacterium for which there is clinical data from the natural products literature, namely MRSA.

5.3
MRSA

S. aureus is a Gram-positive cocci that colonizes the epithelial surfaces of humans [18, 19]. Colonization by *S. aureus* usually does not normally lead to infections, unless the bacterium or their extracellular products breach the epithelial layer, in which case serious infection can result [20]. *S. aureus* has numerous cell surface virulence factors (such as protein A and clumping factor) and secretes exotoxins, which allow the bacteria to cause a myriad of infections. Diseases ranging from relatively benign furuncles and subcutaneous abscesses to scalded skin syndrome, sepsis, necrotizing pneumonia and toxic shock syndrome have been attributed to these organisms [19]. While no single cell surface virulence factor has been shown to be uniquely required for mucous membrane attachment, once colonization occurs, numerous secreted exotoxins, including the pyrogenic toxin superantigens and exfoliative toxins, definitively cause serious human disease. Other secreted exotoxins, such as the four hemolysins and Panton–Valentine leukocidin have also been suggested to contribute to significant illnesses [19].

5.3.1
Natural Products for MRSA: *In Vitro* Data

Over 2770 species of plant are reported to be active against *S. aureus* [15, 16]. Beyond this, there is a growing body of data demonstrating that specific plant extracts may be of benefit for the treatment of MSRA [15, 21–31]. For example, a 50% ethanol extract the dried fruits of *Terminalia chebula* Retz. inhibited the growth of MRSA, with a minimum inhibitory concentration (MIC) of 31.3 μg/ml [31]. This plant is native to many countries of Southeast Asia and has been used in traditional medicine to treat upper respiratory tract infections. The primary constituents of the fruit are hydrolysable tannins and components thereof, including chebulagic acid, chebulinic acid, corilagin, gallic acid, punicalagin, terchebulin and terminalic acid [15].

Another plant, *Melaleuca alternifolia*, commonly referred to as tea tree oil (TTO), is well known for its traditional use as an antimicrobial agent [32]. In one study, 66 isolates of *S. aureus* were susceptible to TTO in both the disk-diffusion and modified broth microdilution assays [21–23]. Of the isolates tested, 64 were methicillin-resistant and 33 were mupirocin-resistant. The MIC and minimum bactericidal concentration were 0.25 and 0.50 μg/ml, respectively, and suggest TTO may be useful in the treatment of MRSA carriage [21]. The major chemical constituents of TTO include 1,8-cineol (4.5–16.5%), terpinen-4-ol (29–45%), γ-terpinene (10–28%) and α-terpineol (2.7–13.0%), and these compounds were shown to reduce the viability of *S. aureus*. Organisms treated with TTO or its chemical constituents at the MIC, or 2 times the MIC, showed a significant loss of tolerance to sodium chloride, had a predisposition to lysis, and altered cell morphology that included the formation of mesosomes and the loss of cytoplasmic contents [22]. The predisposition to lysis, the loss of 260 nm-absorbing material, the loss of tolerance to sodium chloride and the altered morphology seen by electron microscopy all suggest that TTO and its components compromise the cytoplasmic membrane [21–23].

In terms of the development of bacterial resistance to TTO, the screening of 100 MRSA isolates revealed no resistance to TTO at a concentration of 2.5% (v/v) [29, 30]. Relatively resistant subpopulations did occur in all five test strains producing isolates with MICs from 0.5 to 16% (v/v) at a frequency of 1.2×10^{-8} to 1×10^{-5}, but this resistance proved to be relatively stable at 3 months. Unfortunately, the presence of a subpopulation of resistant mutants in isolates unexposed to TTO suggests that exposure may eventually lead to selection of resistance. Stepwise induction of resistance was possible on solid media with a minimum 4-fold increase in MIC and one strain attaining a stable MIC of 16% (v/v), which is greater than the 5–10% (v/v) concentration advocated in topical formulations [29]. The stepwise induction of low-level resistance to TTO is similar to that seen for mupirocin in 1985, with induction of a stable MIC to mupirocin of 40 mg/l [26]. TTO is widely used in nonprescription preparations, frequently at lower concentrations. Therefore, TTO appears to be a viable alternative for study as a topical agent to eradicate MRSA colonization [29].

5.3.2
Natural Products for MRSA: Clinical Data

There are currently only two published clinical trials evaluating the efficacy of TTO for decolonization of MRSA. The first clinical trial was a pilot study comparing the use of a combination of 4% TTO nasal ointment and 5% TTO body wash (intervention care) with a standard 2% mupirocin nasal ointment and triclosan body wash (routine care) for eradication of MRSA [33]. A total of 30 in-patients infected or colonized with MRSA were recruited and randomly assigned to be treated with intervention care or routine care for a minimum of 3 days. Infected patients also received intravenous vancomycin and all participants were screened for continued MRSA carriage 48 and 96 h after the cessation of experimental treatment. Only 18 patients completed the trial – 10 in the routine care group and eight in the intervention care group. More patients in the intervention care than in the routine care group cleared the MRSA (5/8 versus 2/10). Two patients in the intervention group received 34 days of treatment and one cleared the infection while the other remained chronically colonized with MRSA. The intergroup differences were not statistically significant. Some of the adverse events reported by the group receiving TTO nasal ointment were swelling of the nasal mucosa and burning. No adverse events were reported from the use of the TTO body wash or mupirocin ointment. Since this was considered a pilot study, it was too small to give any conclusive results about the use of TTO for MRSA decolonization [33].

The second study was a randomized, controlled clinical trial that evaluated the efficacy of TTO products compared to standard treatments for decolonization of MRSA [34]. Two hundred and twenty-four patients were randomized to receive either 10% TTO cream applied to the nares 3 times daily for 5 days, 5% TTO body wash used daily for 5 days, 10% TTO cream applied to wounds or ulcers (TT) or 2% mupirocin nasal ointment applied to the nares 3 times daily for 5 days, chlorhexidine gluconate 4% body wash used daily for 5 days, silver sulfadiazine 1% cream applied to wounds or ulcers daily for 5 days (ST). Prior to beginning treatment all subjects were swabbed for MRSA detection. Swabs were the obtained for each subject 14 days after completion of treatment. Persistently positive MRSA cultures were considered as failure. Post-treatment positive cultures for nasal carriage were analyzed separately. Of the 110 subjects randomized to the TT group, 64 (59%) remained positive for MRSA post-treatment. Of the 114 patients who received ST, 58 (51%) subjects were still persistently positive for MRSA after completing treatment. There was no significant difference in treatment outcomes ($P = 0.0286$). In the ST group, 74 subjects had positive nasal carriage of MRSA and on day 14 only 16 subjects remained persistently positive. Of the 76 subjects in the TT group with positive MRSA nasal carriage, 40 remained positive at the end of treatment. Mupirocin was significantly more effected than TTO for eradicating MRSA nasal carriage ($P = 0–0001$). There were no reports of adverse events in either treatment group in this study [34]. This study demonstrated that TTO can be considered as an option for MRSA decolonization, as it was effective and safe. However, the optimal concentration to be used in topical products should be evaluated for MRSA eradication.

5.4
Natural Products for MDR Microorganisms

Due to the increasing global problem of MDR organisms, in particular MRSA, the need to identify new and preferably inexpensive treatments is urgent. Not only are many of our commonly used medications derived from plants or plant products, demonstrating a history of efficacy, but these products once developed will be less expensive than synthetic counterparts. The abundance of supporting *in vitro* data available that suggests plants and other natural products can be useful sources of new and novel treatments for MDR organisms if developed. However, while there is an abundance of *in vitro* studies demonstrating the potential use of natural products as antimicrobials, data from animal models and clinical trials are very limited, and few have been published. Where there are clinical data (e.g. in the case of TTO), the results are very promising and suggest that TTO may be useful in preventing MRSA decolonization. Further data from larger-scale studies are needed, especially evaluations for MRSA eradication.

Although *in vitro* antimicrobial data for natural products are limited in their application, these data do serve as starting points for future investigations of these plants and extracts in animals and eventually in humans. The sheer number of plant species with purported antimicrobial and MDR pump-inhibiting activities demonstrates that natural products are a potentially valuable source of new agents for infectious disease, particularly for MDR organisms, where soon no therapy will exist. For future research priorities, a greater emphasis on *in vivo* animal studies and clinical trials in humans is needed to appropriately evaluate natural products for human use.

References

1 World Health Organization (1998) *World Health Report*, World Health Organization, Geneva.

2 World Health Organization (2003) *World Health Report*, World Health Organization, Geneva.

3 Pinner, R., Teutsch, S., Simonsen, L., Klug, L., Graber, J., Clarke, M. and Berkelman, R. (1996) *Journal of the American Medical Association*, **275**, 189–193.

4 Iwu, M., Duncan, A.R. and Okunji, C.O. (1999), in *New antimicrobials of Plant Origin. Perspectives on New Crops and New Uses* (ed. J. Janick), ASHS Press, Alexandria, VA, pp. 457–462.

5 Fauci, A. (1998) *Emerging Infectious Diseases*, **4**, 374–378.

6 Spink, W.W. and Ferris, V. (1945) *Science*, **102**, 102–221.

7 Cowan, M.M. (1999) *Clinical Microbiology Reviews*, **12**, 564–582.

8 Hsieh, Y.C., Hsueh, P.R., Lu, C.Y., Lee, P.I., Lee, C.Y. and Huang, L.M. (2004) *Clinical Infectious Diseases*, **38**, 830–835.

9 Witte, W., Cuny, C., Strommenger, B., Braulke, C. and Heuck, D. (2004) *Euro Surveillance*, **9**, 1–2.

10 Mahady, G.B. (2001) *Journal of Nutrition*, **131** (3s), 1120S–1123.

11 Talalay, P. and Talalay, P. (2001) *Academic Medicine*, **76**, 238–247.

12 Farnsworth, N.R. and Morris, R.W. (1976) *American Journal of Pharmaceutical Education*, **148**, 46–52.

13 Rates, S.M.K. (2001) *Toxicomanies*, **39**, 603–613.

14 Borris, R.P. (1996) *Journal of Ethnopharmacology*, **51**, 29–38.

15 Mahady, G.B. (2005) *Current Pharmaceutical Design*, **19**, 2405–2427.

16 Farnsworth, N.R. (2008) *NAPRALERT Database*, University of Illinois at Chicago, Chicago, IL.

17 Martin, K. and Ernst, E. (2003) *Journal of Antimicrobial Chemotherapy*, **51**, 241–246.

18 Raad, I., Alrahwan, A. and Rolston, K. (1998) *Clinical Infectious Diseases*, **26**, 1182–1187.

19 Tenover, F.C. and Gaynes, R.P. (2000), The Epidemiology of Staphylococcus Infections in *Gram-positive Pathogens* (eds V.A. Fischetti, R.P. Novick, J.J. Ferretti, D.A. Portnoy and J.I. Rood), ASM Press, Washington, DC, pp. 526–534.

20 Yarwood, J.M. and Schlievert, P.M. (2003) *Journal of Clinical Investigation*, **112**, 1620–1625.

21 Carson, C.F., Hammer, K.A. and Riley, T.V. (2006) *Clinical Microbiology Reviews*, **19**, 50–62.

22 Carson, C.F., Mee, B.J. and Riley, T.J. (2002) *Antimicrobial Agents and Chemotherapy*, **46**, 1914–1920.

23 Carson, C.F., Cookson, B.D., Farrelly, H.D. and Riley, T.V. (1995) *Journal of Antimicrobial Chemotherapy*, **35**, 421–424.

24 Czech, E., Kneifel, W. and Kopp, B. (2001) *Planta Medica*, **67**, 263–269.

25 Darwish, R.M., Aburjai, T., Al-Khalil, S. and Mahafzah, A. (2002) *Journal of Ethnopharmacology*, **79**, 359–364.

26 Elsom, G.K.F. and Hide, D. (1999) *Journal of Antimicrobial Chemotherapy*, **43**, 427–428.

27 Lee, C.K., Kim, H., Moon, K.H. and Shin, K.H. (1998) *Archives of Pharmacological Research*, **21**, 62–66.

28 May, J., Chana, C.H., Kinga, A., Williams, L. and Frencha, G.L. (2000) *Journal of Antimicrobial Chemotherapy*, **45**, 639–643.

29 Nelson, R.R.S. (2000) *Journal of Antimicrobial Chemotherapy*, **45**, 549–550.

30 Nelson, R.R.S. (1997) *Journal of Antimicrobial Chemotherapy*, **40**, 305–306.

31 Sato, Y. (1997) *Biological and Pharmaceutical Bulletin*, **20**, 401–404.

32 Mahady, G.B., Fong, H.H.S. and Farnsworth, N.R. (2007) *Fructus chebulae*, in *WHO Monographs on Selected Medicinal Plants, Volume 3*, World Health Organization, Geneva.

33 Caelli, M., Porteous, J., Carson, C.F., Heller, R. and Riley, T.V. (2000) *Journal of Hospital Infection*, **46**, 236–237.

34 Dryden, M.S., Dailly, S. and Crouch, M. (2004) *Journal of Hospital Infection*, **56**, 283–286.

6
West African Plants and Related Phytocompounds with Anti-Multidrug-Resistance Activity

Koné Mamidou Witabouna and Kamanzi Atindehou Kagoyire

Abstract

Diseases due to multidrug resistant (MDR) bacteria have become a major health problem in developing as well as developed countries. Therapy of bacterial infections is a frequent problem due to the emergence of bacterial strains resistant to numerous antibiotics. In many west African countries, the prevalence of MDR bacteria is increasing despite the measures of control. This increase of MDR bacteria necessitates the search for new effective alternatives to replace antibacterial agents currently in use. In this instance, the plant kingdom is undoubtedly a valuable source for new antibiotics. Herbal medicine has been shown to have a genuine utility and rural population depens on it as primary health care. About 80% of the population in West Africa still use traditional medicine for treating diseases, both microbial and non-microbial origins. Ethnobotanical studies on west African medicinal plants has reasonably been done resulting into a collection of information which show that African medicinal plants could be source of leads compounds. A wide range of those medicinal plants have been studied for antibacterial activity against various bacteria, but few have included MDR bacteria. This chapter provides an overview of the West African plants and related phytocompounds with anti-MDR activity. An attempt has been made to summarize the information in order to highlight those promising medicinal plant in the treatment of diseases caused by multidrug resistant bacteria. The most promising are *Erythrina senegalensis*, *Erythrina vogelii*, *Terminalia avicennioides*, *Momordica balsamina*, *Combretum paniculatum*, *Morinda lucida*, *Ocimum gratissimum* and *Trema guineensis*. These plant species show a strong anti-MDR activity on MRSA, vancomycin resistant enterococci (VRE), *Escherichia coli* ESBL, *Klebsiella pneumoniae* ESBL and MDR-*Salmonella typhi* strains. The phytochemicals responsible for this activity are flavonoids, isoflavonoids, naphtoquinones, alkaloids, sesquiterpenoids, phenolic compounds and xanthones. Precise active principles, such as isowighteone, vogelin B and vogelin C with anti-MDR activity, have been isolated from *Erythrina vogelii*.

The importance of MDR bacteria makes these data of special interest since many medicinal plants are used in traditional medicinal in places of West Africa where medical services are not in position to provide health care.

New Strategies Combating Bacterial Infection. Edited by Iqbal Ahmad and Farrukh Aqil
Copyright © 2009 WILEY-VCH Verlag GmbH & Co. KGaA, Weinheim
ISBN: 978-3-527-32206-0

6.1
Introduction

The emergence of bacterial strains with multiple resistances to antibiotics has become a major health problem worldwide. In Africa, it is one of the main concerns for various reasons. As a fact, populations in Africa are more exposed to sicknesses due to bacteria. First, the climate, often humid, favors the increase of bacteria. In addition, most of the population's precarious living conditions and its ignorance of basic hygienic rules contribute to infections. These hardships are worsened with the acquired immunodeficiency syndrome pandemic – favorable ground for the development of all sorts of infections.

Concerning treatment with antibiotics, few patients follow their treatment to the end, often due to lack of money or even ignorance because few people know that antibiotic therapy not taken to the end leads to the development of resistance. During the last decade, the prevalence of MDR bacteria has increased despite the measures of control [1–4]. Facing this problem of multiple resistance, it is a must to find available, on an ongoing basis, active molecules against bacteria multiresistant to antibiotic treatment.

African flora still includes unexplored plant species that could help as a basis for the development of new active molecules against bacterial strains multi-resistant to a course of antibiotics. Several investigations were conducted on the use of African medicinal plants, particularly in West Africa [5–12]. Several plants were identified, particularly the ones used in the treatment of various diseases of the respiratory system (pneumonia, angina, cough, pharyngitis), of the urogenital system (various venereal diseases, cystitis and urinary infection, etc.), pathology of the digestive tract (gastroenteritis, diarrhea, dysentery, food poisoning) and skin diseases (scabies, furuncles, various dermatitis). Bacteria such as *Pneumococcus*, *Staphylococcus*, *Enterococcus* and even *Pseudomonas* are involved in various diseases. Furthermore, many of them include strains multiresistant to antibiotics.

Among the plants found through investigations able to treat bacterial diseases that we have just mentioned, several species have been tested *in vitro* against MDR bacteria [13–15]. Furthermore, the research conducted is scattered among various laboratories without cooperative agreements. The research is not conducted with a coordinated program with a clear reachable goal. The same situation prevails with phytochemical investigations, particularly rare, and only few of them ended with the isolation of pure molecules; even in these cases, few pure molecules have then been evaluated for their antibacterial activities.

No opportunity for a solution should be neglected since the emergence of bacterial strains multiresistant to antibiotics has become an acute problem worldwide. As a consequence, traditional knowledge of African medicinal plants is worth serious exploration. In this chapter, we propose a selection of plants used in West Africa in the treatment of infectious diseases mentioned above, which affect respiratory, digestive, urogenital and skin systems. For particular species whose antibacterial activities have been tested, we give a synthesis of the results through a chart. The chemical composition of rare studied species is given, followed by a discussion concerning the connection between the observed biological activity and the chemical composition of the observed plant species.

6.2
MDR Bacteria in West Africa

Bacterial infections are widespread in West Africa as well as in most of the tropical countries.

Among the bacteria including strains resistant to antibiotic treatment, there are the staphylococci with *Staphylococcus aureus*, enterobacteria (*Escherichia coli*), pneumococcus (*Streptococcus pneumoniae*), mycobacteria (*Mycobacterium tuberculosis*) and *Pseudomonas* (*Pseudomonas aeruginosa*).

6.2.1
Staphylococci

The staphylococci are bacteria implicated in various diseases attacking different internal and external organs of the human body. The strains of staphylococci most frequently belong to *S. aureus*, *Staphylococcus epidermis* and *Staphylococcus saprophyticus*. These bacteria are known to provoke boils, whitlow, abscesses, conjunctivitis, ear infection, endometriosis, salpingitis, pneumonia, meningitis, osteomyelitis, acute cystitis and food poisoning. All these cutaneous, genital or respiratory infections are likely to become complicated and result in septicemia. In West Africa, the treatment of these infections is a frequent problem due to the high prevalence of MDR bacteria [2]. In Abidjan (Ivory Coast), the prevalence of methicillin-resistant *S. aureus* (MRSA) carriage among health care personal is high; such personal constitute an infectious risk for the hospitalized patients who are so exposed to nosocomial infections caused by MRSA. On 269 *S. aureus* isolates, 38.7% were MRSA strains. The carriage rate of MRSA in the population was 17.8% [1]. Furthermore, a variable proportion of MRSA expressed resistance to other families of antibiotics such as aminoglycosides (77.6%), rifampicin (8.8%), fluoroquinolones (34.1%) and vancomycin (5.9%) [16]. A high rate (72%) of MRSA bacteremia, in particular among nosocomial bacteremia, was found in Dakar (Senegal) [3]. For Nigeria, 88.2% of MRSA isolates associated with infections and 11.8% colonizing strains were reported in a study carried out at the University of Ilorin Teaching Hospital [4]. Wound infections seen in the University College Hospital, Ibadan, were mostly caused by MRSA [17].

6.2.2
Enterobacteria

These bacteria comprise *E. coli*, *Klebsiella pneumoniae* and others. *E. coli* is responsible for urinary infections, enterohemorragic, hemorragic colic and bloody diarrhea, due to the consumption of infected food. Extended-spectrum β-lactamase (ESβL)-producing isolates were described for many West African countries. In Ivory Coast, a rate of 10.3% was reported by the Pasteur Institute of Ivory Coast (data unpublished).

6.2.3
Pneumococci (*S. pneumoniae*)

The pneumococci are widespread bacteria and exist in different types. These bacteria are responsible for current severe ear, nose and throat infections, meningitis, bacteremia, septicemia, pneumonia, and sometimes arthritis. *S. pneumoniae* is transmitted from the nose and throat of an infected person to another through kissing, coughing, sneezing and saliva. The occurrence of resistant strains of pneumococci or reduced sensitivity to many antibiotics [18, 19] curbs effective control of diseases caused by *S. pneumoniae*. To study the level of resistance and compare the situations in different cities, a prospective study was carried out in Abidjan (Ivory Coast), Casablanca (Morocco), Dakar (Senegal) and Tunis (Tunisia), from 1996 to 1997. Multiple resistances were more frequent in penicillin-nonsusceptible isolates than in penicillin-susceptible isolates [18].

6.2.4
Pseudomonas

The most common of these bacteria is *P. aeruginosa* which can be found in lung, genital, digestive tract, skin and central nervous system infections. This bacteria is also responsible for bone, ear and eyes diseases. MDR *P. aeruginosa* were reported in a study carried out in Ivory Coast at the University Hospital Center of Cocody (Abidjan). This study revealed that 3% of the strains were MDR [20].

6.2.5
Mycobacteria

Mycobacteria such as *M. tuberculosis* are found in pulmonary infections and tuberculosis. On 79 patients enrolled in the departments of pulmonary disease of two general hospitals in Abidjan (CHU of Cocody and Treichville) and a tuberculosis outpatient clinic, the rate of MDR tuberculosis was 79%. Among MDR tuberculosis patients, those aged of 20–40 were the most concerned group (72%) with a clear male predominant rate (sex ratio: 3) [21]. Simultaneous rifampicin and isoniazide resistance was also reported in patients hospitalized for tuberculosis in the Respiratory Disease Unit of the Treichville University Hospital in Abidjan [22]. Some cases of MDR tubercle bacilli were described in Sierra Leone for the retreatment patients suffering from pulmonary infection [23].

6.3
Plants Used in Bacterial Disease Treatment in West Africa

Ethnobotanical studies on West African medicinal plants have been performed, resulting into a collection of information about traditional medicine practices. Information gathered is numerous, and concerns both microbial and nonmicrobial diseases. Many of the recorded plants and herbal remedies have been used over the years by African traditional medical practitioners to treat a number of the bacterial diseases listed above (Section 6.2) when referring to ethnomedical surveys carried out in West African countries [5–12, 24–38]. A compilation of such plants which are in use for the treatment of different bacterial ailments in West Africa is presented in Table 6.1. From ancient times, plant material has been used for the treatment of boils,

Table 6.1 Traditional uses of West African plants in the treatment of bacterial diseases.

Plant species and family	Part used	Indications
Acalypha wilkesiana Müll. Arg. (Euphorbiaceae)	leaves	dermatological and gastrointestinal disorders [14]; antimycotic, antibacterial, wound dressing [8]
Acanthospermum hispidum DC. (Asteraceae)	leaves	typhoid fever, bacillary dysentery [32]
Adansonia digitata L. (Bombacaceae)	leaves, stem bark, seed oil	childhood diarrhea [33]; diarrhea, dysentery, typhoid fever, bacillary dysentery [7]
Adenia lobata Engl. (Passifloraceae)	leaves, roots	eye infections [36]
Agelaea trifolia (Lam.) Gilg ex. G. Schellenb (Connaraceae)	leaves, stem bark	dysentery [36]
Ageratum conizoides L. (Asteraceae)	leaves	eye infections [36]
Albizia adianthifolia W. F.Wight (Mimosaceae)	leaves, stem bark	childhood diarrhea [36]; gonorrhea [37]
Albizia ferruginea (Guill. and Perr.) Benth. (Mimosaceae)	stem	diarrhea, dysentery [7]
Alchornea cordifolia Müll. Arg. (Euphorbiaceae)	stem and root bark	venereal diseases [33]; ameboid dysentery [36]; mouth infections [31]; cough, postdelivery infections, chronic gonorrhea [37]
Anchomanes difformis Engl. (Araceae)	leaves	cough [8, 12]; venereal diseases [36]; sore throat [37]
Annona senegalensis Pers. (Annonaceae)	leaves, stem and root bark	antiseptic, wounds, leprosy [25]; gastroenteritis, childhood diarrhea, dysentery [12]; cough [33]
Asparagus africanus Lam. (Asparagaceae)	stem, leaves	cough, genital wounds [38]
Bobgunnia madagascariensis (Desv.) J. H. Kirkbr. and Wiersema (Ceasalpiniaceae)	roots	syphilis, leprosy, gastritis, rheumatism, abscesses, diarrhea, wounds, ear infections [10]

(Continued)

Table 6.1 (*Continued*)

Plant species and family	Part used	Indications
Bombax buonopozense P. Beauv. (Bombacaceae)	stem, leaves, seeds	*E. coli* cystitis, gastroenteritis and diarrhea, antiseptic [32]
Bridelia ferruginea Benth. (Euphorbiaceae)	leaves, roots	dysentery [32]; gastrointestinal pain and disorders [7]; cough [8]
Buchholzia coriacea Engl. (Capparidaceae)	stem and root bark, seeds	otitis [12]; cold, venereal diseases [36]; eye infections, back and ear pain, smallpox [10]
Carapa procera DC. (Meliaceae)	stem bark, seed oil	luxation, edema [6]; skin diseases, rheumatism, bacterial infections, ophthalmia [8]
Ceiba pentandra Gaertn. (Bombacaceae)	stem and root bark, sap	cough, fever [33]; diarrhea, dysentery [7]
Cercestis afzelii Schott (Araceae)	leaves	boils [37]
Cissus populnea Guill. and Perr. (Vitaceae)	roots	infected wounds, boils [13]
Cola nitida Schott and Endl. (Sterculiaceae)	stem bark	boils [37]
Combretum paniculatum Vent. (Combretaceae)	not specified	microbial and nonmicrobial diseases [40]
Crossopteryx febrifuga Benth. (Rubiaceae)	stem bark	diarrhea, dysentery [7]
Dalbergiella welwitschii Baker f. (Fabaceae)	leaves	stomach ache, rheumatism [25]
Dioscorea minutiflora Engl. (Dioscoreaceae)	leaves, tubers	white stain on cornea [37]
Enantia polycarpa Engl. and Diels (Annonaceae)	stem bark	wound, leprosy, diarrhea [25]; old wounds [36]
Erythrina senegalensis DC. (Fabaceae)	stem bark, roots	dry cough, gonorrhea, stomach ache [13]; dysentery, cough [8]
Erythrina vogelii Hook. f. (Fabaceae)	root bark	venereal diseases, gonorrhea, jaundice [8, 64]; gonorrhea [10]
Erythrococca anomala Prain (Euphorbiaceae)	leaves	miningitis [12]; cough, wound, headache [36]
Erythrophleum ivorense A. Chev. (Caesalpiniaceae)	leaves	dermatosis [37]; dermatosis and scabies, boils [36]; smallpox [10]
Ficus exasperata Vahl. (Moraceae)	young leaves	cough [6]; diarrhea with blood, cough [36]; cough, venereal diseases [37]
Ficus sur Forssk. (Moraceae)	young leaves	diarrhea [12]; dysentery [37]

Table 6.1 (*Continued*)

Plant species and family	Part used	Indications
Flueggea virosa (Willd.) Voigt (Euphorbiaceae)	leaves, stem	bronchitis, cough, diarrhea, abdominal pain, edema [8]; dysentery [25]
Geophila obvallata (Schumch.) F. Didr. (Rubiaceae)	leaves	gastrointestinal pain, otitis [25]
Garcinia kola Heckel (Clusiaceae)	leaves	cough, inflammation of the respiratory tract [8]
Harrisonia abyssinica Oliv. (Simaroubaceae)	leaves, roots	scabies, varicella, stomach pain [30]; diarrhea, dysentery [7]
Harungana madagascariensis Poir. (Hypericaceae)	leaves, roots, stem bark	cutaneous parasitose, edema, dysentery [25]; asthma, dysentery [10]
Heliotropium indicum L. (Boraginaceae)	leaves	buccal candidosis [12], eye lotion, wound dressing [8];
Hoslundia opposita Vahl (Lamiaceae)	leaves	edema, rheumatism, ophthalmia, dermatosis, venereal diseases [25]
Irvingia gabonensis (Aubry-Lecomte ex. O'Rorke) Baill. (Irvingiaceae)	stem bark	dermatosis, scabies [36]; diarrhea, dysentery [7]
Jatropha curcas L. (Euphorbiaceae)	sap, leaves, stem, roots	toothache [36]; gastritis, stomach ache [95]; diarrhea, dysentery [7]
Keetia hispida (Benth.) Bridson (Rubiaceae)	leaves	respiratory diseases [13]
Kalanchoe crenata (Andrews) Haw. (Crassulaceae)	leaves	otitis, eye inflammation [95]
Kigelia africana (Lam.) Benth. (Bignoniaceae)	stem bark, roots, fruits	venereal diseases, cough [36]; dysentery, gonorrhea, syphilis, cough, rheumatism [10]
Khaya ivorensis A. Chev. (Meliaceae)	roots	diarrhea, dysentery [7]
Khaya senegalensis A. Juss. (Meliaceae)	stem, fruits	gastrointestinal disorders, dermatosis, vomiting, wounds [10]
Lannea acida A. Rich. (Anacardiaceae)	stem bark	diarrhea, stomach ache, gonorrhea, rheumatism [13]
Mallotus oppositifolius Müll. Arg. (Euphorbiaceae)	leaves, whole plant	testicle furuncle [37]; diarrhea, dysentery [7]
Momordica balsamina L. (Cucurbitaceae)	not specified	microbial and nonmicrobial diseases [40]

(*Continued*)

Table 6.1 (*Continued*)

Plant species and family	Part used	Indications
Manniophyton fulvum Müll. Arg. (Euphorbiaceae)	leaves	dry cough, bronchitis [37]
Mansonia altissima A. Chev. (Sterculiaceae)	stem bark	phlegmy cough with respiratory difficulties [37]; tuberculosis, leprosy [36]
Mareya micrantha (Benth.) Müll. Arg. (Euphorbiaceae)	leaves	sore throat, old wounds, gonorrhea [37]
Mezoneuron benthamianum Baill. (Ceasalpiniaceae)	leaves, roots, sap	gum, ear and eye pain [36]; diarrhea, dysentery [7]
Microglossa pyrifolia (Lam.) Kuntze (Asteraceae)	leaves, roots	umbilical tetanus [12]; conjunctivitis, gonorrhea [37]
Morinda lucida Benth. (Rubiaceae)	roots	stomach ache [36]; stomach ache, conjunctivitis [13]
Motandra guineensis (Thonn.) A. DC. (Apocynaceae)	leaves	venereal diseases, mouth wounds [36]
Nauclea latifolia Sm. (Rubiaceae)	leaves, roots and stem bark	inflammation, rheumatism, scabies, skin diseases [95]; throat ache [36], diarrhea, dysentery [7]
Newbouldia laevis (P. Beauv.) Seem. (Bignoniaceae)	stem	bloody diarrhea, stomach wound, head and throat ache [36]; diarrhea, colitis, gastroenteritis [12]
Ocimum gratissimum L. (Lamiaceae)	leaves	diarrhea, stomach ache [14]; cold [8]
Pachypodanthium staudtii Engl. and Diels (Annonaceae)	stem bark	bronchitis, gastrointestinal disorders [25]; scabies [36]
Paullinia pinnata L. (Sapindaceae)	leaves or whole plant	bloody diarrhea, wound [36]; pain or gastrointestinal disorders, diarrhea dysentery [7, 10]
Pentadesma butyracea Sabine (Clusiaceae)	root and stem bark	diarrhea, dysentery [7]
Psidium guajava L. (Myrtaceae)	leaves	diarrhea, cough, cutaneous irritation, healing [95]; diarrhea, dysentery [7, 37]
Piliostigma thonningii (Schumach.) Milne-Redh. (Caesalpiniaceae)	leaves, stem	diarrhea, dysentery [7]; cough, wounds [38]
Phyllanthus discoides Müll. Arg. (Euphorbiaceae)	stem bark	stomach ache [14]

Table 6.1 (*Continued*)

Plant species and family	Part used	Indications
Pothomorphe umbellata Miq. (Piperaceae)	stem, leaves	smallpox [25]
Pycnanthus angolensis (Werw.) Warb. (Myristicaceae)	leaves, stem bark, sap	gingivitis/stomatitis, tooth and gum ache [37]
Pseudocedrela kotschyi (Schweinfurt) Harms (Anacardiaceae)	roots, stem	gastrointesinal disorders [7]; tooth and gum care [38]
Rauvolfia vomitoria Afzel. (Apocynaceae)	leaves, stem, roots	varicella, gonorrhea [10]
Saba senegalensis (A. DC.) Pichon (Apocynaceae)	root bark, latex	serious pulmonary disease [6]; ophtalmia, respiratory problems, tuberculosis, food poisoning [10]
Scoparia dulcis L. (Scrophulariaceae)	leaves	antidiarrhoea, antivomiting, venereal diseases treatment, stomace diseases [96]; fever, diarrhea [33]; cough, varicella [36]
Spathodea campanulata P. Beauv. (Bignoniaceae)	stem, leaves, flowers	healing, cough, skin diseases, syphilitic chronic wounds, edema, [10]; venereal diseases, wounds [36]; gingivitis, stomatitis [37]
Spondias mombin Jacq. (Anacardiaceae)	leaves	stomach ache, eye diseases, headache, wounds, tooth decay [6]
Terminalia avicennioides Guill. and Perr. (Combretaceae)	leaves	cough [36]; dental caries, skin infections [14]
Tieghemella heckelii Pierre ex. A. Chev. (Sapotaceae)	stem bark	acute gonorrhea [12]
Thonningia sanguinea Vahl (Balanophoraceae)	flowers, roots, stem	skin diseases, dysentery, sore throat, bronchial asthma [8]
Trema guineensis (Thonn.) Ficalho (Ulmaceae)	stem bark	cough [8, 36, 37]; tuberculosis [36]
Trichilia monadelpha (Thonn.) J. de Wilde. (Meliaceae)	root bark	gonorrhea, dermatosis [37]; gastrointestinal pain and disorders [7]
Uapaca togoensis Pax. (Euphorbaiceae)	leaves	vomiting, rheumatism, fever [13]
Vernonia nigritiana Oliv. and Hiern. (Asteraceae)	roots	diarrhea, dysentery [7]

(*Continued*)

Table 6.1 (*Continued*)

Plant species and family	Part used	Indications
Waltheria lanceolata R. Br. ex. Mast. (Sterculiaceae)	roots	diarrhea, vomiting, cough [13]
Ximenia americana L. (Olacaceae)	roots	mouth wounds, diarrhea, rheumatism [13]
Zanthoxylum zanthoxyloides (Lam.) B. Zepernick and Timler. (Rutaceae)	leaves, stem bark	toothache, sore throat [6]; scabies [37]; antiseptic, antiinflamatory [32]

abscesses, wounds, respiratory tract diseases, gastrointestinal disorders, ear, throat and urinary infections, and some of the most common diseases caused by bacteria. The pharmacological preparations are mostly decoction; pounded fresh plants or powders of dried or burned plant parts, primarily leaves, roots and stem barks. Water or alcohol (ethanol or palm wine) are the main solvents used in the preparation of traditional remedies. Furthermore, most of the remedies are prepared from a single plant species rather than a mixture of plants. The main problem of West African traditional medicine is the lack of standardization of most of the remedies. Most of the plant species in use are well distributed throughout West Africa, but some differences in uses could be observed in the therapeutic indications and the receipts from one country to another or in the same country from one ethnic group to another.

Many of these ethnobotanical surveys have been published or are available on PHARMEL 2 [12] and PRELUDE databases.

6.4
West African Medicinal Plants with Activity against MDR Bacteria

Antibacterial screening in order to evaluate the efficiency of West African medicinal plants on bacteria has been carried out in different countries. However, those including MDR bacteria are still scarce, fragmented or unpublished. The study of West African medicinal plants for their activity against MDR bacteria is relatively recent, probably due to the emergence of bacteria resistant to numerous antibiotics. It was not possible to draw up an exhaustive list of all the plants with anti-MDR activity. Only the 37 plants with sufficient documentation are reported for bactericidal or bacteriostatic effects against various MDR bacteria such as MRSA and MDR *Salmonella typhi* (Table 6.2).

Thirty plants were reported to be effective against MRSA strains. These are for example *Erythrina vogelii, Potomorphe umbellata* [39], *Erythrina senegalensis, Bobgunnia madagascariensis, Waltheria lanceolata* [13], *Terminalia avicennioides, Phyllanthus discoideus, Ocimum gratissimum, Acalypha wilkesiana* [14] and *Thonningia sanguinea* [15]. While on MDR-*S. typhi* strains, six plant species were found to be active: *T. avicennioides, Momordica balsamina, Combretum paniculatum, Trema guineensis,*

Table 6.2 West African plants with anti-MDR activity.

Plant species	Family	Plant parts	Target	MIC (µg/ml)	References
Acalypha wilkesiana	Euphorbiaceae	leaves	MRSA	24	[14]
Ageratum conizoides	Asteraceae	leaves	MRSA	43	[14]
Anchomanes difformis	Araceae	leaves	MRSA	375	[39]
Bobgunnia madagascariensis	Caesalpiniaceae	roots	MRSA, MLSB	23	[13]
Bridelia ferruginea	Euphorbiaceae	leaves	MRSA	30.6	[14]
Carapa procera	Meliaceae	stem bark	MRSA, MLSB	94	[39]
Cercestis afzelii	Araceae	leaves	MRSA, MLSB	188	[39]
Cissus populnea	Vitaceae	roots	MRSA, MLSB	188	[13]
Cola nitida	Sterculiaceae	stem bark	MRSA, MLSB	94	[39]
Combretum paniculatum	Combretaceae		MDR Salmonella typhi	9.60–14	[40]
Dalbergiella welwitschii	Fabaceae	stem bark	MRSA, MLSB	188	[39]
Dioscorea minutiflora	Dioscoreaceae	leaves	MRSA, MLSB	188	[39]
Erythrina senegalensis	Fabaceae	roots	MRSA, MLSB	12, 47	[13]
Erythrina vogelii	Fabaceae	roots	MRSA, MLSB	12, 6	[39]
Erytrophleum ivorense	Caesalpiniaceae	leaves	MRSA, MLSB	188, 375	[39]
Ficus sur	Moraceae	stem bark	MRSA, MLSB	188	[39]
Garcinia kola	Clusiaceae	leaves	MRSA, MLSB	47, 94	[39]
Harungana madagascariensis	Hypericaceae	leaves, stem bark	MRSA, MLSB	94	[39]
Heliotropium indicum	Boraginaceae	leaves	MLSB	188	[39]
Keetia hispida	Rubiaceae	leaves	MDR enterococci	94	[13]
Lannea acida	Anacardiaceae	roots	MRSA, MLSB	188	[13]
Momordica balsamina	Cucurbitaceae		MDR Salmonella typhi	9.60–14	[40]
			MDR Shigella spp.	3	[41]
Monodora myristica	Annonaceae		MRSA, Escherichia coli ESβL		[81]

(Continued)

Table 6.2 (*Continued*)

Plant species	Family	Plant parts	Target	MIC (µg/ml)	References
Morinda lucida	Rubiaceae		MDR *Salmonella typhi*	9.60	[40]
Ocimum canum	Lamiaceae	leaves	MRSA, *Escherichia coli* ESβL		[81]
			Klebsiella pneumoniae ELSB		[81]
Ocimum gratissimum	Lamiaceae	leaves	MRSA	22.3	[81]
			MDR *Shigella* spp.	3	[41]
			MDR *Salmonella typhi*	40	[40]
Phyllanthus discoides	Euphorbiaceae	stem bark	MRSA	20.5	[14]
Pothomorphe umbellata	Piperaceae	leaves and stem	MRSA, MLSB	94	[39]
		roots	MRSA, MLSB	188–94	[39]
Pycnanthus angolensis	Myristicaceae	stem bark	MRSA, MLSB	375	[39]
Spondias mombin	Anacardiaceae	leaves	MRSA, MLSB	94	[39]
Terminalia avicennioides	Combretaceae	stem bark	MRSA	18.2	[14]
			MDR *Salmonella typhi*	9.60–14	[40]
			MDR *Shigella* spp.	3	[41]
Tieghemella heckelii	Sapotaceae	stem bark	MRSA, MLSB	94	[39]
Thonningia sanguinea	Balanophoraceae	leaves	MRSA	31.25	[15]
			Escherichia coli ESβL	6250	[15]
			Klebsiella pneumoniae ESβL	6250	[15]
Trema guineensis	Ulmaceae	stem bark	MRSA, MLSB	94	[40]
			MDR *Salmonella typhi*	9.60–14	[40]
Uapaca togoensis	Euphorbaiceae	leaves	MDR enterococci	94	[13]
Waltheria lanceolata	Sterculiaceae	roots	MRSA, MLSB	47	[13]
Ximenia americana	Olacaceae	roots	MDR enterococci	94	[13]

MLSB = macrolides, lincosamides, streptrogramines B.

Morinda lucida and *Ocimum gratissimum* [40]. *T. sanguinea* is reported for its antibacterial activity against ESβL *E. coli* and ESβL *Klebsiella pneumoniae* [15]. *E. vogelii* [39], *E. senegalensis*, *B. madagascariensis*, *W. lanceolata*, *Keetia hispida* and *Uapaca togoensis* [13] were effective against MDR enterococci. *O. gratissimum* and *T. avicennoides* markedly inhibited the growth of MDR *Shigella* species such as *Shigella dysenteriae*, *Shigella flexneri*, *Shigella sonnei* and *Shigella boydii*. Except *S. flexneri*, an *O. gratissimum/T. avicennoides* concoction exhibited a higher shigello-cidal property than other extracts in *S. dysenteriae*, *S. sonnei* and *S. boydii*. *O. gratissimum* showed a greater shigellocidal effect against the *S. flexneri* isolates, while extracts of *M. balsamina* possessed low shigellocidal potential [41].

It is important to appreciate that there are some West African plants that possess marked anti-MDR activity. These results offer a scientific basis for the traditional uses of herbal remedies of the active plants against bacterial diseases, in particular MDR-associated diseases. All the plant extracts tested and active on MDR bacteria were obtained using ethanol or water. Those solvents are the most commonly used for the preparation of the receipts in West African traditional medicine practices. Ethanol extracts exhibited greater anti-MDR potency than the corresponding aqueous extracts at the same concentrations. This could be explained by the nature of the bioactive compounds which are enhanced in the presence of ethanol [14].

Many of the anti-MDR plants such as *Anchomanes difformis* and *M. balsamina* [10] have analgesic and antipyretic effects, which may complete the antibacterial property. In case of bacterial diseases, fever and pain are some of the symptoms.

6.5
Metabolites Isolated from West African Medicinal Plants Effective on MDR Bacteria

As a result of the recent interest in the plant kingdom as a potential source of new drugs, strategies for the fractionation of plant extracts based on biological activity rather than on a particular class of compound have been developed [42]. West African medicinal plants were investigated, and yielded isoflavones, pterocarpans, essential oil, saponins, anthrones, flavonoids, anthraquinones and alkaloids. It has been documented that tannins, saponins, alkaloids [43] and flavonoids [44] are plant metabolites well known for antimicrobial activity. In addition, anti-MDR properties were reported for alkaloids [45], tannins, flavonoids [46], naphthoquinones [47], sesquiterpenoids [48], xanthones [49, 50] and anthraquinones [14]. The mechanism or mode of action of antibacterial agents in plants used for the treatment of bacterial diseases vary. The wide range of structures of those constituents which appear to be active anti-MDR principles suggests different sites of action within the body. Most of the pure compounds isolated from the promising plants were not assayed for their antibacterial potency. Furthermore, phytochemical studies are lacking for some of the plants for which antibacterial activity has been proven. The compounds reported for several plants effective on bacteria and which provide justification for their anti-MDR activity are described in this section.

6.5.1
B. madagascariensis (Caesalpiniaceae) syn: *Swartzia madagascariensis*

The main constituents of fruits, seeds and pods are catechin tannins, saponins [51] and triterpene saponins which were found to be glucuronides of oleanolic acid (1) and gypsogenin (2) [52]. Isoflavonoids such as pterocarpine, homopterocarpine and demethylpterocarpine were also isolated from the stem wood [53, 54]. They derived from the basal compound pterocarpane, and belong to three structure types [pterocarpane (3), pterocarpene (4) and coumestane (5) groups]. None of these compounds was tested for its antibacterial activity. The effect of *B. madagascariensis* on bacteria was described for the first time by Koné *et al.* [13]. Saponins or isoflavonoids may be responsible for its anti-MDR activity. This plant is reported to be toxic as well as some of the pure compounds (medicarpine) isolated from different plant parts [55]. See Figure 6.1.

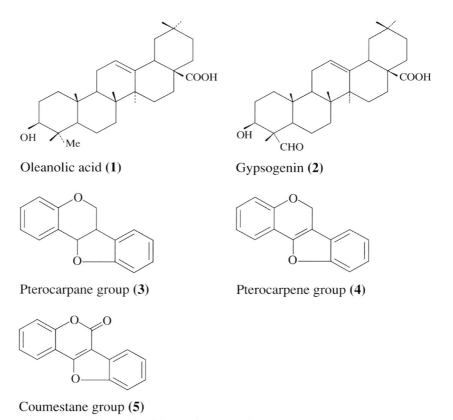

Oleanolic acid (**1**)

Gypsogenin (**2**)

Pterocarpane group (**3**)

Pterocarpene group (**4**)

Coumestane group (**5**)

Figure 6.1 Metabolites isolated from *Bobgunnia madagascariensis*.

6.5.2
E. senegalensis and *E. vogelii* (Fabaceae)

This genus *Erythrina* is widely found in the tropical and subtropical areas of the world. Many of the species are used in traditional medicine for the treatment of microbial diseases [44]. Two species, *E. senegalensis* [14] and *E. vogelii* [39], are reported to exhibit anti-MDR activity on MRSA. The stem bark and roots of these two plant species contain ring B prenylated isoflavonoids named vogelin A–J (**6–15**) for *E. vogelii* [56–58], and erysenegalensin F–M (**16–19**) and senegalensin (**20**) for *E. senegalensis* [59–63], in addition with other compounds such as isowighteone (**21**), isolupalbigenin (**22**), 1-methoxyphaseollidin (**23**), ulexone (**24**), 6,8- diprenyl-genistein, 8-prenylluteone, warangalone, scandenone, auriculatin, 2,3-dihydroau-riculatin and carpachromene. Only isowighteone, vogelin B, vogelin C [64] and isolupalbigenin were tested for their antibacterial activity and were effective on MRSA with, respectively, minimum inhibitory concentrations of 25, 12.5 and 1.6 µg/ml. The nontested compounds could also contribute to the anti-MDR properties of *E. senegalensis* and *E. vogelii*. Many of the purified compounds are isoflavonoids and related compounds that are known to have antibacterial properties [44]. The antibacterial activity of flavonoids is attributed to the presence of phenolic hydroxyls as inhibitors of the microbial enzymes [65]. In addition, the prenylated groups render them more lipophilic, inducing antimicrobial activity within interactions with cell membranes. *Erythrina* species from Ivory Coast contain traces of alkaloids, especially in the seeds, and are slightly toxic [25]. See Figure 6.2.

6.5.3
Erythrophleum ivorense (Caesalpiniaceae)

This plant is rich in alkaloids which belong to the cassaine (**25**), 19-hydroxycassaine (**26**), erythrophlamine (**27**), cassamine (**28**), norcassaine (**29**), norcassamidine (**30**) and norcassamidide (**31**) groups. The stem bark contains cassaine, cassaidine, coumidine, 3-(3-methylcrotonyl)-cassaine, 19-hydroxycassaine, erythrophlamine, cassamine, cassamidine, erythrophleguine, 3-(3-methylcrotonyl)-norcassaine (ivorine), norcassa-midine, norcassamine, norerythrosuamine, dehydro-norerythrosuamine, norcassa-mide, erythrophlamide and norcassaide [66–69]. The antibacterial properties of *E. ivorense* are due to these alkaloids. The stem bark of the plant was used as a criminal poison by Dan and Guéré from the west of Ivory Coast [10]. Therefore, traditional remedies are comprehensibly prescribed in topical application. See Figure 6.3.

6.5.4
Garcinia kola (Clusiaceae)

Kolanone, a polyisoprenylated benzophenone with antimicrobial properties, has been isolated from the fruit [70]. Antibacterial biflavonoids were also elucidated from

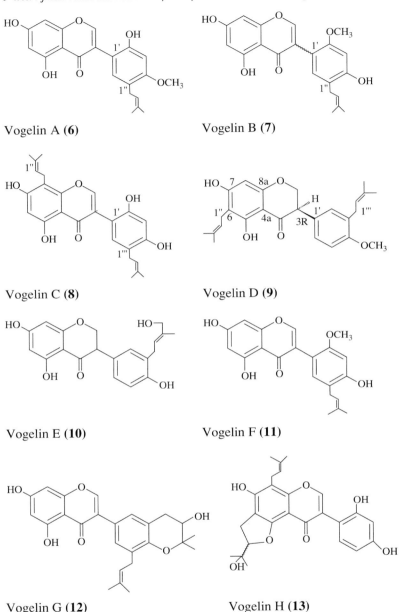

Vogelin A (**6**)

Vogelin B (**7**)

Vogelin C (**8**)

Vogelin D (**9**)

Vogelin E (**10**)

Vogelin F (**11**)

Vogelin G (**12**)

Vogelin H (**13**)

Figure 6.2 Metabolites isolated from *Erythrina senegalensis* and *E. vogelii*.

the seeds [71]. Biflavonoids have been isolated, together with polyprenylated xanthones, from the related species *Garcinia scortechinii* with anti-MRSA activity, Kolanone and biflavonoids may probably be responsible for the anti-MRSA activity of *G. kola*.

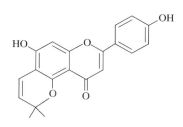

Vogelin I (**14**)

Vogelin J (**15**)

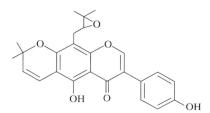

Erysenegalensin F (**16**)

Erysenegalensin G (**17**)

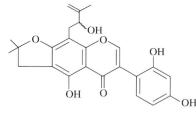

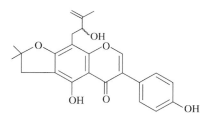

Erysenegalensin L (**18**)

Erysenegalensin M (**19**)

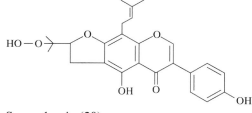

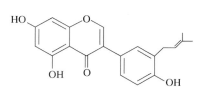

Senegalensin (**20**)

Isowighteone (**21**)

Figure 6.2 (*Continued*)

6.5.5
Harungana madagascariensis (Hypericaceae)

The leaves were investigated, and yielded various polyphenols, catechins (epicatechin, gallic acid, protocatechic), procyanidins, flavonoids such as astilbine (**32**),

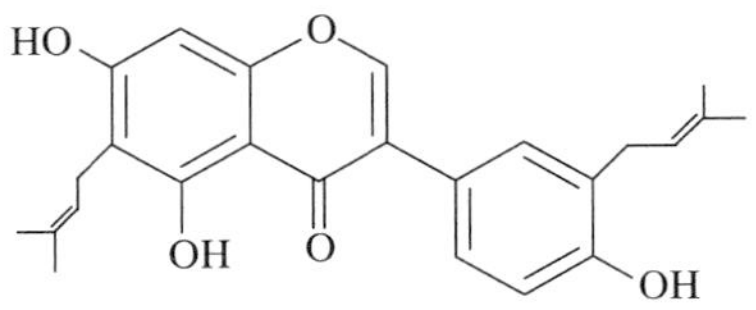

Isolupalbigenin (**22**) 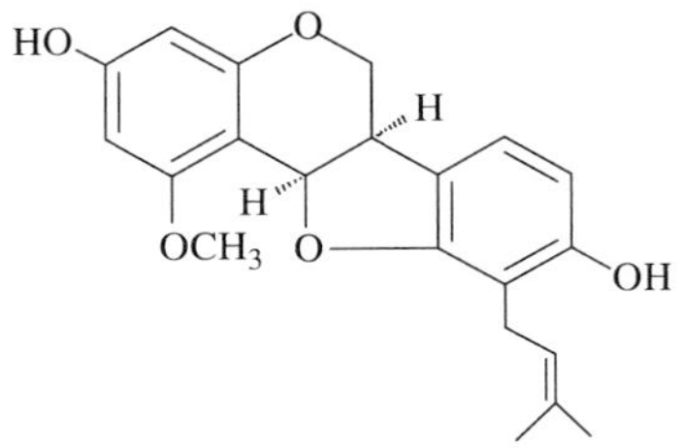 1-methoxyphaseolludin (**23**)

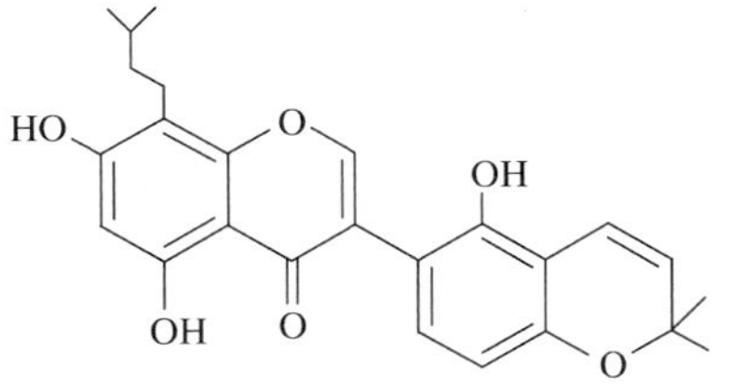

Ulexone (**24**)

Figure 6.2 *(Continued)*

quercetin (**33**), flavonosides (3-rhamnoside, 3-arabinoside and 3-xyloside of quercetin), naphthodianthrone derivatives such as hypericin (**34**) and pseudohypericin (**35**), madagascarin (**36**), β-sitosterol, *p*-coumaric acid, α-pinene, and anthraquinones such as emodine (**37**) and anthrones [72–74]. From the stem bark, pigments [physcion (**38**), euxanthone (**39**), chrysophanic acid, harunganin (**40**), madagascin (**41**), madagascin anthrone (**42**), harongin anthrone (**43**)] and triterpenoids (betulinic acid and friedelin) were isolated [75, 76]. From the roots [74], prenylated anthrones like harunganol A and B (**44** and **45**) and anthraquinones (madagascin, chrysophanol (**46**) and vismiaquinone) were found. Quercetin reported to inhibit the growth of bacteria [77] and anthraquinones [14] could induce the anti-MDR activity of *H. madagascariensis*. See Figure 6.4.

6.5.6
O. gratissimum and *Ocimun canum* (Lamiaceae)

Ocimum species contain volatile oils together with various other compounds. Thymol (**47**), *p*-cymene, eugenol (**48**) and hydrocarbons including γ-terpinene [78], alkaloids, tannins, saponins, flavonoids (trace), anthraquinones, and reducing and nonreducing carbohydrates [38] were isolated from *O. gratissimum*. The essential oil of *O. canum* was characterized by its high content of *trans*-methyl cinnamate, thymol, eugenol and a large amount of hydrocarbons including *p*-cymene [79]. The volatile oil (rich in thymol) exhibits a relatively strong antimicrobial effect [79–81]. According to Akinyemi *et al.* [14], the presence of anthraquinones contributed to the anti-MRSA activity observed for *O. gratissimum* from Nigeria. The anti-MDR activity of *Ocimum*

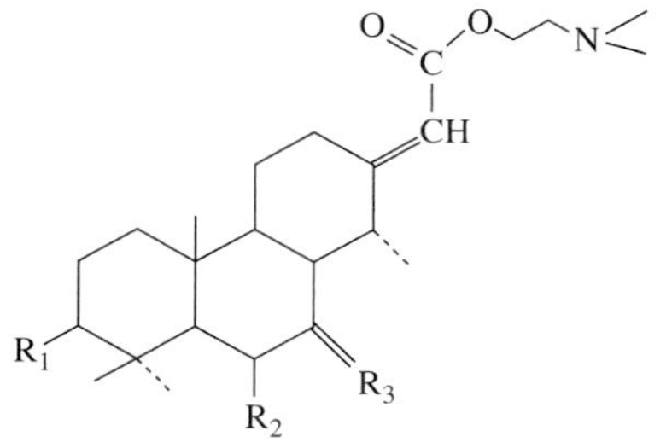

Cassaine group (25)

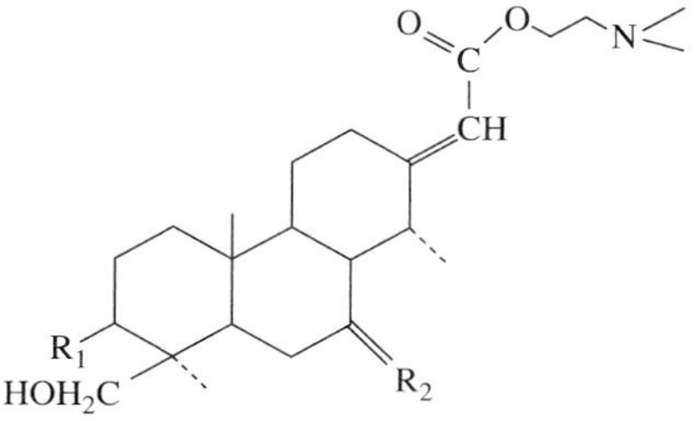

19-Hydroxycassaine group (26)

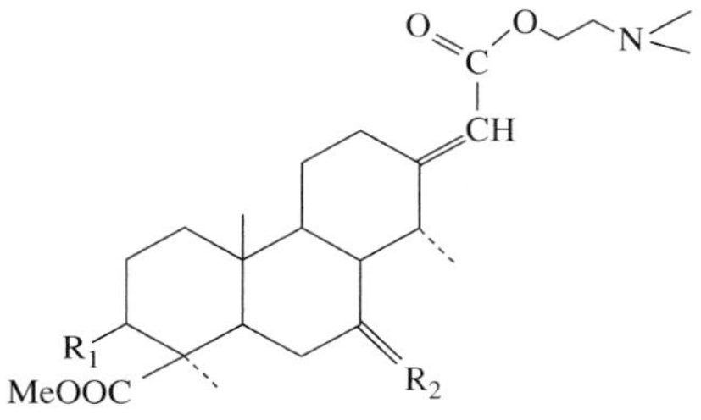

Erythrophlamine group (27)

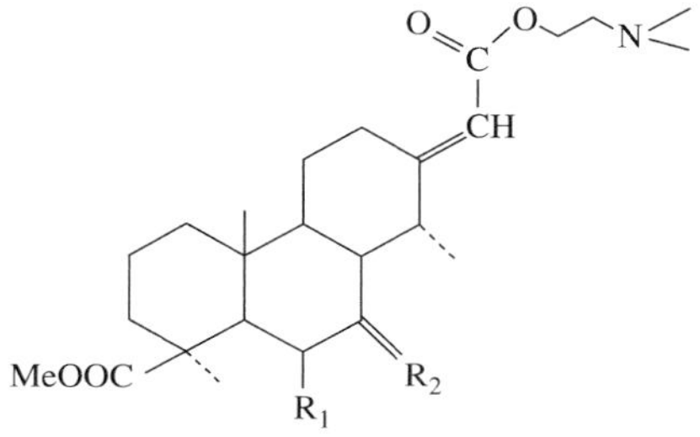

Cassamine group (28)

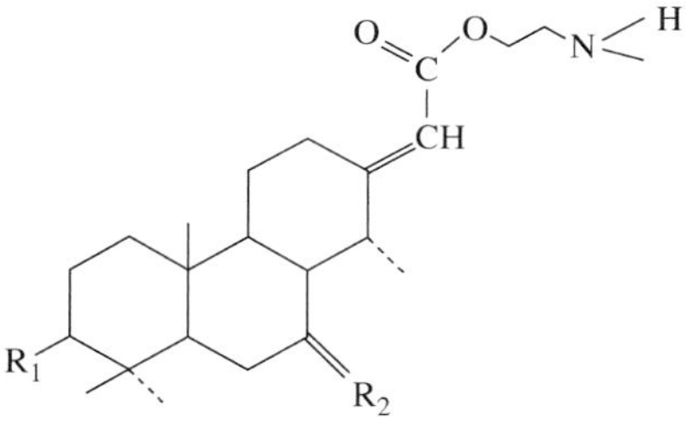

Norcassaine group (29)

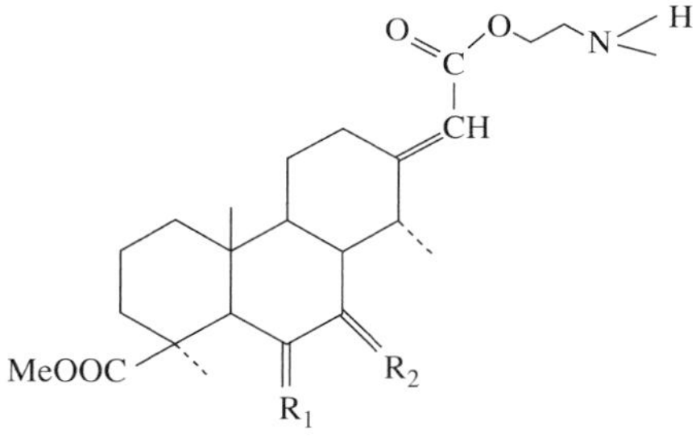

Norcassamidine group (30)

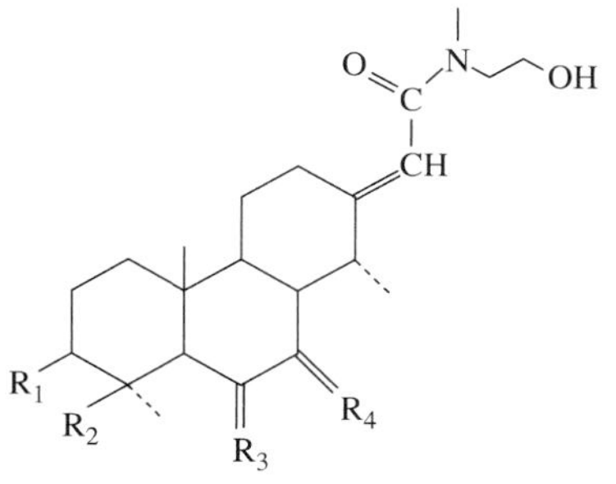

Norcassamidide group (31)

Figure 6.3 Metabolites isolated from *Erythropleum ivorense*.

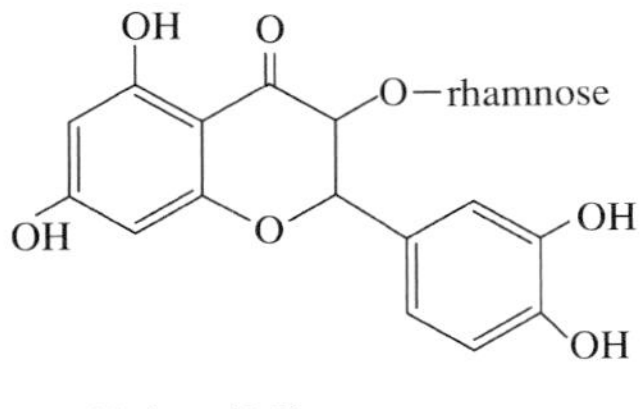

Astilbine (**32**)

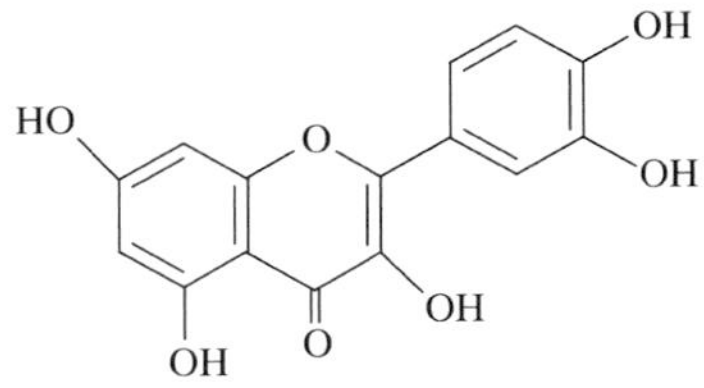

Quercetin (**33**)

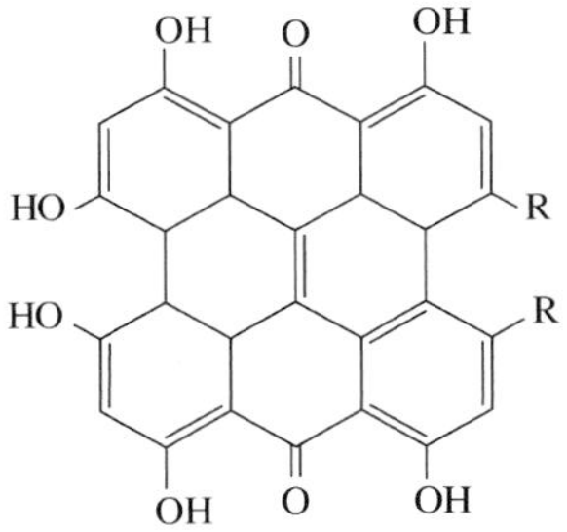

Hypericin (R = Me) (**34**)
Pseudohypericin (R = CH$_2$OH) (**35**)

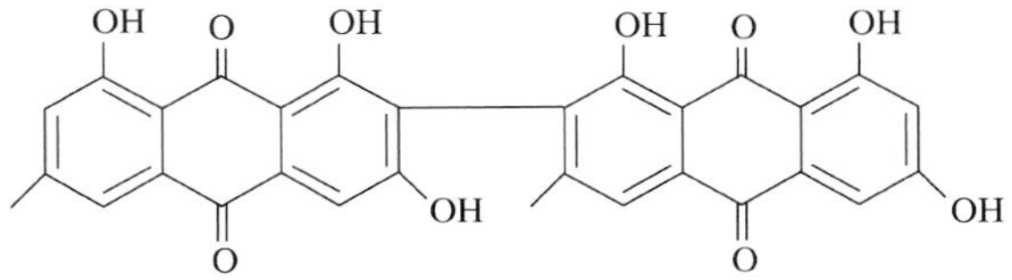

Madagascarin (**36**)

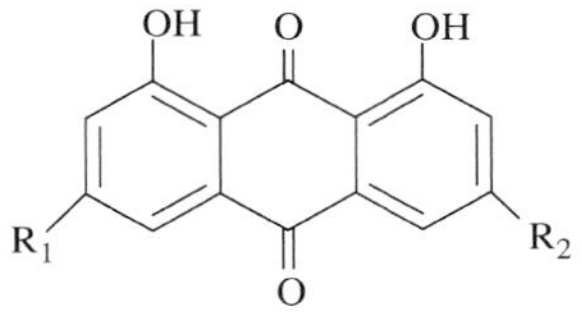

Emodine (R$_1$ = Me ; R$_2$ = OH) (**37**)
Physcion (R$_1$ = Me ; R$_2$ = OMe) (**38**)

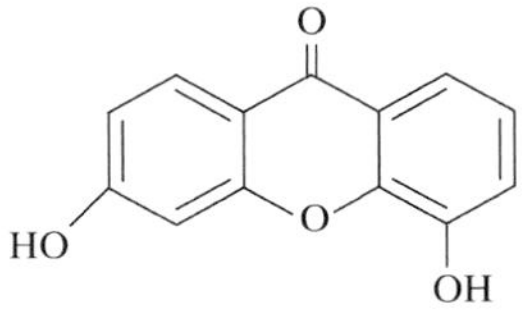

Euxanthone (**39**)

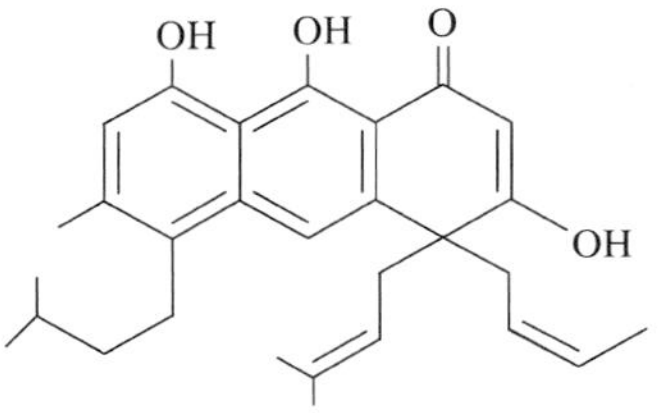

Harunganin (**40**)

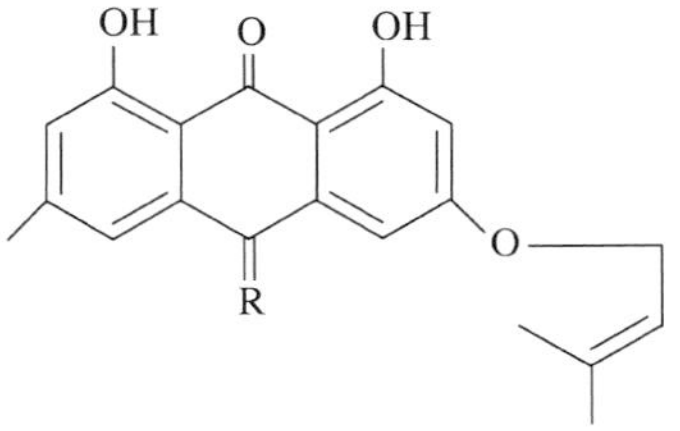

Madagascin (R = O) (**41**)
Madagascinanthrone (R = H) (**42**)

Figure 6.4 Metabolites isolated from *Harungana madagascariensis*.

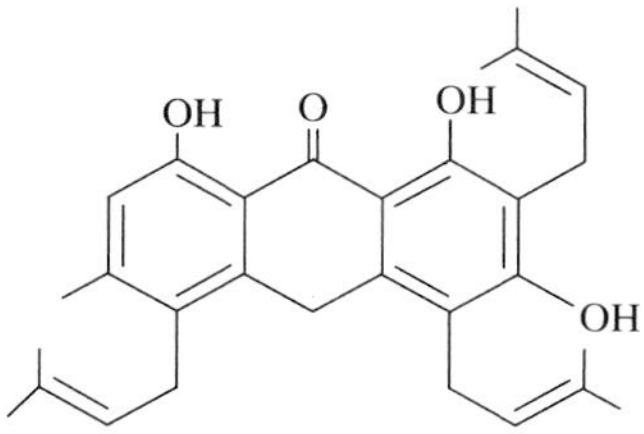

Haronginanthrone (**43**)

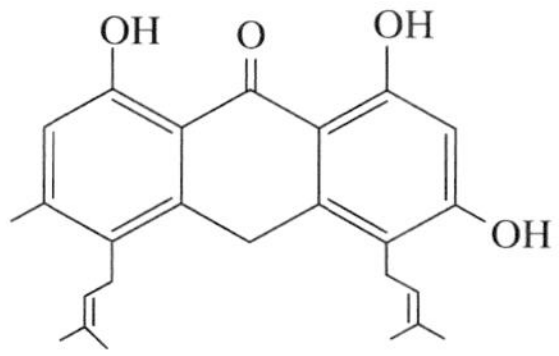

Harunganol A (**44**)

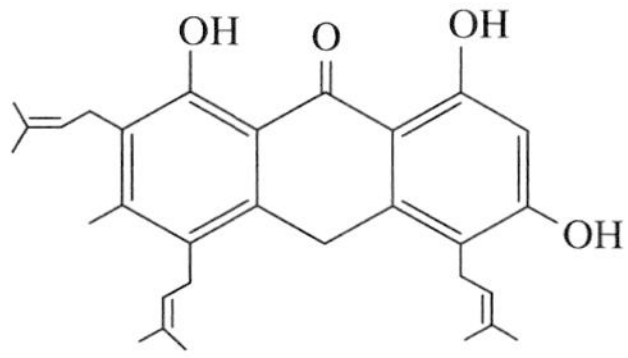

Harunganol B (**45**)

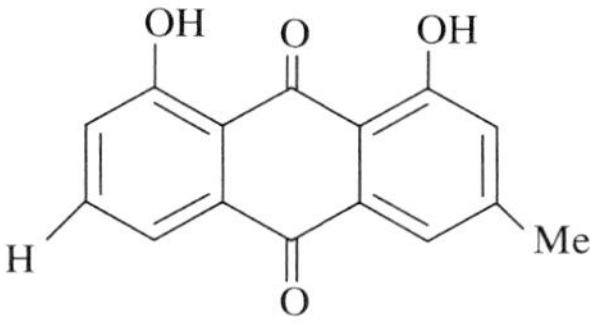

Crysophanol (**46**)

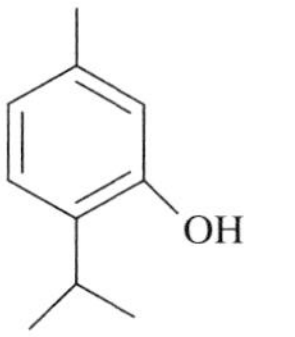

Thymol (**47**)

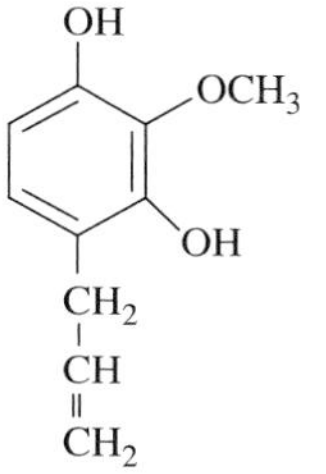

Eugenol (**48**)

Figure 6.5 Metabolites isolated from *Harungana madagascariensis* (43–46) and *Ocimum* species (47, 48).

species may be based on the presence of essential oils and anthraquinones. See Figure 6.5.

6.5.7
M. balsamina (Cucurbitaceae)

Phenylpropanoids esters, named verbascoside and calceolarioside E, were isolated from the aerial parts in addition to rosmarinic acid (**49**) [82]. The presence of flavonoids, cucurbitacins and saponins was also reported [28]. The phytochemicals present in the dried powdered extract of the fruit pulp include resins, alkaloids, tannins, flavonoids, saponins, carbohydrates, steroids and terpenes [83]. The anti-MDR activity could be due to the presence of either rosmarinic acid [84] and phenylpropanoids [82], known for their antibacterial properties, or the other groups of compounds. See Figure 6.6.

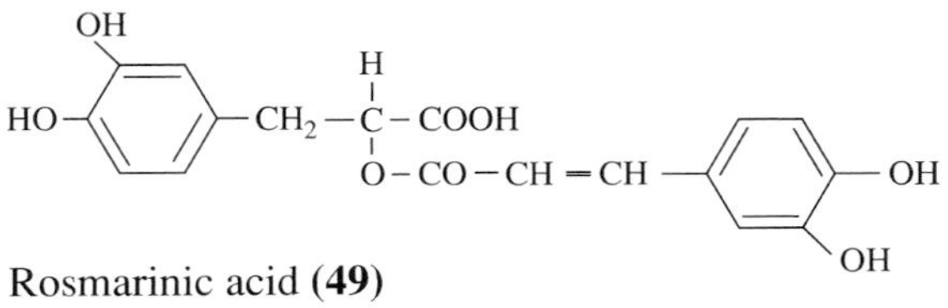

OH
|
HO⟍⟋⟍⟋⟍— CH₂ — C̲ — COOH
H
O — CO — CH = CH —⟍⟋⟍⟋⟍— OH
OH

Rosmarinic acid (**49**)

Figure 6.6 Metabolites isolated from *Momordica balsamina*.

6.5.8
M. lucida (Rubiaceae)

Morinda species are rich in anthraquinones [85], which are known to be antibacterial agents. Alkaloids, quinines [25] and flavonoids glycosides (quercetin-3-*O*-rhamoside, quercetin-3-*O*-rutinoside, kaempferol-3-*O*-rhamoside, kaempferol-3-*O*-rutinoside, luteolin-7-*O*-glucoside, apigenin-7-*O*-glucoside and crysoeriol-7-*O*-neohesperidoside) were isolated from *Morinda* species including *Morinda morindoides* [18].

6.5.9
P. umbellata (Piperaceae)

Piperaceae species are known to contain terpenoids, phenols, phenolic esters, ethers and acids [86], pyrrolidine alkaloids, volatile oil, and lignans [42]. In a search for antioxidant activity of natural products, the catechol derivative named 4-Nerolidyl-catechol (**50**) was isolated from the roots of *P. umbellata*. The same compound was found in the leaves of *Pothomorphe peltata* [87]. Phenolic acids are known for their antibacterial properties [88]. See Figure 6.7.

6.5.10
S. mombin (Anacardiaceae)

The leaves and stems were found to contain series of 6-alkenyl-salicylic acids. These phenolic acids were shown to have pronounced antibacterial activity [88]. The salicylic acid (**51**) is a known standard antibacterial agent. See Figure 6.7.

6.5.11
Other Plant Species

A. wilkesiana [14], *A. difformis* [89], *Ageratum conizoides*, *Bridelia ferruginea*, [25, 14], *Phyllanthus discoides*, *Terminalia avicennoides* [14], *Tieghemella heckelii* [8], *T. sanguinea* [90, 91] and *U. togoensis* [25] were investigated for their global chemical composition. Some of these plants gave positive reactions to alkaloids, tannins, flavonoids, coumarins, phenolic esters, saponines, triterpenes and anthraquinones.

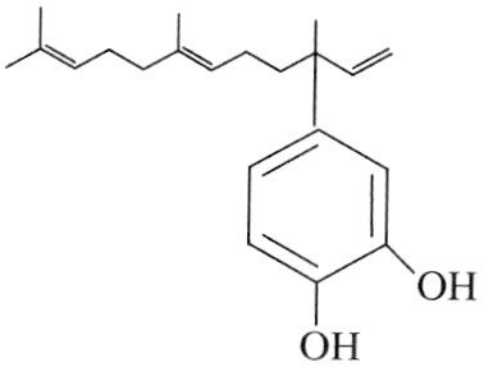

4-Nerolidylcatechol (**50**) 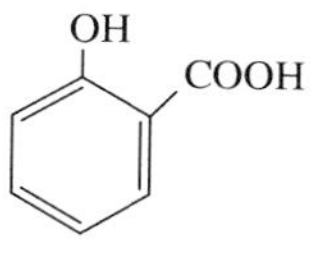   Salycilic acid (**51**)

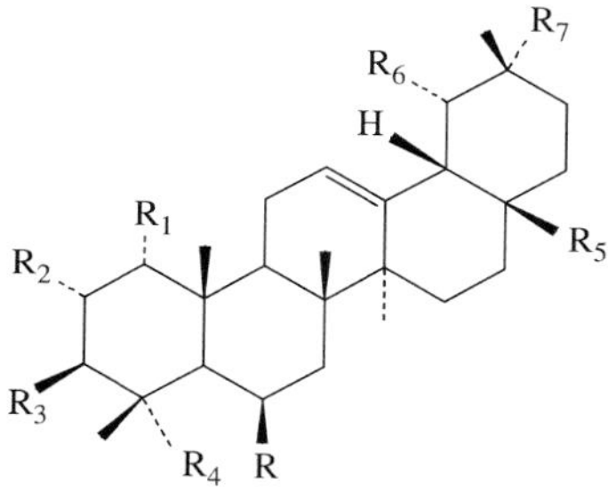

29-carboxy-1α-hydroxyoleanes group (**52**)

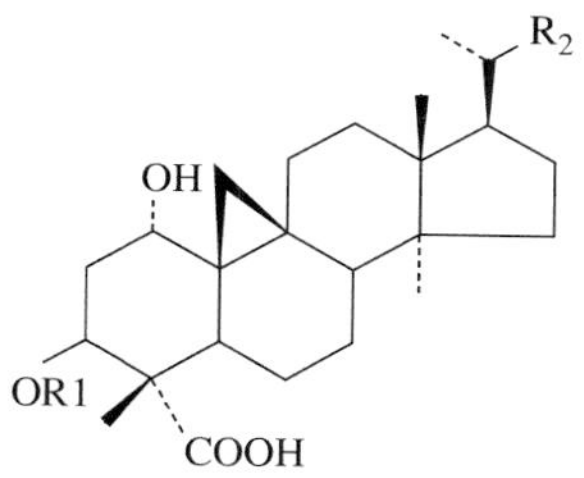

30-carboxy-1α-hydroxyoleanes group (**53**)

Figure 6.7 Metabolites isolated from *Pothomorphe umbellata* and *P. peltata*.

Most of these groups of compounds, in particular alkaloids, tannins, flavonoids, coumarins and anthraquinones, could be responsible for the antibacterial activity and therefore the anti-MDR effect. No information was available on *C. paniculatum* chemistry. However, from *Combretum* species, triterpenoids have been isolated that almost exclusively belong to the 29-carboxy-1α-hydroxyoleane (**52**) and 30-carboxy-1α-hydroxycycloartane (**53**) types [92]. Cycloartane-type triterpenes were already reported for their anti-MDR activity against vancomycin-resistant enterococci and methicillin-resistant staphylococci [93]. See Figure 6.7.

Ficus species contain tannins [25], alkaloids [94], anthracene glycosides, flavonoids, sterols, triterpenes and volatile oils [10] which could be responsible for the anti-MDR activity of *Ficus sur*.

The results indicate that natural products from West African medicinal plants can be effective potential candidates for the development of new strategies to treat MDR bacterial infections.

6.6
Conclusions

A cursory glance at various research on West African plant resources shows that West African plants have a potential for the treatment of diseases due to bacteria, even those connected with MDR bacteria. Ethnomedical investigations counted an important number of plants used by the populations to treat diseases due to bacteria. Although it certainly is an important work, it does not seem exhaustive. More focused investigations, targeting only the fight against bacterial multiresistance, could allow at this level a selection of the most known plants; such a project should allow the registration of other unexplored and very efficient plants. Programs of antibacterial screening were undertaken in various West African countries, but most of them did not include MDR bacteria. To our knowledge, only 37 plants have been tested for their action on MDR bacteria such as MRSA, MDR *S. typhi* and ESβL *E. coli*. Generally, tests were made on crude extracts. The observed antibacterial power could be the result of a synergy among several elements of the plants rather than the result of just one molecule.

Chemical research on active plants is often limited to the global composition, due to the lack of appropriate equipment. The rare active plants on which serious chemical investigation has been made (*Erythrina senegalensis, E. vogelii, H. madagascarienis*, etc.) to obtain the pure compounds have not been systematically tested for their eventual antibacterial action. Only vogelin B and C, isowighteone, and isolupalbigenin are known for their action on MRSA bacterial strains, and macrolides, lincosamides and streptrogramines B on resistant strains. Most Fabaceae, particularly species of *Erythrina* genera, include similar compounds for which their antibacterial power should be tested on MDR strains.

The toxicity of West African medicinal plants has been little studied. Even if their use over generations could suggest the absence of acute toxicity, it is important to evaluate their medical risk.

To make the best possible use of West African plants, coordinated work with precise objectives is necessary within each country and within the region to aim (i) at the discovery of new active molecules for the pharmaceutical industry and (ii) the development of efficient phytomedicines to fight against diseases connected with MDR bacteria. Towards this goal, the most promising species should be the subjects of botanical, pharmacological, chemical and toxicological plant studies. The areas of distribution of the most promising species should be known and protected if necessary in order to perpetuate their use. To reach this goal, West African countries should be aware of the stakes and invest more in research. The training of multidisciplinary and regional teams should be encouraged to promote teamwork towards the global objective of the efficient use of West African plant resources in the fight against the emergence of bacteria strains multiresistant to antibiotics.

References

1 Akoua Koffi, C., Dje, K., Toure, R., Guessennd, N., Acho, B., Faye Kette, H., Loukou, Y.G. and Dosso, M. (2004) *Dakar Médical*, **49**, 70–74.

2 Keasah, C., Odugbmi, T., Ben Redjeb, S., Boye, C.S. and Dosso, M. CB the members of Palm Project (1998) Prevalence of methicillin-resistant *Staphylococcus aureus* in eight African hospitals and Malta, in *38th ICAAC, poster E.093.*

3 Seydi, M., Sow, A.I., Soumare, M., Diallo, H.M., Hatim, B., Tine, R., Diop, B.M. and Sow, P.S. (2004) *Medecine et Maladies Infectieuses*, **34**, 210–215.

4 Taiwo, S.S., Bamidele, M., Omonigbehin, E.A., Akinsinde, K.A., Smith, S.I., Onile, B.A. and Olowe, A.O. (2005) *West African Journal of Medicine*, **24**, 100–106.

5 Kerharo, J. and Bouquet, A. (1950) *Plantes médicinales et toxiques de la Côte d'Ivoire-Haute Volta*, Vigot Frères, Paris.

6 Adjanohoun, E., Aké Assi, L., Florent, J.J., Guinko, S., Koumaré, M., Ahyi, A.M.R. and Raynal, J. (1980) *Médecine traditionnelle et Pharmacopée. Contribution aux études ethnobotaniques et floristiques au Mali*, ACCT, Paris.

7 Abbiw, D.K. (1990) *Useful Plants of Ghana. West African Uses of Wild and Cultivated Plants*, Royal Botanic Gardens, London.

8 Iwu, M.M. (1993) *Handbook of African Medicinal Plants*, CRC Press, London.

9 Bellomaria, B. and Kacou, P. (1995) *Fitoterapia*, **LXV**, 117–141.

10 Neuwinger, H.D. (1996) *African Ethnobotany Poisons and drugs. Chemistry–Pharmacology–Toxicology*, Chapman & Hall, London.

11 Koné, M., Atindehou Kamanzi, K. and Traoré, D. (2002) *Annales de Botanique de l'Afrique de l'Ouest*, **2**, 13–21.

12 Pharmel 2 (1992) *Banque de données de Médecines Traditionnelles et Pharmacopées des pays d'Afrique et de l'Océan Indien*, ACCT, Paris.

13 Koné, W.M., Kamanzi Atindehou, K., Terreaux, C., Hostettmann, K., Traoré, D. and Dosso, M. (2004) *Journal of Ethnopharmacology*, **93**, 43–49.

14 Akinyemi, K.O., Oladapo, O., Okwara, C.E., Ibe, C.C. and Fasure, K.A. (2005) *BMC Complementary and Alternative Medicine*, **5**, 6.

15 N'guessan, J.D., Dinzedi, M.R., Guessennd, N., Coulibaly, A., Dosso, M., Djaman, A.J. and Guede-Guina, F. (2007) *Tropical Journal of Pharmaceutical Research*, **6**, 779–783.

16 Akoua-Koffi, C., Guessennd, N., Gbonon, V., Faye-Kette, H. and Dosso, M. (2004) *Medecine et Maladies Infectieuses*, **34**, 132–136.

17 Okesola, A.O., Oni, A.A. and Bakare, R.A. (1999) *African Journal of Medicine and Medical Sciences*, **28**, 55–57.

18 Benbachir, M., Benredjeb, S., Boye, C.S., Dosso, M., Belabbes, H., Kamoun, A., Kaire, O. and Elmdaghri, N. (2001) *Antimicrobial Agents and Chemotherapy*, **45**, 627–629.

19 Kacou-N'Douba, A., Bouzid, S.A., Guessennd, K.N., Kouassi-M'Bengue, A.A., Faye-Kette, A.Y. and Dosso, M. (2001) *Annals of Tropical Paediatrics*, **21**, 149–154.

20 Aka, J., Dosso, M. and Michel, G. (1987) *Medicina Tropical*, **47**, 53–59.

21 Kouassi, B., Horo, K., N'douba, K.A., Koffi, N., Ngom, A., Aka-Danguy, E. and Dosso, M. (2004) *Bulletin de la Société de Pathologie Exotique*, **97**, 336–337.

22 Ouedraogo, M., Ouedraogo, S.M., Diagbouga, S., Coulibaly, G., Achi, V., Domoua, K., N'Dathz, M. and Yapi, A. (2000) *Revue des Maladies Respiratoires*, **17**, 477–480.

23 Sticht-Groh, V., Rüsch-Gerdes, S., Boillot, F. and Remillieux, M. (1993) *Tubercle and Lung Disease*, **74**, 411–412.

24 Irvine, F.R. (1961) *Woody Plants of Ghana: With Special Reference to their Uses*, Oxford University Press, London.

25 Bouquet, A. and Debray, M. (1974) *Plantes médicinales de la Côte d'Ivoire. Mémoires ORSTOM 32*, ORSTOM, Paris.

26 Adjanohoun, E. and Aké Assi, L. (1970) *Plantes pharmaceutiques de Côte d'Ivoire.* Ministère de la Recherche Scientifique de Côte d'Ivoire, Abidjan.

27 Adjanohoun, E. and Aké Assi, L. (1979) *Contribution au recensement des plantes médicinales de Côte d'Ivoire,* Centre National de Floristique, Abidjan.

28 Adjanohoun, E.J., Adjakidje, V., Ahyi, M.R.A., Aké Assi, L., Akoegninou, A., d'Almeida, J., Apovo, F., Boukef, K., Chadare, M., Cusset, G., Dramane, K., Eyme, J., Gassita, J.N., Gbaguidi, N., Goudote, E., Guinko, S., Houngnon, P., Issa, L.O., Kéita, A., Kiniffo, H.V., Koné-Bamba, D., Musampa, A., Seyya, N., Saadou, M., Sodogandji, Th., de Souza, S., Tchabi, A., Zinsou Dossa, C., and Zohoun, Th. 1989. *Contribution aux études ethnobotaniques et floristiques en République Populaire du Bénin,* ACCT, Paris.

29 Burkill, H.M. (1985) *The Useful Plants of West Tropical Africa. Families A–D,* Royal Botanical Gardens, London.

30 Oliver-Bever, B.E.P. (1986) *Medicinal Plants in Tropical West Africa,* Cambridge University Press, London.

31 Vangah-Manda, M.O. (1986) Contribution à la connaissance des plantes médicinales utilisées par les ethnies Akans de la région littorale de la Côte-d'Ivoire, Thèse 3e Cycle, Université Nationale.

32 De Souza, S. (1988) *Flore du Bénin. Tome 3. Noms des plantes dans les langues nationales Béninoises,* Université Nationale du Bénin, Cotonou.

33 Aké Assi, L. and Guinko, S. (1991) Plantes utilisées dans la médecine traditionnelle en Afrique de l'Ouest, Roche, Basel.

34 Koné, M.W., Atindehou, K.K., Téré, H. and Traoré, D. (2002) *Bioterre, Revue Internationale des Sciences de la vie et de laterre, Spécial,* 30–36.

35 Sofowora, A. (1996) *Plantes médicinales et médecine traditionnelle d'Afrique,* Karthala-Académie Suisse des Sciences Naturelles, Paris.

36 Tra-Bi, F.H. (1997) Utilisation des plantes, par l'homme, dans le Haut-Sassandra et de Scio, en Côte-d'Ivoire, Thèse 3e cycle, Université de Cocody, Abidjan.

37 Weiss, C. (1997) *Ethnobotanische und pharmakologische Studen zu Arzneipflanzen der traditionellen Medizin der Elfenbeinküste,* Universität Basel, Basel.

38 Koné, M.W. (1998) Evaluation de l'activité antibactérienne des plantes utilisées en médecine traditionnelle dans la région de Ferkessédougou, Côte-d'Ivoire, Mémoire DEA, Université de Cocody, Abidjan.

39 Kamanzi Atindehou, K., Koné, M., Terreaux, C., Traoré, D., Hostettmann, K. and Dosso, M. (2002) *Phytotherapy Research,* **16,** 497–502.

40 Akinyemi, K.O., Mendie, U.E., Smith, S.T., Oyefolu, A.O. and Coker, A.O. (2005) *Journal of Herbal Pharmacotherapy,* **5,** 45–60.

41 Iwalokun, B.A., Gbenle, G.O., Adewole, T.A. and Akinsinde, K.A. (2001) *Journal of Health, Population and Nutrition,* **19,** 331–335.

42 Evans, W.C. (2002) *Trease and Evans: Pharmacognosy,* 15th edn, Saunders, Edinburgh.

43 Tschesche, R. (1971) Advances in the chemistry of antibiotics substances from higher plants: pharmacognosy and Phytochemistry, in 1st International Congress Proceeding of Murich.

44 Mitscher, L.A., Drake, S., Gollapudi, S.R. and Okwate, S.K. (1987) *Journal of Natural Products,* **50,** 1025–1040.

45 Zhang, Z., ElSohly, H.N., Jacob, M.R., Pasco, D.S., Walker, L.A. and Clark, A.M. (2002) *Journal of Natural Products,* **65,** 856–859.

46 Lin, R.D., Chin, Y.P. and Lee, M.H. (2005) *Phytotherapy Research,* **19,** 612–617.

47 Machado, T.B., Pinto, A.V., Pinto, M.C., Leal, I.C., Silva, M.G., Amaral, A.C., Kuster, R.M. and Netto-dos Santos, K.R. (2003) *International Journal of Antimicrobial Agents,* **21,** 279–284.

48 Kubo, I., Muroi, H., Kubo, A., Chauderie, S.K., Sanchez, Y. and Ogura, T. (1994) *Planta Medica,* **60,** 218–221.

49 Rukachaisirikul, V., Tadpetch, K., Watthanaphanit, A., Saengsanae, N. and Phongpaichit, S. (2005) *Journal of Natural Products*, **68**, 1218–1221.

50 Sukpondma, Y., Rukachaisirikul, V. and Phongpaichit, S. (2005) *Journal of Natural Products*, **68**, 1010–1017.

51 Beauquesne L. (1947) *Annales Pharmaceutiques Francaises*, **5** 470–483.

52 Borel, C. and Hostettmann, K. (1987) *Helvetica Chimica Acta*, **70**, 570–576.

53 Bezanger-Beauquesne, L. and Pinkas, M. (1967) *Comptes Rendus Hebdomadaires des Seances de l'Academie des Sciences D*, **264**, 401–403.

54 Harper, S.H., Kemp, A.D., Underwood, W.G.E. and Campbell, R.V. (1969) *Journal of the Chemical Society (C)*, 1109–1116.

55 Chapuis, J.C., Sordat, B. and Hostettmann, K. (1988) *Journal of Ethnopharmacology*, **23**, 273–284.

56 Atindehou, K.K., Queiroz, E.F., Terreaux, C., Traoré, D. and Hostettmann, K. (2002) *Planta Medica*, **68**, 181–182.

57 Queiroz, E.F., Atindehou, K.K., Terreaux, C., Antus, S. and Hostettmann, K. (2002) *Journal of Natural Products*, **65**, 403–406.

58 Kamdem Waffo, A.F., Coombes, P.H., Mulholland, D.A., Nkengfack, A.E. and Fomum, Z.T. (2006) *Phytochemistry*, **67**, 459–463.

59 Tanee Fomum, Z., Ayafor, J.F., Ifeadike, P.N., Nkengfack, A.E. and Wandji, J. (1986) *Planta Medica*, **52**, 341.

60 Taylor, R.B., Corley, D.G., Tempesta, M.S., Fomum, Z.T., Ayafor, J.F., Wandji, J. and Ifeadike, P.N. (1986) *Journal of Natural Products*, **49**, 670–673.

61 Wandji, J., Fomum, Z.T., Tillequin, F., Skaltsounis, A.L. and Koch, M. (1994) *Planta Medica*, **60**, 178–180.

62 Wandji, J., Nkengfack, A.E., Fomum, Z.T., Ubillas, R., Killday, K.B. and Tempesta, M.S. (1990) *Journal of Natural Products*, **53**, 1425–1429.

63 Tanaka, H., Doi, M., Etoh, H., Watanabe, N., Shimizu, H., Hirata, M., Ahmad, M., Qurashi, I. and Khan, M.R. (2001) *Journal of Natural Products*, **64**, 1336–1340.

64 Kamanzi Atindehou, K. (2002) Plantes médicinales de Côte d'Ivoire: Investigations phytochimiques guidées par des essais biologiques. Mémoire de Doctorat d'Etat ès-Sciences, Biologie Végétale, Université de Cocody, Abidjan.

65 Barron, D. and Ibrahim, R.K. (1996) *Phytochemistry*, **43**, 921–982.

66 Cronlund, A. and Sandberg, F. (1971) *Acta Pharmaceutica Suecica*, **8**, 351–360.

67 Cronlund, A. (1973) *Planta Medica*, **24**, 371–374.

68 Cronlund, A. (1973) *Acta Pharmaceutica Suecica*, **10**, 507–514.

69 Cronlund, A. (1976) *Planta Medica*, **29**, 123–128.

70 Hussain, R.A., Owegby, A.G., Parimoo, P. and Waterman, P.G. (1982) *Planta Medica*, **44**, 78–81.

71 Han, Q.B., Lee, S.F., Qiao, C.F., He, Z.D., Song, J.Z., Sun, H.D. and Xu, H.X. (2005) *Chemical and Pharmaceutical Bulletin*, **53**, 1034–1036.

72 Buckley, D.G., Ritchie, E., Taylor, W.C. and Young, L.M. (1972) *Australian Journal of Chemistry*, **29**, 1523–1533.

73 Baldi, A., Gehrmann, B., Romani, A. and Vincieri, F.F. (1992) *Planta Medica*, **58**, 691.

74 Iinuma, M., Tosa, H., Ito, T., Tanaka, T. and Aqil, M. (1995) *Phytochemistry*, **40**, 267–270.

75 Stout, G.H., Alden, R.A., Kraut, J. and High, D.F. (1962) *Journal of the American Chemical Society*, **84**, 2653–2654.

76 Ritchie, E. and Taylor, W.C. (1964) *Tetrahedron Letters*, **23**, 1431–1436.

77 El-Gammal, A.A. and Mansour, R.M. (1986) *Zentralblatt fuer Bakteriologie*, **141**, 561–565.

78 Ntezurubanza, L., Scheffer, J.J. and Svendsen, A.B. (1987) *Planta Medica*, **53**, 421–423.

79 Martins, A.P., Salgueiro, L.R., Vila, R., Tomi, F., Cañigueral, S., Casanova, J., da Cunha, A.P. and Adzet, T. (1999) *Planta Medica*, **65**, 187–189.

80 Sofowora, A. (2002) Plants in Africa traditional medicine-an overview, in *Trease and Evans Pharmacognosy*, 15th edn, Saunders, Edinburgh. pp. 488–496.

81 Oussou, K.R., Kanko, C., Guessennd, N., Yolou, S., Koukoua, G., Dosso, M., N'Guessan, Y.T., Figueredo, G. and Chalchat, J.-C. (2004) *Comptes Rendus Chimie*, **7**, 1081–1086.

82 De Tommasi, N., De Simone, F., De Feo, V. and Pizza, C. (1991) *Planta Medica*, **57**, 201.

83 Bot, Y.S., Mgbojikwe, L.O., Chika, N., Alash'le, A. and Jelpe, D.D.D. (2007) *African Journal of Biotechnology*, **6**, 47–52.

84 Lamaison, J.L., Petitjean-Freytet, C., Durand, F. and Carnat, A.P. (1991) *Fitoterapia*, **52**, 166–171.

85 Kerharo, J. and Bouquet, A. (1947) *Revue de Botanique Appliquée et d'Agriculture Tropicale*, **27**, 418–420.

86 Barros, S.B.M., Teixeira, D.S., Aznar, A.E., Moreira, J.R., Ishii, I. and Freitas, C.D. (1996) *Ciência e Cultura*, **48**, 114–116.

87 Desmarchelier, C., Barros, S., Repetto, M., Latorre, L.R., Kuto, M., Coussio, J. and Ciccia, G. (1997) *Planta Medica*, **63**, 561–563.

88 Corthout, J., Pieters, L., Claeys, M., Geerts, S., Vanden Berghe, D. and Vlietinck, A. (1994) *Planta Medica*, **60**, 460–463.

89 Tchiakpe, L., Balansard, G., Bernard, P. and Placidi, M. (1980) *Herba Hungarica*, **96**, 55–63.

90 Cabalion, P., Fournet, A., Mangeney, P. and Bouquet, A. (1980) *Fitoterapia*, **51**, 89–111 .

91 Angenot, L. (1970) *Plantes Médicinales Phytothérapie*, **4**, 263–278.

92 Rogers, C.B. and Verotta, L. (1996) Chemistry and biological properties of the African Combretaceae, in 1st International Symposium Proceedings of the International Organization for Chemistry and Development.

93 Gutierrez-Lugo, M.T., Singh, M.P., Maiese, W.M. and Timmermann, B.N. (2002) *Journal of Natural Products*, **65**, 872–875.

94 Oates, J.F., Swain, T. and Zantovska, J. (1977) *Biochemical Systematics and Ecology*, **5**, 317–321.

95 NGala Malume, T.J., Paulus, J., Kabeya, M., Nlandu, L. and Kisika, K. (1995) *Revue de Médecine et Pharmacopée Africaines*, **9**, 9–14.

96 Rasolondratovo, B., Manjary, F., Rabemanantsoa, C., Rasoanaivo, P. and Ratsimamanga-Urveg, S. (1995) *Revue de Médecine et Pharmacopée Africaines*, **9**, 135–146.

7

Essential Oils and New Antimicrobial Strategies

Sabulal Baby and Varughese George

Abstract

Essential oils are complex mixtures of plant secondary metabolites, composed mostly of terpenoids. They show considerable structural diversity with even nitrogen- and sulfur-containing compounds as their constituents. In plant systems they act, according to the design, as defense compounds against microbes, herbivores and other ecological factors. Essential oils have applications as antimicrobials, food preservatives, flavours and cosmetics. They are also used in clinical applications and aromatherapy. This article describes new essential oil based antimicrobial strategies, particularly giving emphasis to essential oil constituents. A range of recent mechanistic studies of essential oil constituents is compiled. The mode of action of essential oils as antimicrobials, specific assay techniques and other recent developments in these areas are also addressed. Further, the evolution of research on essential oils, biosynthesis extraction techniques, chemical analysis, physical parameters, olfactory evaluation and other basic concepts are also described.

7.1
Introduction: Essential Oils

There are an estimated 250 000–500 000 species of plants on Earth [1]. Humans use a small percentage of these plants for food, medicine and other applications. At present, about 100 000 plant secondary metabolites have been isolated and characterized by systematic scientific studies [2]. This is a very small portion of their total count in the plant kingdom. Secondary metabolites play significant roles in plant defense against herbivores, microbes, insects etc. [3]. They are responsible for plant flavours and pigments. Secondary metabolites also show a range of other bioactivities. Over the years, various secondary metabolites (alkaloids, quinones, terpenoids, lectins, flavonoids, phenolics, polyphenols, tannins, coumarins ect.), classified based on their structures, have been studied for their antimicrobial properties [1, 4].

New Strategies Combating Bacterial Infection. Edited by Iqbal Ahmad and Farrukh Aqil
Copyright © 2009 WILEY-VCH Verlag GmbH & Co. KGaA, Weinheim
ISBN: 978-3-527-32206-0

Essential oils, a group of plant secondary metabolites, have been known since antiquity to possess biological activities. They are also known as 'volatile' or 'ethereal' oils. The term 'essential oil' was derived from the Latin phrase '*Quinta essentia*', meaning 'the effective component of a drug', coined in the sixteenth century by the Swiss physician and alchemist, Paracelsus von Hohenheim [5]. Essential oils were regularly used in ancient Egypt, China, India, Rome, Greece and the Middle East as perfumes, deodorants, food flavours, preservatives, pharmaceuticals and embalming antiseptics. The oil of turpentine was mentioned by Roman and Greek historians. Hippocrates, the Greek physician, wrote in 400 BC, 'The way to health is to have an aromatic bath and scented massage every day'. Distillation as a method of extracting essential oils was used in India, Egypt and Persia more than 2000 years ago [5, 6]. In the fourteenth century, distillation was developed in Spain and France to produce concentrated essences of rosemary and sage. Essential oils were mentioned in early publications such as *The Book of Nurture* in 1430. By 1550, lavender oil was being produced in France as a trading commodity. By this time, flavours and aromas were being distilled from an increasing number of new plant sources. Also, scientists and physicians began studies on the physicochemical and medicinal properties of essential oils during this period.

Essential oils comprise volatile chemical constituents responsible for the characteristic odor in many plants. There are nearly 4500 individual constituents known to occur in essential oils and the chemical diversity of these constituents does not allow a common structural definition for them. Essential oils have monoterpenes, sesquiterpenes and their derivatives as their predominant constituents. They also have aliphatic hydrocarbons, acids, alcohols, aldehydes, acyclic esters and lactones, nitrogen- and sulfur-containing compounds, coumarins, phenylpropanoids etc. as their constituents. The terpenoids in essential oils originate by the conversion of their biosynthetic precursors in plants to hemi- (C_5), mono- (C_{10}), sesqui- (C_{15}) and diterpenes (C_{20}) by terpene synthases [7]. Over 1000 monoterpenes and 3000 sesquiterpenes are currently known [8]. Essential oil constituents are small molecules with molecular weights up to 350 Da.

Labiatae, Apiaceae, Pinaceae, Zingiberaceae, Rutaceae, Compositae, Myrtaceae, Myristicaceae, Fabaceae, Lauraceae, Cupressaceae and Asteraceae are examples of essential oil-bearing plant families. Essential oils are found in the flowers, buds, fruits, peel, leaves, bark, wood, roots, seeds and other plant parts. The yield and composition of essential oils vary widely depending on factors such as season, location, cultivation technique, soil structure and climate. In some plants a single component dominates their essential oils (> 40%), whereas in others relatively small percentages of several components constitute their oils [9, 10]. Trace components in essential oils are also significant in deciding their flavour, odor and biological activities.

Essential oils are utilized in the food (55%), perfumery and cosmetics (20%), pharmaceutical industries (< 10%), and for the isolation of active constituents (15%) [11]. For example, essential oils rich in linalool and linalyl acetate find prominent use in cosmetic and soap industries [12]. The most common methods employed for the isolation of essential oils from plant specimens are hydrodistillation, steam distillation, enfleurage technique, supercritical extraction and microwave extraction. Headspace sampling is another technique for the isolation of trace amounts of volatile constituents from difficult sample matrices, preferably for gas chromatographic analysis. Quality control is critical for essential oils, especially in the cosmetic and flavour industries. The physical parameters for quality control are specific gravity, refractive index, specific rotation, solubility, color and appearance of the oil. Chemical and olfactory evaluations are also critical steps in essential oil quality control [13, 14].

Essential oils show inhibitory effects against bacteria, fungi and viruses [15]. They also show potent antimicrobial effects against a variety of human pathogens and food spoilage microorganisms [16]. Essential oils protect plants from microbial attack, and help in pollination by attracting bees and flies [17, 18]. These oils also have significant applications in pharmaceutical preparations [19]. Volatile oils also play an important role in aromatherapy on the basis of their therapeutic value, and due to the influence of their fragrance on human thoughts and emotions [20–22].

This chapter describes the role of essential oils and their constituents as antimicrobials, their mode of action, protocols for antimicrobial assays, and strategies for the development of new essential oil-based antimicrobials. In addition, basic concepts and practices in essential oil research (biosynthesis, extraction, chemical analysis, physical parameters, olfactory evaluation etc.) are also briefly outlined here.

7.2
Biosynthetic Origin

Essential oils have been investigated in innumerable studies on their chemical, physical and medicinal properties since the sixteenth century. As a result, significant advancements were brought into these fronts. The latter part of the twentieth century witnessed a surge in biochemical studies to unravel the synthesis of primary and secondary metabolites in plant systems. Primary metabolites (DNA, RNA, α-amino acids, etc.) are critical to the metabolic functioning of the plant, whereas secondary metabolites (alkaloids, steroids, terpenoids, coumarins, etc.) help in its functioning and survival. Plant terpenoids, the dominant group in essential oils, are the most functionally diverse class of compounds in plants [23]. Terpenoids function as electron carriers, mediators of polysaccharide synthesis, hormones, photosynthetic pigments, membrane components, pollinators, herbivore repellents etc. [24–26].

Research in plant terpenoid biosynthesis is an area dominated by the group of Rod Croteau at the Institute of Biological Chemistry, Washington State University in the United States and research groups at the Max Planck Institute for Chemical Ecology in Germany [27]. In plants, mono-, sesqui- and diterpenes are synthesized from their parental skeletal types by the actions of various terpene synthases [28–30]. These terpenoid biosynthetic pathways are conveniently divided into various stages. The first stage involves the synthesis of isopentenyl diphosphate, isomerization to dimethylallyl diphosphate, prenyltransferase-catalyzed condensation of these two C_5 units to geranyl diphosphate and the subsequent 19-4 additions of isopentenyl diphosphate to generate farnesyl diphosphate and geranylgeranyl diphosphate. In the second stage, the prenyl diphosphates undergo a range of cyclizations to produce the parent skeletons of monoterpenes (C_{10}) from geranyl diphosphate, sesquiterpenes (C_{15}) from farnesyl diphosphate and diterpenes (C_{20}) from geranylgeranyl diphosphate. These transformations are catalyzed by terpenoid synthases (cyclases). In the last stage, a variety of redox modifications of the parent skeletal types produce the various terpenoid metabolites of plant essential oils, turpentines and resins [29].

Terpene synthases are capable of forming multiple products from a single substrate [31]. All these terpene synthases are related to each other, forming a large family of proteins [29, 32]. Since terpenoid biosynthetic pathways are very complex, purification of terpene synthases and isolation of gene probes from the target proteins are difficult tasks using classical biochemical approaches. Thus, functional genomic approaches directed toward the characterization of genes involved in the formation of essential oil constituents have also been introduced recently [33]. Since essential oils have commercial applications, the steps involved in their biosynthesis and secretion are also targets for genetic engineering. Biosynthetic pathways of other essential oil constituents, such as aliphatic hydrocarbons, acids, alcohols, aldehydes, coumarins, acyclic compounds (esters and lactones), nitrogen- and sulfur-containing compounds, are relatively less explored as compared to terpenoids.

7.3
Extraction of Essential Oils

Extraction of volatile oils from plants, in various forms, began in antiquity and extraction techniques have undergone constant refinements. The most common methods are steam distillation, hydrodistillation, enfleurage technique and super-critical extraction. The extraction technique is a careful choice based on the nature of the plant specimen and thermal lability of the target oil constituents [34]. The principles and methodologies of selected essential oil isolation techniques are briefly described here.

7.3.1
Steam Distillation

Steam distillation is done by passing dry steam through the plant specimen, whereby the steam volatile aromatic molecules are volatilized, later condensed and collected in receivers. Steam distillation has been in use for essential oil extraction for many years. Recent examples from the literature are extraction of oils from *Rosmarinus officinalis* (rosemary) [35] and *Satureja fruticosa* [36]. Hydrosteam distillation is carried out when the perfumery plant material is vulnerable to direct steam. In this technique, the plant material is supported on a screen or a perforated grid inserted at some distance above the bottom of the still. Then, distillation is carried out with low-pressure steam replacing the volatiles from the intact plant material by osmotic action.

7.3.2
Hydrodistillation

Hydrodistillation involves boiling the plant material with water and the volatiles in the plant specimen are carried away along with the steam, which is condensed by the circulation of cold water and collected over water. Hydrodistillation using the Clevenger-type apparatus is the most commonly used technique for laboratory-scale extraction of essential oils [37–39].

7.3.3
Enfleurage Technique

Enfleurage technique is used to extract traces of volatile oils from very delicate plant parts such as flowers. Flower petals are placed on glass plates layered with odorless vegetable or animal fat, which will absorb its essential oil. After the fat has absorbed as much of the essential oil as possible, in a few hours or days, the depleted petals are removed and replaced with fresh ones. This is continued until the fat becomes saturated with the essential oil. Then, the essential oil is extracted from the fat by the addition of alcohol [6, 40].

7.3.4
Supercritical Fluid Extraction

Supercritical fluid extraction, introduced in the 1980s, uses CO_2 under extremely high pressure to extract essential oils. This is an energy-efficient, environmentally friendly process [41]. The plant specimen is placed in a stainless steel tank and CO_2 is injected into this tank. Under high pressure, CO_2 turns into a liquid and acts as a solvent to extract the essential oil from the plant specimen. When the pressure is decreased, CO_2 returns to the gaseous state leaving only the volatile oil and no other residues behind. These CO_2 extractions at lower temperatures are gentler on plant samples compared to steam distillation, and result in fresher, cleaner and crisper oils, which smell similar to the natural plant aromas. Studies have shown that CO_2 extraction produces very potent essential oils with good therapeutic effects. Further, it produces higher yields, and makes some materials (e.g. gums, resins) easier to handle. Many essential oils that cannot be extracted by steam distillation are obtainable with CO_2 extraction. Moyler [41] reported a comparative study of steam distilled and CO_2-extracted volatile oils of celery seed, hop, clove bud, juniper berry, mace and ginger by using free-choice odor profiling and gas chromatographic fingerprint analysis [41].

7.3.5
Microwave Extraction

Solvent-free microwave extraction (SFME) is another recently developed green technology in which a combination of microwave heating and dry distillation at atmospheric pressure results in the extraction of essential oils from plant samples.

SFME results in substantial savings in time, energy, solvents and plant material. The oil obtained by microwave extraction can be directly used for analysis without further solvent exchange or trace water removal. Lucchesi *et al.* [42] compared the extraction of essential oils from basil, garden mint and thyme by single-stage SFME and conventional hydrodistillation. The essential oils extracted by SFME for 30 min were similar in yield and quality to those obtained by hydrodistillation for 4.5 h. This study also observed that SFME yields an essential oil with higher amounts of oxygenated compounds [42]. Wang *et al.* [43] isolated essential oils by SFME, improved SFME, microwave-assisted hydrodistillation and conventional hydrodistillation from dried *Cuminum cyminum* (cumin) and *Zanthoxylum bungeanum*. The chemical profiles of these oils were compared by gas chromatography-mass spectrometry (GC-MS) and no major changes were observed between them [43]. Other recent examples are microwave extraction of *Nigella sativa* seeds [39], *Carum ajowan* (ajowan), *C. cyminum* and *Illicium anisatum* (star anise) [44].

7.3.6
Headspace Sampling Techniques

Headspace sampling is a microextraction technique for the isolation of volatile constituents from plant matrices. In this technique, isolated volatile constituents are directly injected into a gas chromatograph and, thus, avoid contamination of the inlet system and the column with semi- or non-volatile residues. A number of variations of headspace sampling techniques such as dynamic headspace (DHS), static headspace (SHS), solid-phase microextraction (SPME), headspace solvent microextraction (HSME), headspace sorptive extraction (HSSE), direct thermal desorption (DTD), purge-and-trap sampling (PTS), solid-phase microextraction-gas chromatography mass Spectrometry (SPME-GC-MS), headspace solid-phase microextraction-gas chromatography (HS-SPME-GC) etc. exist [45–50]. As an example, HSME is a rapid technique for the extraction and preconcentration of the volatile components of a plant sample. Extraction is by suspension of a microliter drop of the selected solvent from the tip of a microsyringe to the headspace of the sample in a sealed vial. After a set extraction time, the microdrop is retracted back into the microsyringe and injected directly into a gas chromatograph. SPME-GC-MS is a similar technique to extract volatiles from plant matrices and to directly inject the volatiles using GC-MS, leading to the identification of its chemical constituents. In headspace sampling techniques, parameters such as nature of the extracting solvent, particle size of the sample, temperatures of the microdrop and sample, sample volume, and the extraction time are to be standardized for optimal results. Headspace sampling coupled with GC and its variations are now widely used in essential oil analysis [46–48, 51, 52].

In addition to volatile oils, plants also synthesize and emit volatile organic compounds (VOCs). VOCs play significant roles in plant–environment interactions, ecology and atmospheric chemistry. The development of static and dynamic techniques for headspace collection of volatiles in combination with portable GC systems has significantly improved the understanding of the chemistry, biosynthesis and ecology of plant VOCs [3, 53–55].

7.4
Storage of Essential Oils

Storage is an important aspect, especially in essential oil-based industries. Essential oils are affected by light, heat and air on long standing [56, 57]. They must be clarified, and also made free from moisture and other impurities before storage. Oxidation, resinification, polymerization, hydrolysis of esters and interaction of functional groups cause darkening of oil and spoilage. Thus, essential oils must be stored in airtight containers in a cool, dark place [58].

7.5
Chemical Analysis of Essential Oils

As mentioned in Section 7.1, essential oils are complex mixtures of aliphatic and aromatic hydrocarbons, mono-, sesqui-, diterpenes and their oxygenated derivatives. Terpenoids are composed of isoprene units. Monoterpenes are made of two and sesquiterpenes by three isoprene units. Some essential oils have no mono- and sesquiterpenes, instead they have aliphatic and aromatic hydrocarbons and their derivatives. Chromatographic techniques such as thin-layer chromatography (TLC), high-performance thin-layer chromatography (HPTLC), high-performance liquid chromatography (HPLC) and column chromatography are used for separation of the fragrant complex into individual components. Techniques such as gas chromatography flame ionization detection (GC-FID), GC-MS, GC-MS-MS etc. result in the separation of the volatile mixture, and subsequent identification and quantification of its individual constituents [59, 60]. Identification of the fragrance constituents may also be accomplished through infrared (IR) and nuclear magnetic resonance (NMR) spectroscopy [61]. IR spectroscopy gives data on the functional groups present in an essential oil constituent. [^{1}H-NMR, ^{13}C-NMR, two-dimensional (2-D)-NMR etc.] is applied to obtain data on the proton and carbon skeletons of essential oil constituents [62].

7.5.1
Chromatographic Techniques in Essential Oil Analysis

Chromatography is the differential migration of solute components in a system of two phases, i.e. mobile phase (liquid or gas) and stationary phase (solid or liquid). The term 'chromatography' originated from two Greek words – 'chromo' (color) and graphy (to write). The Russian botanist Mikhail Tswett discovered chromatography in 1906 by separating pigments in plant leaves [63]. Major chromatographic techniques are column chromatography, paper chromatography, TLC, HPTLC, GC, HPLC and the recently evolved hyphenated techniques. TLC, HPTLC, GC, GC-MS, HPLC-ultraviolet (UV) spectroscopy and LC-MS-MS are widely used in essential oil analysis [64].

7.5.1.1 **TLC/HPTLC**

N.A. Izmailov and M.S. Schraiber laid the foundation for TLC in 1938 [65]. In 1958, Egon Stahl refined this method and developed it into a technique of general application [66]. TLC is now widely used in natural product chemistry and synthetic organic chemistry [67]. In TLC, a uniform thin layer of a solid adsorbent held on a glass plate forms the stationary phase. A mobile phase (solvent) runs over separating the components of a sample (extract, essential oil, etc.) applied onto the stationary phase. Detection of compounds on a TLC plate is usually carried out by spraying with a suitable reagent. Examples of commonly used spray reagents are ceric sulfate, vanilline sulfuric acid, Dragendorff's reagent etc. The separated compounds can be detected under UV light at 254 or 365 nm. The movement of any substance relative to the solvent front in a given chromatographic system is constant and characteristic of the substance, under reproducible experimental conditions. This parameter, R_f value, is defined as $R_f =$ [distance moved by the component]/[distance moved by the solvent front]. TLC is widely used for the quick profiling of essential oils.

HPTLC is a major advancement of the principles of TLC. It is very useful for the qualitative as well as quantitative analysis of essential oils and their constituents [68]. The basic differences between conventional TLC and HPTLC are in the particle and pore sizes of the sorbents. Precoated plates are available for HPTLC in which silica gel of very fine particle size is used. The smaller particle size of the stationary phase helps in better resolution and sensitivity, and requires a shorter time. HPTLC needs lower amounts of concentrated sample (extract, essential oil, etc.) and the size of the sample spot is in the range of 1 mm in diameter. Examples of sample spotting techniques are contact spotting, pre-adsorbent or concentrating zones and chemical focusing. The Linomat sample injection system of Camag (Switzerland) uses the programmed multiple development for spotting of samples, in which the amount of sample to be sprayed on the plate is programmed. Linear development is the most commonly used technique in HPTLC. The qualitative and quantitative estimation of the components in a developed plate is achieved by densitometric scanning. HPTLC is very useful in obtaining fingerprint patterns of herbal formulations, extracts, essential oils and in detection of adulteration. Reich and Schibli [68] suggested toluene : ethyl acetate (95 : 5, 85 : 15), chloroform and dichloromethane as suitable mobile phases for screening of essential oils in HPTLC [68].

7.5.1.2 **GLC**

GLC was introduced as an analytical tool by A.J.P. Martin and A.T. James in 1952 [69]. It is a gas–liquid partition mode of chromatography. The main variables in GLC are the nature of the stationary phase of the column and the temperature of operation. These are varied according to the polarity and volatility of the compounds being separated. Some very polar biomolecules cannot be volatilized without the loss of their chemical identity. In such cases, suitable volatile derivatives that are stable for GC investigations are prepared [70].

The GLC apparatus is sophisticated compared to TLC or paper chromatography. It has four main components: column, heater, gas flow and detection. The GC column is a long narrow tube usually made of metal or glass in the form of a coil to conserve space. In capillary columns, the stationary phase is held as a thin film adhering to the column walls. Highly reproducible capillary columns with poly-siloxane (BP1, BP5, DB5, HP5, HP5-MS, BPX5, etc.)- and polyethylene glycol (BP20, BP21)-based stationary phases are used for GC separations now. HP5-MS, cross-linked 5% phenyl methyl siloxane column has the following dimensions: 30 m (length) $\times$ 0.32 mm (inner diameter) $\times$ 0.25 μm (film thickness). DB5, HP5 and HP5-MS are nonpolar capillary columns suitable for essential oil analysis. In packed columns, the column contains an inert supporting material (chromosorb W, celite, etc.) carrying the stationary phase (5–15% silicone oil) absorbed on it. In GC, the heater provides heat to the column progressively from 50 to 350 °C at a standard rate and holds the temperature at a desired point for a specified period. The temperature of the column inlet is separately controlled so that the sample can be rapidly vaporized as it is passed on to the column. The sample dissolved in a suitable solvent is injected by a hypodermic syringe into the inlet port through a rubber septum. The gas flow consists of an inert carrier gas such as nitrogen or helium. Separation of the compound on the column depends on passing this gas through at a controlled rate. Detection is usually based on either flame ionization or electron capture. FID, commonly used for essential oil analysis, requires hydrogen gas to be added to the gas mixture and to be burnt off in the detector. The detection device is linked to a potentiometric recorder, which produces the results of the separation as a series of peaks of varying intensity.

In GLC, retention time (R_t) is the time required for elution of the compound. R_t is expressed relative to a standard compound that is added to the sample (essential oil) or the solvent used for dissolving the sample. GLC provides both quantitative and qualitative data. Area of a peak in the GLC trace is directly related to the concentration of the component in the sample. The percentage of each component is determined from the area under each peak and expressed as proportion of total area, provided that the detector is equally sensitive to each of the eluted components of the volatile sample. In a simple two component mixture, the percentage of A = [area of peak A/(area of peak A + peak B)] $\times$ 100.

7.5.1.3 GC-MS

GC-MS is the most useful analytical technique for essential oils, and it provides the characteristic molecular ion peak and fragment ion peaks from a separated com-pound in an essential oil [71]. It is a hyphenated technique in which gas chromato-graph is set up so that the separated components are subjected to mass spectral analysis. GC provides an effective resolution of the individual components in an essential oil. Identification of these separated compounds is accomplished through their characteristic molecular fingerprints, mass spectra [60]. Any compound that passes through a gas chromatograph is converted to ions in the mass spectrometer. The characteristic nature of a mass spectrum makes the mass spectrometer a very specific GC detector. In GC-MS, little sample and relatively high volumes of carrier gas are present in the effluent from the GC unit. Another incompatibility between the units is that the gas chromatograph operates at high pressure and the mass spectrometer operates under high vacuum. The interface unit of GC-MS helps to operate both gas chromatograph and mass spectrometer without degrading the performance of either unit.

The main advantages of a mass spectrometer as a detector for GC are its increased sensitivity and its specificity in identifying unknowns. In GC-MS, essential oils or other complex mixtures can be subjected to a total ion chromatogram mode, which records the entire mass spectra, or to a selected ion monitoring mode, in which the intensities of specific ions only are recorded. Further, individual component spectra are compared with known spectra in in-built databases (WILEY, NIST, Flavour and Fragrance Database etc.) and then further interpreted for identification of the constituent. Coinjection of oil with authentic standards is another GC-based technique for the identification of essential oil constituents. The relative retention index (RRI), determined by coinjection of sample with *n*-alkanes, is another parameter for confirming the identity of oil constituents [8, 72, 73].

$$\text{RRI} = 100i[(E_x - H_n)/(H_{n+i} - H_n)] + 100n$$

where E_x is the retention time of the compound (oil constituent), H_n and H_{n+i} are retention times of reference hydrocarbons with n and $n+i$ carbon atoms, respectively.

GC-MS also has certain limitations. In some cases, a target compound cannot be quantified at a desired level due to interference from the other signals of the sample or an unknown compound cannot be identified due to a lack of good fragmentation patterns. Thus, the technique of isolating a single ion called a 'precursor ion' from a mass spectral fragmentation and subjecting this isolated ion to undergo a collision induced dissociation in order to obtain its mass spectrum was introduced using multiple quadrupole mass filters in tandem. This tandem MS technique is called MS-MS because it can produce a secondary mass spectrum of an ion, resulting from a primary fragmentation event. GC-MS-MS has a triple quadrupole mass spectrometer, which links two stages of mass analysis. GC-MS-MS and GC-MS-MS-MS techniques are now used in essential oil analysis [74] as well as in doping analysis, forensic science, drug testing, pesticide testing in food, biomarker analysis in oil exploration etc. Patchoulic oil, the volatile oil of *Pogostemon cablin*, is widely used in the cosmetic and oral hygiene industries. It is also used as an indicator for the quality assessment of dried *P. cablin*. However, the complexity of its herbal constituents makes it difficult for using conventional GC for analytical purpose. Zhao *et al.* [75] subjected patchoulic oil to GC-MS-MS analysis, and established an accurate and reproducible technique for the quality assessment of the dried patchouli herb based on its patchoulic alcohol content [75].

7.5.2
Spectroscopic Techniques

Isolation, characterization and bioactivity studies of individual essential oil constituents gained priority since the latter part of the twentieth century. For example, essential oil constituents (carvacrol, citronellol etc.) were tested as food preservatives. Column chromatography, preparative TLC and HPLC-based techniques lead to the isolation of individual constituents from essential oils. In most cases, isolation is limited to major oil constituents. ^{1}H-NMR, ^{13}C-NMR, 2 D-NMR, IR and MS techniques provide data on the structural features of isolated oil constituents, leading to their characterization [59, 76–79]. Further, advanced Fourier transform-IR techniques are now used for the classification and quantification of essential oils and chemotaxonomy of essential oil bearing plants [80, 81].

7.6
Physical Parameters

Essential oils are complex mixtures, and variations in oil compositions are often caused by climatic influences, genetic effects, type and development of the plant organ from which oils are isolated, and culture conditions of the plant. The quality of an essential oil is assessed by its physicochemical characteristics. Physical parameters for essential oil quality control are specific gravity, refractive index, congealing point, freezing point, melting point, specific rotation, solubility, colour and appearance, acid, ester, iodine, carbonyl and sap values etc.

7.6.1
Specific Gravity

Specific gravity is the ratio of weight of an essential oil to that of water at 20 °C. It is characteristic of the unique composition of each essential oil under specific conditions. Specific gravity is used as a parameter to check oil quality and adulteration.

7.6.2
Refractive Index

Refractive index measures the rate at which light passing through the sample (essential oil) is refracted. The speed and number of degrees at which the oil refract light is characteristic of the chemical constituents of the oil. The passage of a ray of light from one medium to another causes a change in its direction called refraction. If it passes from a less dense to a more dense medium, as from air to water, it is refracted towards the normal so that the angle of refraction r is less than the angle of incidence i. The refractive index n of the second medium with respect to the first is given by $n = \sin i / \sin r$. The refractive index of a liquid can be determined to a high degree of accuracy using a digital refractometer. It is a characteristic property of a liquid, and it varies with the temperature and wavelength of light used. The wavelength of the D-line of the sodium spectrum is generally used for standard measurements. The symbol $^{20}n_D$ indicates that the refractive index has been determined at 20 °C using the D-line of sodium as the source of light. Since refractive index is a ratio, it has no unit. Refractive index is an important criterion of the purity of essential oils [82].

7.6.3
Specific Rotation

Optically active constituents in essential oils influence the direction and degree to which light rays bend as they pass through the oil. A beam of ordinary light consists of electromagnetic waves oscillating in many planes. When passed through a polarizer only waves oscillating in a single plane pass through. The emerging beam of light is 'plane polarized', having oscillations in a single plane. When plane polarized light is passed through certain organic compounds, its plane of polarization is rotated. A compound that can rotate the plane of polarization of plane polarized light is 'optically active'. A compound that rotates the plane polarized light to the left (anticlockwise) is 'laevo rotatory' and to the right (clockwise) is 'dextro rotatory'. By convention, rotation to the left is given a minus sign (–) and rotation to the right is given a plus (+) sign. Optical rotation is measured with a polarimeter. The polarimeter consists of a monochromatic light source, polarizer, sample cell, a second polarizer (analyzer) and a light detector. The analyzer is oriented 90° to the polarizer so that no light reaches the detector. When an optically active substance is present in the beam, it rotates the polarization of the light reaching the analyzer so that there is a component that reaches the detector. The angle that the analyzer must be rotated to return to the minimum detector signal is the optical rotation, α.

The rotation of plane polarized light is an intrinsic property of an optically active molecule. When a polarized beam of light is passed through the solution of an optically active compound, the angle of rotation depends on the number of optically active molecules encountered by it. Thus, optical rotation (α) is proportional to the concentration and length of the sample solution. It is expressed as $[\alpha]_\lambda^t = \alpha/lc$, where $[\alpha]_\lambda^t$ is the specific rotation at temperature t (°C) and at wavelength λ (nm) of the light used for the measurement. If the D-line of Na (589 nm) is used at 20 °C, specific rotation is denoted as $[\alpha]_D^{20}$. α is the observed angle of rotation (degrees), l is the length of sample solution (in dm) and c is the concentration of the sample solution (g/ml). Thus, the specific rotation of a substance is defined as the observed angle of rotation at a concentration of 1 g/ml and path length of 1 dm.

Specific rotation is another physical parameter for the quality control of essential oils. In a recent study, Sugawara *et al.* isolated (R)-(−)-linalool with specific rotation of $[\alpha]_D = -15.1°$ from lavender oil, (S)-(+)-linalool with $[\alpha]_D = +17.4°$ from coriander oil and (R,S)-(±)-linalool with $[\alpha]_D = 0°$ from commercial linalool by repeated flash column chromatography. The odor distinctiveness of these optically active linalools on inhalation by humans was also examined in terms of sensory tests [83].

7.7
Olfactory Evaluation

Olfactory evaluation is a critical step in essential oil quality control. For most commercial purposes, a professional perfumer or aroma chemist carries out olfactory evaluation by smelling the aroma of oils or oil products. Electronic noses (Z-nose, E-nose etc.) have been introduced recently to assist in olfactory evaluation of essential oils and their flavour and cosmetic products [84]. A combination of physical, chemical and olfactory evaluation could lead to the quality assessment of essential oils, especially for commercial purposes. Examples of recent olfactory studies on volatile oils are Ngassoum *et al.* [85], Kondoh *et al.* [86] and Sugawara *et al.* [87].

7.8
Essential Oils as Antimicrobials

Essential oils are biosynthesized by plants and their bioactivities act as a chemical barrier against plant diseases. For example, plant pathogens could easily penetrate wound sites in plants caused by herbivores. However, wounds at essential oil-bearing plant tissues result in the rupture of oil glands, causing the oil to flow over the wound. The antimicrobial activities of the oil could defend the attack of the pathogenic organisms from entering the plant system. In 1676, Antony van Leeuwenhoek first recorded the antimicrobial activity of a 'spice' by observing a decline in the number and activity of 'animalcules' in a sample of well water on addition of pepper [88, 89]. Antimicrobial activities of essential oils are now utilized in food preservation, disinfection, antiseptics and therapeutics as well as they play a role in aromatherapy, flavour and cosmetics. Burt [58] listed commercial products of essential oils such as food preservatives (DMC Base Natural, Protecta One, Protecta Two, etc.), antiseptics, dental root canal sealers and feed supplements [58]. Essential oils are clinically used in dermatology, respiratory infections, asthma, urology, sleep and nervous disorders, cardiac, and vascular diseases. They are also used for laxatives, immunomodulating drugs, erosive gastritis, colds and coughs, depression, panic disorder, jet lag, gastrointestinal diseases etc. [90].

Over the years the *in vitro* antimicrobial properties of essential oils, their constituents and their mechanisms of action have been studied and reviewed by various authors [15, 58, 91–94]. Further, the activities of essential oils against a wide range of organisms such as food-spoiling organisms, food-poisoning organisms, mycotoxigenic filamentous fungi, animal and plant viruses were tested [95–101]. However, poor absorption of essential oils and their constituents in the human intestine compared with synthetic antimicrobial drugs limits their clinical applications [102, 103]. Clinical efficiency of essential oil-based drugs depends critically on the systemic availability of the bioactive molecules in the target organs. Thus, data on drug delivery systems, absorption (dermal, pulmonary or intestinal), distribution, stability, metabolism and pharmacokinetics of oils and oil constituents are critical in correlating the *in vitro* and *in vivo* observations [104, 105]. Bioavailability and pharmacokinetics data are also important in assessing the safety of essential oil-based drugs [104]. A few existing pharmacokinetic studies on intravenous administration found that oil constituents are relatively quickly eliminated in humans. Kohlert *et al.* [104] emphasized the need for bioavailability and pharmacokinetics studies of essential oil constituents in biological matrices, particularly in human systems [104]. Certain strategies now adopted to improve the efficacy of essential oils are the combined use of different oils for potential synergistic effects, and the combination of essential oils with oil constituents and synthetic antimicrobials [103, 106].

The current interest in the use of essential oil constituents as natural antimicrobial agents was initiated in the 1980s mainly due to the change in consumer attitudes towards the use of synthetic preservative agents [107–110]. Toxic side-effects, interaction of synthetic antibiotics with other drugs, over prescription, misuse, funds required for the development of new synthetic antimicrobial agents etc. are the other major concerns which led the scientific community to search for natural antimicrobials [15, 111]. Microbial resistance against synthetic antimicrobials is another reason to the enhanced search for natural alternatives [112]. For example, resistance of *Staphylococcus aureus* to methicillin increased from near zero to approx. 70% in Japan and the Republic of Korea, 40% in Belgium, 30% in the United Kingdom and 28% in the United States in 10–15 years by 1998 [112]. Similarly, the resistance rate of *Streptococcus pneumoniae* was 7% in Germany, 9.5% in Iceland, 25% in Romania, 44% in Spain and 58% in Hungary [113].

The chemical composition and, thus, the physical properties and biological potency of essential oils are dependent on factors such as plant part used, extraction method, season of harvest, genotype, chemotype, geographical origin and ecological conditions [114–121]. Thus, the consistency in chemical and physical properties of an essential oil even from the same plant specimen is to be ascertained before its end use. Owing to these reasons, individual components of essential oils are now gaining priority over essential oils in the development of antimicrobials. For example, carvacrol, citronellol, eugenol, geraniol and limonene etc. were tested as food preservatives. Thus, essential oil constituents and their synthetic analogues could reduce the uncertainty associated with the parental oils and they are more dependable in terms of reproducibility, safety and economical viability.

7.9
Antimicrobial Activity: Mode of Action

The mode of action of essential oils is an area of interest to microbiologists, chemists and physicians. Considerable literature is available dealing with the mechanism of action of essential oils against pathogenic bacteria, fungi and other microorganisms. Most of these studies agree on the concept of disruption of microbial cell membranes by constituent molecules in essential oils. This 'disruption' results in membrane expansion, increased membrane fluidity and permeability, disturbance of membrane-embedded proteins, inhibition of respiration and alteration of ion transport processes. The lipophilicity of oil constituents, lipid composition of bacterial membranes and their net surface charge are the major factors deciding the membrane permeability of oil constituents. Oil constituents might also cross the cell membranes, penetrating the interior of the cell and interacting with intracellular sites critical for antibacterial activity [122]. Several active molecules in an essential oil might have a synergistic effect in this disruption of microbial cell membranes. At certain dosages, the volatile oils saturate the membranes and show effects similar to those of local anesthetics. Biological activity of an essential oil is a combined effect of both their active and inactive constituents. The inactive constituents play their role by influencing resorption, rate of reactions and bioavailability of the active molecules.

Over the years numerous studies have dealt with the antimicrobial activities of essential oils and attempted the mechanistic details of these activities. A few recent examples of such studies are cited here. Cox *et al.* [123] studied the action of *Melaleuca alternifolia* (tea tree) oil against *Escherichia coli*, *Staphylococcus aureus* and *Candida albicans* cells [123]. This study showed enhanced permeability of bacterial cytoplasmic and yeast plasma membranes at minimum inhibitory levels of the tea tree oil. Minimum inhibitory levels of tea tree oil also increased the uptake of the nucleic acid stain propidium iodide, to which the cell membrane is normally impermeable. They also observed leakage of potassium ions immediately upon adding tea tree oil to these microbial suspensions. Thus, it is evident from this study that the antimicrobial activity of tea tree oil results from its ability to disrupt the permeability barrier of microbial membrane structures. The mode of action of tea tree oil is similar against *E. coli*, *S. aureus* and *C. albicans*, and to that of other broad-spectrum, membrane-active disinfectants and preservatives [123, 124].

de Billerbeck *et al.* [125] studied the growth inhibitory effect of *Cymbopogon nardus* var. *nardus* essential oil on *Aspergillus niger* (Van Tieghem) mycelium on agar medium [125]. The mycelium growth was completely inhibited at 800 mg/l and this concentration was found to be lethal under the test conditions. *C. nardus* oil at 400 mg/l caused growth inhibition of 80% after 4 days of incubation and a delay in conidiation of 4 days compared with the control. Further, in this study, transmission electron microscopic observations were carried out to determine the ultrastructural modifications of *A. niger* hyphae after treatment with *C. nardus* oil. These observations found marked thinning in the hyphal diameter and the hyphal wall. de Billerbeck *et al.* [125] suggested that these modifications in the cytological structure of *A. niger* are due to the interference of *C. nardus* oil with the enzymes responsible for wall synthesis, which disturb normal growth. In addition, *C. nardus* oil caused plasma membrane disruption and mitochondrial structure disorganization. These findings indicate the possibility of exploiting *C. nardus* oil as an effective inhibitor of biodegrading and storage-contaminating fungi [125].

As the search for natural antimicrobials is gaining priority of late, a shift in the approach of antimicrobial activity studies on essential oils has also been observed. An increasing number of studies are now describing the antimicrobial activities of essential oils with reference to their major active constituent(s). One example is the antifungal activity studies of the essential oil of *Thymus pulegioides* against several clinical isolates and authentic type strains of *Aspergillus*, and five clinical strains of dermatophyte fungi by Pinto *et al.* [126]. *T. pulegioides* oil exhibited significant antifungal activity against clinically relevant fungi, mainly due to lesion formation in the cytoplasmic membrane and a considerable reduction of the ergosterol content. The antifungal activity of the oil was attributed to carvacrol and thymol, the major constituents in *T. pulegioides* oil identified by GC-MS analysis. This study recommended *T. pulegioides* oil for further investigations leading to its clinical applications [126].

In another study, Lambert *et al.* [127] determined the minimum inhibitory concentration (MIC) of oregano (*Origanum campactum*) essential oil and its major components, thymol and carvacrol, against *Pseudomonas aeruginosa* and *S. aureus* based on turbidimetric growth data [127]. Inhibition profiles of a range of concentrations of each of the oil, thymol, carvacrol and carvacrol/thymol mixtures were determined. The data suggested that mixtures of carvacrol and thymol gave an additive effect, and that the overall inhibition by oregano oil can be attributed mainly to the additive antimicrobial action of these two constituents. Further, addition of low amounts of each additive increased permeability of cells to the fluorescent nuclear stain ethidium bromide, dissipated pH gradients as indicated by the fluorescent probe (carboxyfluorescein diacetate succinimidyl ester) and caused leakage of inorganic ions. Mixing carvacrol and thymol at proper amounts led to the total inhibition. This inhibition is due to damage in membrane integrity, which further affects pH homeostasis and equilibrium of inorganic ions. Since certain carvacrol/thymol combinations were found to provide as high inhibition as oregano oil with a smaller flavour impact, Lambert *et al.* suggested further studies on the extent and mode of action of these oil constituents leading to the application of them as natural preservatives in foods [127].

Further, the antimicrobial screening and mechanistic studies targeted specifically on potent essential oil constituents are on the ascent now. Carvacrol, the major constituent in essential oils of oreganum (60–70%) and thyme (45%), is one of the most investigated essential oil constituents in a mechanistic perspective [128–130]. Ultee *et al.* studied the mechanism of action and structural requirements of carvacrol against the food-borne pathogen *Bacillus cereus*. This study showed that the hydroxyl group of carvacrol and the presence of a system of delocalized electrons are important structural features in its antimicrobial activity (Figure 7.1) [128]. *p*-Cymene, the biosynthetic precursor of carvacrol, showed weak antimicrobial activity possibly due to the absence of the phenolic hydroxyl group (Figure 7.1) [128]. On the mechanism of action, Ultee *et al.* [128] hypothesized that carvacrol destabilizes the cytoplasmic membrane and acts as a proton exchanger, thereby reducing the pH gradient across the cytoplasmic membrane. This leads to the collapse of the proton motive force and depletion of the ATP pool, eventually leading to cell death. This study suggested that carvacrol could be applied both as an antimicrobial agent and as a flavouring compound in products associated with outbreaks of *B. cereus* [129]. Recently, Veldhuizen *et al.* [131] investigated the structure–activity relationship of carvacrol by comparing the activities of its structural analogs. This study inferred that the hydroxyl group of carvacrol itself is not essential for its antimicrobial activity [131]. They also observed decrease in the antimicrobial activity of carvacrol on the removal of its

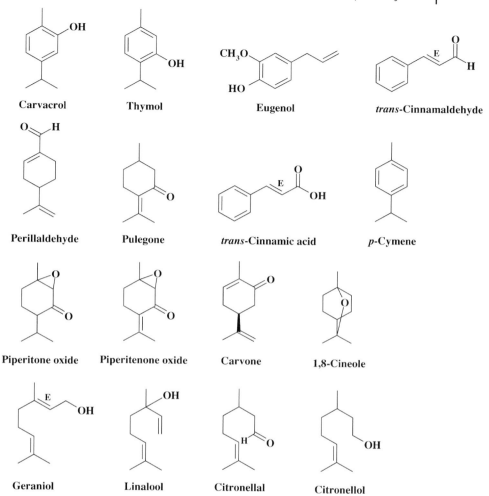

Figure 7.1 Examples of antimicrobial essential oil constituents and related compounds.

aliphatic ring substituents (Figure 7.1) [131]. Further, Knowles *et al.* [132], investigated the effects of carvacrol on dual-species biofilms formed by *S. aureus* and *Salmonella enterica* serovar Typhimurium with a constant-depth film fermentor [132]. This study demonstrated the inhibitory effects of carvacrol on a dual-species biofilm at various stages of maturation. Pulse and continuous exposure studies showed that carvacrol was as effective as commercial-grade sanitizing agents against biofilms [132].

Gill and Holley [89] studied the mechanism of antibacterial activity of cinnamaldehyde against *Listeria monocytogenes*, and eugenol against *L. monocytogenes* and *Lactobacillus sakei* [89]. Eugenol (5 mM) and cinnamaldehyde (30 mM) were bactericidal to *L. monocytogenes* in broth media at 20 °C. Eugenol (6 mM) was bactericidal to

L. sakei, but treatment with 0.5 M cinnamaldehyde had no significant effect. In this study, the cellular and extracellular ATP levels in HEPES buffer were measured at 20 °C to study the role of interference with energy generation in the mechanism of action. Rapid inhibition of the energy metabolism of *L. monocytogenes* and *L. sakei* was observed when the cells are exposed to bactericidal concentrations of eugenol and cinnamaldehyde. The suggested mechanism of inhibition of energy generation is inhibition of glucose uptake or its utilization and effects on membrane permeability [89]. Ali *et al.* [133] studied the antimicrobial activities of eugenol and cinnamaldehyde against the human gastric pathogen *Helicobacter pylori* [133]. Both eugenol and cinnamaldehyde inhibited the growth of all the 30 *H. pylori* strains tested, at a concentration of 2 µg/ml, at 9 and 12 h of incubation, respectively. At acidic pH, increased activity was observed for both these compounds. Further, *H. pylori* did not develop any resistance towards these compounds even after 10 passages grown at sub-inhibitory concentrations. These results indicate that eugenol and cinnamaldehyde may prevent *H. pylori* growth *in vitro*, without acquiring any resistance [133].

Similar mechanistic studies on antimicrobial activity of essential oil constituents are currently on the rise [58, 64, 89, 128, 129, 132, 134–137]. Table 7.1 is a compilation of recent antimicrobial activity studies on essential oil constituents, test methods, significant results and related references. The most studied essential oil constituents, in a mechanistic perspective, are carvacrol, thymol, eugenol, geraniol, perillaldehyde, cinnamaldehyde etc. The chemical structures of these antimicrobial oil constituents and related compounds are shown in Figure 7.1. In a structure–activity correlation, oil constituents with phenolic hydroxyl groups (e.g. carvacrol and thymol) were found to have the broadest spectrum of activity against the Gram-negative and Gram-positive bacteria and fungi. Ultee *et al.* [129] have shown that the phenolic ring with the delocalized electrons of carvacrol is a structural requirement for its activity, as supported by the lack of activity in menthol with no delocalized electrons [129]. The activities of carvacrol (hydroxyl group *ortho* to methyl) and thymol (hydroxyl group *meta* to methyl) were observed to be comparable against *B. cereus*, *S. aureus* and *P. aeruginosa* (Figure 7.1) [129, 138]. Thus, it appears that the relative position of the hydroxyl group on the phenolic ring does not strongly influence the degree of antibacterial activity. Contrary to previous reports, a recent mechanistic study of carvacrol and its structural analogs showed that the hydroxyl group of carvacrol itself is not essential for its antimicrobial activity, but it renders special features that add to the antimicrobial mode of action of carvacrol [131]. Acyclic (non-phenolic) alcohols such as geraniol and (±)-linalool are relatively less active compared to these phenolic constituents. These observations on structure–activity relationships could assist in the search for natural as well as synthetic analogs of these phenolic compounds, with potential antimicrobial applications. The hydrophobicity of oil constituents enables them to penetrate the lipid bilayer of the cell membrane and this permeability leads to leakage of cell contents such as cytoplasmic constituents, disruption of proton motive force, electron flow, active transport and coagulation of cell contents. Some mechanistic studies have also found interaction between essential oil constituents and proteins embedded in the cytoplasmic membrane. These interactions could distort the lipid–protein interaction or result in direct interaction between the oil

Table 7.1 Antimicrobial activity studies on essential oil constituents.

Constituent(s)	Microorganism(s) or disease(s)	Assay(s)	Results	References
1. Carvacrol	*Bacillus cereus* IFR-NL94-25	Determination of antibacterial activity, monitoring viability of *B. cereus* cells, influence of pH, phase transition temperature	Carvacrol showed a dose-dependent growth inhibition of *B. cereus*, total growth inhibition was observed at 0.75 mmol/l and above; membrane fluidity was the factor influencing the bactericidal activity of carvacrol	[134]
2. Carvacrol	*Bacillus cereus* IFR-NL94-25	Determination of viability of *B. cereus* cells, intra- and extracellular ATP concentrations, influence of carvacrol on the membrane potential	Carvacrol interacts with *B. cereus* cell membranes by changing its permeability for cations like H^+ and K^+ and dissipation of ion gradients leading to impairment of essential cell processes, thus, to cell death	[128]
3. Carvacrol	*Bacillus cereus* IFR-NL94-25	Determination of partition coefficient of carvacrol, intracellular pH measurements, influence of carvacrol and cymene on membrane potential, antibacterial activity	Antibacterial activity studies of thymol, cymene, menthol, carvacrol methyl ester and so on showed that the hydroxyl group of carvacrol and presence of a delocalized phenolic electron system are important for the activity of carvacrol	[129]
4. Carvacrol, its structural analogs	*Escherichia coli* ATCC 25922, *Staphylococcus aureus* ATCC 6538	Determination of bacterial growth curves in LB containing carvacrol and its analogs, cell death time and effect of compounds on the membrane potential	The activity of carvacrol was compared to its structural analogs to find the structural requirements for its antimicrobial activity; this study showed that the hydroxyl group of carvacrol is not essential for its activity, but it does have features that add to the antimicrobial mode of action of carvacrol	[131]

(Continued)

Table 7.1 (*Continued*)

Constituent(s)	Microorganism(s) or disease(s)	Assay(s)	Results	References
5. Carvacrol	*Staphylococcus aureus* NCTC 10788, *Salmonella enterica* serovar Typhimurium NCTC 74	Effects of carvacrol on dual-species biofilms formed by *S. aureus* and *S. enterica* serovar Typhimurium were investigated with a constant-depth film fermentor	Cryosectional studies detected viable *S. aureus* and *S. enterica* serovar Typhimurium at depths of 320 and 180 μm from the biofilm surface, respectively; carvacrol pulses (1.0 mmol/h) inhibited *S. aureus* by 2.5 log colony-forming units per biofilm during the early stages of film formation; carvacrol is an effective natural means to control dual-species biofilm formation	[132]
6. Carvacrol, nisin (antibacterial polycyclic peptide)	*Listeria monocytogenes* Scott A, *Bacillus cereus* IFR-NL94–25	*L. monocytogenes* and *B. cereus* were exposed to 96 combinations of nisin and carvacrol and optical densities were monitored, determination of the combined effect of nisin and carvacrol on the viable count of *L. monocytogenes* and *B. cereus*	*B. cereus* was more sensitive towards nisin than *L. monocytogenes* and the inhibitory effect of nisin was stronger towards cells cultivated and exposed at 8 °C than towards cells cultivated and exposed at 20 °C; combination of nisin with sublethal doses of carvacrol resulted in an enhanced reduction in the viable count of both *L. monocytogenes* and *B. cereus*, indicating synergy between nisin and carvacrol; related studies were reported by Periago and Moezelaar, 2001 [144] and Periago *et al.* [145].	[143–145]
7. Carvacrol, thymol	*Escherichia coli* O157:H7	Determination of MIC and minimum bactericidal concentration, checkboard assay for synergism and antagonism	Carvacrol and thymol displayed bacteriostatic and bactericidal properties with MICs of 1.2 mmol/l and were additive in combination; they were effective in preventing the growth of *E. coli* O157: H7 in liquid foods, especially when used in combination with a stabilizer or in ethanol solution	[64]

8. Carvacrol, eugenol	*Candida albicans* strain 1E 111PV515 isolated from the vaginal secretions of a woman with acute vaginitis	Determination of MIC	*In vitro* MIC of carvacrol and eugenol against the *C. albicans* were 10^3 mg/l and 2×10^3 mg/l, respectively; these two naturally occurring antifungal agents are promising drugs for the treatment and prevention of vaginal candidiasis	[130]
9. Carvacrol, (+)-carvone, thymol, *trans*-cinnamaldehyde	*Escherichia coli* ATCC 35150 O157:H7, *Salmonella typhimurium* ATCC 13311, *Photobacterium leiognathi* ATCC 33469	Growth inhibition tests, photobacter toxicity test, uptake of 1-N-phenylnaphthylamine, bacteriolysis, lipopolysaccharide and protein release, determination of intra- and extracellular ATP	Carvacrol, thymol and *trans*-cinnamaldehyde inhibited *E. coli* and *S. typhimurium* at 1–3 mM; *trans*-Cinnamaldehyde was most inhibitory towards *P. leiognathi*; carvacrol and thymol disintegrated the outer membrane, released outer membrane-associated material from the cells to the external medium, decreased the intracellular ATP pool of *E. coli* and increased extracellular ATP	[146]
10. Thymol, carvacrol	*Staphylococcus aureus* ATCC 6538, *Pseudomonas aeruginosa* ATCC 2730	Determination of MIC, collection of turbidimetric growth data	Combinations of carvacrol and thymol provided total inhibition; The mechanism of action is damage in bacterial membrane integrity, which affects pH homeostasis and equilibrium of inorganic ions	[100]
11. Thymol, eugenol	*Escherichia coli* APL 87/1, *Bacillus subtilis* APL 87/35	Determination of MIC and minimal lethal concentration (MLC)	MICs and MLCs were identical for clove oil and eugenol and oregano oil and thymol; Clove and oregano oils exerted bactericidal effects due to these major phenolic constituents causing damage in bacterial envelopes	[147]
12. Thymol, eugenol	*Saccharomyces cerevisiae* SB36-85	Determination of the MIC and minimum fungicidal concentration, treatment of yeast cells with thymol and eugenol, counting viable cells, scanning electron microscope observations	Scanning electron microscopy data showed alteration of both membrane and cell wall of the yeast on treatment with thymol and eugenol	[148]

(Continued)

Table 7.1 (*Continued*)

Constituent(s)	Microorganism(s) or disease(s)	Assay(s)	Results	References
13. Thymol, eugenol, *trans*-cinnamaldehyde, linalool	Influence on total microbial count (bacteria, yeasts and moulds) in air	Determination of the total airborne microbial count in the testing room, using an air sampler	All tested compounds showed a reduction in total microbial count and total count of yeasts and moulds in the air at all concentrations; total microbial count reduction: thymol, 69.50%, eugenol, 69.40%, *trans*-cinnamaldehyde, 65.93%, linalool - 69.92%; total count reduction of yeasts and moulds: thymol, 50.60%, eugenol, 58.31%, *trans*-cinnamaldehyde, 46.19%, linalool, 50.25%.	[149]
14. Thymol, *p*-cymene, estragol, linalool, carvacrol	*Shigella sonnei* CIP 82.49, *Shigella flexneri* CIP 82.48	Agar well diffusion assay	Thymol and carvacrol showed inhibition of *Shigella* spp.; decontamination of lettuce by thymol and carvacrol at 0.5 and 1.0% v/v was evaluated; this study resulted in inconsistent antimicrobial data and hampered sensoric properties of lettuce	[150]
15. Thymol, carvacrol, eugenol, *trans*-cinnamic acid	*Escherichia coli*, *Salmonella enterica* Typhimurium LT2	Determination of MIC by checkerboard assay	Thymol was most effective with the lowest MIC values against *S. enterica* (1.0 mmol/l) and *E. coli* (1.2 mmol/l); after thymol, the order of antimicrobial activity was carvacrol > eugenol > cinnamic acid	[151]
16. Eugenol, cinnamaldehyde	29 indigenous and one standard strain of *Helicobacter pylori* ATCC 26695, *one strain of Escherichia coli* NCIM 2089	Disc diffusion susceptibility test, determination of MIC by agar dilution method	Eugenol and cinnamaldehyde inhibited the growth of 30 *H. pylori* strains at a concentration of 2 μg/ml, without acquiring any resistance towards these compounds even after 10 passages grown at subinhibitory concentrations	[133]

17. Eugenol encapsulated in water-soluble micellar nonionic surfactant solution	*Escherichia coli* O157:H7 (H1730, F4546, 932, E0019), *Listeria monocytogenes* (Scott A, 101, 108, 310)	Determination of MIC by microbroth dilution assay	Eugenol encapsulated in surfactant micelles inhibited *E. coli* and *L. monocytogenes* at pH 5, 6 and 7; inhibition of *L. monocytogenes* and *E. coli* O157: H7 decreased with increasing pH; the MIC was 0.2, 0.5 and 0.5% at pH 5, 6 and 7, respectively.	[152]
18. Eugenol, cinnamaldehyde	*Listeria monocytogenes* strain (a somatic serotype 1 meat plant isolate), *Lactobacillus sakei* strain (isolated from spoiled cured meats)	Determination of bactericidal concentrations by broth dilution technique	5 mM eugenol and 30 mM cinnamaldehyde were bactericidal to *L. monocytogenes* in broth media at 20 °C; 6 mM eugenol was bactericidal to *L. sakei*, but treatment with 0.5 M cinnamaldehyde had no significant effect	[89]
19. Cinnamaldehyde	Bacteria: Gram-negative (7 isolates), Gram-positive (1), fungi: yeasts (4), filamentous molds (4), dermatophytes (3)	Determination of MIC by agar dilution method	MICs of cinnamaldehyde for bacteria: 75–600 μg/ml, yeasts: 100–450 μg/ml, filamentous fungi: 75–150 μg/ml, dermatophytes: 18.8–37.5 μg/ml; this study suggested that broad-spectrum antibiotic activities of the essential oil of the Chinese medicinal herb *Cinnamomum cassia* are due to its major constituent, cinnamaldehyde (85%)	[153]
20. Fifteen essential oil constituents	*Haemophilus influenzae* ATCC 33391, *Staphylococcus pyogenes* ATCC 12344, *Streptococcus pneumoniae* IP-692, *Streptococcus pneumoniae* PRC-53, *Staphylococcus aureus* FDA 209P JC-1, *Escherichia coli* NIHJ JC-2	Antibacterial activity of essential oil constituents against respiratory tract pathogens by gaseous contact, determination of minimum inhibitory dose (MID)	The MIDs of these essential oil constituents were close to those of their parent oils; cinnamaldehyde showed the highest activity with a MID < 6.25 mg/l air against all tested strains; thymol showed activity comparable to that of cinnamaldehyde; among terpene alcohols, geraniol showed the highest activity	[154]

(Continued)

Table 7.1 (*Continued*)

Constituent(s)	Microorganism(s) or disease(s)	Assay(s)	Results	References
21. Citral, *trans*-cinnamaldehyde, (−)-perillaldehyde, (−)-citronellal, eugenol, carvacrol	Influence of microbial count in air	Vaporization of test compounds with an air washer	The highest reduction of germ count was observed for (−)-perillaldehyde (53%) and eugenol (13%)	[155]
22. Geraniol, (*R*)-(−)-linalool, terpineol, *γ*-terpinene, 1,8-cineole	Influence on airborne microbes on vaporization	Vaporization of test compounds with an air washer	Terpineol and 1,8-cineole showed the highest antimicrobial activities with average reduction of germ count 68 and 64%, respectively	[156]
23. Twenty natural identical essential oil constituents	*Escherichia coli* ATCC 11775, *Staphylococcus aureus* ATCC 25923, *Bacillus cereus* ATCC 11778, yeast *Candida albicans* ATCC 10231	Disc diffusion assay, determination of MIC by *p*-iodonitrotetrazolium violet microplate method	Carvacrol (MIC 1.66–13.3 mM) and geraniol (MIC 19.5–51.9 mM) showed the broadest spectrum of antimicrobial activity	[157]
24. Geraniol, nerol, their synthetic derivatives	Bacteria: Gram-positive (one ATCC strain, one clinical isolate); Gram-negative (five ATCC strains and clinical isolates); yeast: *Candida albicans* ATCC 10231	Agar diffusion disk method, determination of MIC by the agar serial tube dilution method	Geranial, geraniol, nerol, aromatic esters as well as geranyl amine and geranyl chloride were found to have high activity against all strains of microorganisms tested	[158]
25. Pulegone	Five Gram-positive bacteria, four Gram-negative bacteria and one fungus (*Candida albicans* ATCC 1023)	Disc diffusion method	Pulegone exhibited significant antibacterial and antifungal activity, particularly against *C. albicans* and *S. typhimurium*	[159]

26. Pulegone, piperitenone oxide, piperitone oxide, carvone, limonene, menthone	19 Gram-positive and Gram-negative bacteria, three fungi	Solid-phase assay; microtitration assays	Pulegone-rich essential oil of *Mentha suaveolens* efficiently inhibited all the microorganisms tested with MICs ranging between 0.69 and 2.77 ppm; pulegone was most effective against the tested microorganisms, followed by piperitenone oxide and piperitone oxide	[78]
27. Linalool, 1,8-cineole, α-pinene, β-pinene, β-caryophyllene, limonene	Three gram-negative bacteria, four gram-positive bacteria and three fungi	Disc diffusion method, determination of MIC by microdilution broth susceptibility assay	Synergy between the six tested and other constituents causes the antimicrobial activity of essential oils from *Salvia* spp. (*S. santolinifolia, S. hydrangea* and *S. mirzayanii*); linalool and 1,8-cineole showed the highest antimicrobial activity among the tested compounds	[160]
28. Linalyl acetate, menthol, thymol	*Staphylococcus aureus* ATCC 6538P, *Escherichia coli* ATCC 15221.	Determination of MIC by microdilution method	The antimicrobial effect of (+)-menthol, thymol and linalyl acetate may be due to a perturbation of the lipid fraction of bacterial plasma membranes, resulting in alterations of membrane permeability and in leakage of intracellular materials	[122]
29. Acetanisole, benzaldehyde, cinnamaldehyde, diacetyl, phenylpropionaldehyde, pyruvaldehyde	*Staphylococcus aureus* WRRC B124	Tested against growth of *S. aureus* in brain heart infusion broth; determination of MIC	The carbonyl compounds were effective antistaphylococcal agents and their use in combination with thermal processing may serve as a new approach to control *S. aureus* growth and other Gram-positive food borne pathogens	[161]

constituents and hydrophobic groups of the cell wall proteins [139]. Certain oil constituents might also cross the cell membranes, penetrating the interior of the cell and interacting with intracellular sites critical for antibacterial activity [122]. Gram-negative bacteria are observed to be less susceptible to the action of essential oils and their constituents, compared to Gram-positive bacteria, due to the presence of the outer membrane surrounding the cell wall, which restricts the diffusion of lipophilic compounds through the lipopolysaccharide covering [58, 140]. Physical conditions such as low pH, low oxygen levels and low temperature enhance the action of oil constituents. Synergism i.e. the combined activity of two or more oil constituents being more than sum of their individual effects has been observed between carvacrol and its precursor *p*-cymene [135], and between cinnamaldehyde and eugenol [141, 142]. Synergy between oil components and certain mild preservation methods has also been observed [58].

7.10
Antimicrobial Assays

As cited previously, there are numerous literature reports on the *in vitro* antimicrobial assays of essential oils. Antimicrobial assays have evolved from disc diffusion techniques introduced in the mid-twentieth century to the recent broth or agar dilution methods. Bacterial and fungal strains for *in vitro* antimicrobial testing are usually obtained from authentic repositories such as the American Type Culture Collection (ATCC). In clinical studies, bacterial and fungal strains are collected and identified from body tissues of individual patients, and their susceptibility against various synthetic and natural antibiotics ascertained by various *in vitro* assay techniques.

7.10.1
Disc Diffusion Assay

Disc or agar diffusion assay is based on the movement of molecules through an agar matrix under controlled conditions. Agar, a polysaccharide obtained from the cell walls of certain seaweeds [162], when allowed to harden forms a matrix permeable to diffusion of small molecules. The movement of a molecule, under controlled conditions, in the agar matrix depends on its concentration. This principle of agar diffusion is used to determine the susceptibility or resistance of a bacterial strain to antimicrobial agents such as synthetic antibiotics, plant extracts and essential oils. Agar diffusion assays are easy, economical, and allow preliminary screening of plant extracts and essential oils against a wide range of microorganisms. However, the accuracy, reproducibility and inter-laboratory precision of disc diffusion assays were found to be poor in most cases. Recently, there have been efforts to standardize disc diffusion assays. Examples are Alderman and Smith [163], National Committee for Clinical Laboratory Standards [164] and Gabhainn *et al.* [165].

A survey of the literature on antimicrobial assays of essential oils reveals that a comparative evaluation of these data is not possible due to the differences in experimental parameters (sample size, incubation times, agar recipes, standards, choice of solvents etc.) adopted by various laboratories. Again, in practice, most antimicrobial assays are optimized for hydrophilic substances [166]. However, essential oils are complex, volatile mixtures insoluble in water. Due to these physicochemical characteristics, in antimicrobial assays, it is difficult to obtain a stable dispersion of essential oils in the aqueous medium and, thus, the uniform diffusion of lipophilic oil components through the agar is not achieved in most cases [167]. Hence, when testing essential oils, it is necessary to incorporate an emulsifier or solvent into the test medium to ensure contact between the test organism and oil constituents. Such agents most commonly used are Tween 80, Tween 20, ethanol and dimethylsulfoxide. Another problem associated with these assays is the determination of the number of viable bacteria remaining after the addition of the oil. Thus, due to these factors, standard and reproducible methods are to be used in testing the *in vitro* antimicrobial activities of essential oils.

Owing to these concerns, attempts have been made to develop specific assays for determining antimicrobial activities of essential oils i.e., disc diffusion, well diffusion, agar dilution and broth dilution assays [168]. This revealed that the disc diffusion, well diffusion and agar dilution assays were unreliable and produced inconsistent results. This inconsistency was due to problems in achieving stable dispersion of the oils, nonuniform diffusion of lipophilic constituents in aqueous media and variation in methods used for determining the number of viable bacteria remaining after the addition of the oil [168]. In the same study, Hood *et al.* [168] attempted the standardization of disc diffusion assay for essential oils [168]. Five essential oils, i.e., *M. alternifolia* (tea tree oil), *Anetholea anisata* (aniseed myrtle oil), *Santalum spicatum* (Western Australian sandalwood oil), *Melaleuca quinquenervia* (cineole type, niaoli oil) and *M. quinquenervia* (linalool-nerolidol type, Australian nerolina oil), and a combination of 11 bacteria, type strains and hospital isolates were used for this assay. The data from this study inferred that disc or well diffusion assay cannot be used for evaluative work. Even its usefulness in the initial screening of antimicrobial activity of essential oils was questioned [168]. Moon *et al.* [169] studied the differences in zone of inhibitions of essential oils obtained with two different agar types – IsoSensitest agar and nutrient agar. The data showed that microbial growth media type can have a significant effect on the size of inhibition zones. However, this effect is not consistent across all microorganisms or essential oils [169].

7.10.2
Broth and Agar Dilution Methods: Determination of MIC

Broth and agar dilution methods are widely used to determine the MIC of essential oils [168, 170]. The MIC is the lowest concentration of an essential oil or a compound capable of inhibiting the growth of the challenging organism. Hood *et al.* [168], optimized a broth dilution method using 0.02% Tween 80 to emulsify the essential oil samples [168]. For this assay, they used the same five essential oils and 11 bacterial strains as in disc diffusion assay (Section 7.10.1). Bacterial culture conditions and controls were also same. The essential oil was emulsified into the aqueous test medium as follows. The oil was added to a sterile Eppendorf tube and 1/10 of the oil's volume of a 10% solution of Tween 80 in water was added and the solution mixed by vortexing. Aqueous solution (e.g. nutrient broth) was then added in 10- to 20-μl aliquots with brief vortexing between each addition. This was continued until the ratio of aqueous solution to oil was 2 : 1 and the final volume 4.5 ml. An overnight bacterial culture (500 μl) was then added to each tube. Three control tubes, the first containing 0.02% Tween 80 in nutrient broth, the second containing 100 μl of canola oil, 0.02% Tween 80 and nutrient broth, and the third containing only nutrient broth were also prepared. The test tubes were then incubated with shaking at 37 °C for 12 h after which each suspension was serially diluted (10-fold) with sterile nutrient broth to a final concentration of 10^{-7}. Then, 500 μl of the 10^{-5}, 10^{-6} and 10^{-7} dilutions were plated out onto nutrient agar plates using an alcohol flamed glass spreader. The plates were then incubated overnight at 37 °C and the bacterial colonies counted. In this assay, *M. alternifolia* and *M. quinquenervia* (cineole type) oils completely inhibited the growth of all tested organisms. Hood *et al.* [168] found this as a reliable method which could be adopted for antimicrobial assays of essential oils. Further, this method provided the most reliable results when evaluated against a range of essential oils and allowed direct comparison of the antibacterial activity of the test oils, irrespective of their viscosity and hydrophobicity [168].

Hammer *et al.* [15] introduced a modified the agar dilution method for the analysis of essential oils [15]. In this study, 52 plant oils and extracts were evaluated for activity against 10 bacterial and fungal strains. A final concentration of 0.5% (v/v) Tween 20 was incorporated into the agar after autoclaving to enhance oil solubility. A series of 2-fold dilutions of the essential oil, ranging from 2 to 0.03% (v/v), was prepared in Mueller Hinton agar with 0.5% (v/v) Tween 20. Plates were dried at 35 °C for 30 min prior to inoculation with 1–2 mL spots containing approximately 10^4 colony-forming units of each organism, using a multipoint replicator. Mueller Hinton agar, with 0.5% (v/v) Tween 20, but no oil, was used as a positive growth control. Inoculated plates were incubated at 35 °C for 48 h. MICs, the lowest concentration of oil inhibiting the visible growth of each organism on the agar plate, were determined after 24 h for the bacteria and after 48 h for *C. albicans*. The presence of one or two colonies was disregarded. Lemon grass, oregano and bay oils inhibited all organisms at concentrations of 2.0% or less (v/v). Six oils (apricot kernel, evening primrose, macadamia, pumpkin, sage and sweet almond) did not inhibit any of the organisms even at the highest concentration (2.0%, v/v). Variable activity was observed for the remaining oils [15].

As mentioned earlier, Tween 80, Tween 20 and ethanol are the usual emulsifiers or solvents recommended to ensure contact between the test organism and the agent. However, these emulsifiers could cause changes in the physicochemical properties of the test system, resulting in variations in the extent of antimicrobial activity. Further, lipophilic molecules such as essential oil constituents could become solubilized within the micelles formed by nonionic surfactants (Tween 20, Tween 80) and are thus partitioned out of the aqueous phase of the suspension. Studies have shown that antimicrobial test molecules solubilized within the micelles do not come into direct contact with the microorganisms and, thus, they do not contribute to the activity [171]. This effect is more at higher concentrations of the surfactant [172]. For example, a reduction in the bioactivity of tea tree oil has been reported in presence of Tween [173]. Ethanol (approx. 5%) has also been reported to have a marked potentiating effect on the activity of some antimicrobial agents [172]. Remmal *et al.* [174] suggested the incorporation of low concentrations of bacteriological agar as a stabilizer of the oil–water mixture, which overcomes these disadvantages [174].

The resazurin MIC assay, introduced by Mann and Markham in 1998, is a unique broth microdilution assay suitable for the assessment of antimicrobial activity and MIC determination of essential oils [167]. This assay employs a chemically and microbially inert stabilizer, and an indicator capable of reliably predicting MIC either visually or instrumentally. Tea tree oil, with 40.1% terpinen-4-ol and 4.6% 1,8-cineole, was used for this assay. The bacteria used were *E. coli* NCTC 8196, *S. aureus* NCTC 4163, methicillin-resistant *S. aureus* (clinical isolate), Group A *Streptococcus* JM12 (clinical isolate), Group B *Streptococcus* (clinical isolate) and *Proteus vulgaris* NCTC 4635. A solution of resazurin sodium salt (0.01%, w/v) was prepared in sterile distilled water. Emulsifiers tested were Tween 20, Tween 80, 2% ethanol and dimethylsulfoxide. Bacteriological agar (0.15%) was dissolved, sterilized by autoclaving and cooled to room temperature before use. The test organisms were stored on nutrient agar slopes, prepared from nutrient broth base and bacteriological agar at 25 °C, except for streptococcal strains, which were stored on Brain Heart Infusion Agar, prepared from Brain Heart Infusion Broth and bacteriological agar. All microorganisms used in MIC assays were twice-passaged 16–18 h cultures grown in either nutrient broth or Brain Heart Infusion Broth. Optical densities of all inocula were measured at 420 nm using a spectro-photometer. Cell densities were estimated from standard curves and confirmed by the pour plate method on Plate Count Agar.

The inocula were diluted to the appropriate cell densities in nutrient broth containing 0.15% (w/v) agar for all test cultures except the streptococci, for which Brain Heart Infusion Broth with 0.15% (w/v) agar was used. Prior to inoculation, test media were melted by steaming and tempered to 37 °C, at which temperature they remained as liquids. Further, the cell concentration necessary to cause reduction of resazurin within 2 h was determined for each of the test organism. Serial 10-fold dilutions of each culture were prepared in prewarmed nutrient broth. Aliquots (1.7 ml) were dispensed into tubes containing 0.2 ml 'sloppy' (0.15%, w/v) agar and 0.1 ml resazurin solution. The tubes were incubated for 2 h at 37 °C, after which time aliquots from adjacent blue (oxidized), mauve and pink (reduced) dilution tubes were tested by the plate count method.

For the MIC assay, serial 2-fold dilutions (0.005–1.25%, v/v) of the test oil were prepared by vortexing the oil in room-temperature sloppy agar. The assay medium described above was then inoculated with the test organism to yield a final cell density around 1 log cycle lower than the cell density required to reduce resazurin (usually 5–6 $\log_{10}$ colony-forming units/ml). The inoculum density was confirmed by plate count. A sterile 96-well microtiter tray with lid was set up with each of the test bacteria ($n = 8$) as follows: column 1–9, 170 µl inoculum plus 20 µl of an oil dilution; column 10, 170 µl inoculum plus 20 µl oil diluent (positive control); column 11 and 12, 170 µl sterile resazurin assay medium plus 20 µl oil diluent (negative control and blank, respectively). Well contents were thoroughly mixed using the micropipettor. Two trays were prepared for each organism and incubated at 37 °C for either 3.5 or 18 h. After incubation, 10 µl of resazurin solution was added to all except column 12, to which 10 µl of distilled water was added. After a second incubation of 2 h at 37 °C, three methods were used to determine the MIC values. First, wells were assessed visually for color change, with the highest dilution remaining blue indicating the MIC. Absorbance was then read at 570 nm using a plate reader blanked against wells containing only assay medium, oil diluent and distilled water. The MIC was indicated by a rise in absorbance at 570 nm. Immediately afterwards, plate counts were carried out on samples from the microwells, to determine whether bacterial numbers correlated with either indicator of MIC. The MIC of tea tree oil was reliably determined using this resazurin microdilution method [167].

MICs of essential oils are also determined by various modifications of broth and agar dilution techniques. Examples are colorimetric broth microdilution and impedimetric methods [175, 176].

7.11
Other Applications of Essential Oils

Essential oils exhibit a wide range of other biological activities such as antioxidant, anti-inflammatory, analgesic, insecticidal, feeding deterrent, larvicidal and pest management [177–182]. Numerous reports on these biological activities of essential oils are available in the literature. Further, recent studies have demonstrated antiviral properties for many essential oils [183–186]. The antiviral activity of an essential oil depends on its chemical composition and on the virus structure. Studies have found that normal cells acquire enhanced resistance to viral penetration on treatment with essential oils. Tea tree, lavender, juniper, melissa, eucalyptus, thyme, palmarosa etc. are examples of essential oils showing antiviral activities. In addition, essential oils have been widely used in aromatherapy since ancient times. In 400 BC, Hippocrates said that aromatic baths and massages are the best ways to health. Aromatherapy is a combination of counseling, massage and odorous substances. It is one of the most actively growing forms of alternative medicine. In aromatherapy, essential oils exert their effect through the olfactory system, which in turn causes brain stimulation, alters emotions and relieves anxiety. However, the mode of action of essential oils in aromatherapy is not very well understood. Essential oils also have a supporting role in aromatherapy due to their antimicrobial activity and other pharmacological effects on human tissues [187–189]. Over 3000 essential oils are known now. Of these, about 10% are commercially used, mainly as flavours and cosmetics. International trade on essential oils and their end-use products is growing. This growth is mostly reflected in the cosmetic and flavour industries [190, 191].

7.12
Toxicity of Essential Oils

Recently, more systematic toxicity studies on essential oils and their constituents have been appearing in the literature [192–195]. As mentioned in Section 7.11, essential oils are widely used in conjunction with massage in aromatherapy. Most commonly used oils in aromatherapy have low systemic toxicity, especially when used at 2% dilution. Lis-Balchin [196] recommended the evaluation of parameters such as adulteration and safety during pregnancy for essential oils, extracts and phytols used by some aromatherapists [196]. The use of tea tree oil, a potential antimicrobial, anti-inflammatory and anticancer agent, over decades suggests that the topical use of the oil is relatively safe, and that adverse events are minor and occasional. However, cases of tea tree oil toxicosis have been reported by veterinarians to the National Animal Poison Control Center, University of Illinois, USA, when the oil was applied dermally to dogs and cats. In most of these cases, the oil was used to treat dermatologic conditions at inappropriate high doses and the typical signs observed were depression, weakness, lack of coordination and muscle tremors [197]. Hammer *et al.* [198] reviewed the toxicity of tea tree oil, and found that it is toxic only if ingested in higher doses and can also cause skin irritation at higher concentrations. Allergic reactions to tea tree oil occur in predisposed individuals, and may be due to the various oxidation products that are formed by exposure of the oil to light and air. Avoiding ingestion of the oil, applying only diluted oil topically and using properly stored oil can minimize adverse reactions of its use. Data from individual components of tea tree oil also suggest that it has the potential to be developmentally toxic if ingested at higher doses. However, tea tree oil and its components are not genotoxic [198].

Smith *et al.* [199] developed a guide to evaluate the safety of essential oils with applications as flavour ingredients [199]. This guide is based on the chemical composition of the essential oil and the variability of the composition of the oil in the product. Chemically identified oil constituents, biosynthesized by common pathways, are clustered into a limited number of groups called 'congeneric groups'. The safety of the intake of each congeneric group from consumption of the essential oil is evaluated based on the data on absorption, metabolism and toxicology of members of the congeneric group. The intake of the group of unidentified constituents is evaluated in the context of consumption of the essential oil as a food and also from the toxicity data on the oil or an oil of similar chemotaxonomy [199]. Further, the toxicity of essential oils can be tested by the brine shrimp (*Artemia salina*) bioassay [200, 201]. Safety and toxicity need to be evaluated before the usage of essential oils and their constituents for flavour, food preservation or medicinal purposes [202].

7.13
Scope for Future Research

Search for natural antimicrobial agents has gained priority since the 1980s mainly due to consumer attitudes against the use of synthetic preservative agents and growing resistance against synthetic antimicrobials. Essential oils have been used as antimicrobials since ancient times. Further, the use of essential oils as antimicrobials has undergone a shift from raw essential oils to potential oil constituents of late. Oil constituents are preferred as antimicrobial drug candidates due to the inconsistency in the physicochemical characteristics of parental oils and also due to difficulties associated with the adsorption of oil-based drugs in biological systems. The antimicrobial properties of potential oil constituents are now being utilized in commercial products such as food preservatives, biofilms, feed supplements, dental root canal sealers and antiseptics [58].

The search for new promising oil constituents and studies on their antimicrobial mode of action are to be identified as priority areas. As mentioned in Section 7.1, a number of essential oil-bearing species, especially in the biodiversity-rich and remote spots on Earth, are yet to be screened for their chemical composition and potential antimicrobial constituents [59, 203]. Biosynthetic studies on potential essential oil constituents, providing data on their enzymatic pathways, could give us suitable tracks towards engineering the enhanced production of these constituents by the source plants. Further, mechanistic studies on the mode of action of active essential oil constituents are gaining increasing priority of late, but still these studies and the target oil constituents are limited. Again, the action of essential oil constituents on microbial membrane components (proteins, phospholipids, etc.) is little understood. Studies on synergism and antagonism of essential oils and their constituents could be utilized to optimize their antimicrobial activities and, thus, to minimize the amount required to achieve the extent of activity. Interaction of volatile oil constituents with other ingredients in their commercial products (food preservatives, cosmetic products, etc.) has also been poorly investigated.

The clinical efficiency of essential oil-based drugs depends on the systemic availability of oil constituents in the target organs. Thus, data on drug delivery systems, absorption, distribution, stability, metabolism and pharmacokinetics of oil constituents are critical in correlating the *in vitro* and *in vivo* observations [104]. These parameters are also important in assessing the safety of essential oil-based drugs. However, only very limited studies are available in the literature on the bioavailability and pharmacokinetics of essential oil constituents in biological matrices, particularly in human systems. The toxicity of potential essential oils and oil constituents need to be systematically addressed. In conclusion, the search for new potential oil constituents, development of standardized extraction and isolation protocols, understanding of their biosynthesis, antimicrobial studies, studies on their mode of action, drug development protocols and safety evaluation could lead to new essential oil-based antimicrobial products.

Acknowledgments

B.S. acknowledges Dr. K.R. Harikumar, Department of Chemistry, University of Toronto, Canada, for providing the required literature. B.S. also thanks various authors who provided reprints of their work.

References

1 Cowan, M.M. (1999) *Clinical Microbiology Reviews*, **12**, 564–582.

2 Goossens, A., Häkkinen, S.T., Laakso, I., Seppänen-Laakso, T., Biondi, S., De Sutter, V., Lammertyn, F., Nuutila, A.M., Soderlund, H., Zabeau, M., Inze, D. and Oksman-Caldentey, K.-M. (2003) *Proceedings of the National Academy of Sciences of the United States of America*, **100**, 8595–8600.

3 Gershenzon, J. (2007) *Proceedings of the National Academy of Sciences of the United States of America*, **104**, 5257–5258.

4 Wallace, R.J. (2004) *Proceedings of the Nutrition Society*, **63**, 621–629.

5 Guenther, E. (1948) *The Essential Oils*, Van Nostrand, New York.

6 Bauer, K., Garbe, D. and Surburg, H. (2001) *Common Fragrance and Flavor Materials: Preparation, Properties and Uses*, Wiley-VCH Verlag GmbH, Weinheim.

7 Bohlmann, J., Meyer-Gauen, G. and Croteau, R. (1998) *Proceedings of the National Academy of Sciences of the United States of America*, **95**, 4126–4133.

8 Adams, R.P. (2007) *Identification of Essential Oil Components by Gas Chromatography/Mass Spectrometry*, 4th edn, Allured, Carol Stream, IL.

9 Lopes, D., Bizzo, H.R., Sa Sobrinho, A.F. and Pereira, M.V.G. (2000) *Journal of Essential Oil Research*, **12**, 705–708.

10 Sabulal, B., Dan, M., Pradeep, N.S., Kurup, R. and George, V. (2007) *Journal of Essential Oil Research*, **19**, 279–281.

11 Pauli, A. (2006) *Medicinal Research Reviews*, **26**, 223–268.

12 McLeod, J. (1994) *Lavender Sweet Lavender*, Kangaroo Press, Kenthurst.

13 Bauer, K., Garbe, D. and Surburg, H. (1997) *Common Fragrance and Flavor Materials*, 3rd edn, VCH, Weinheim.

14 Boelens, M.H. (1997) *Perfumer and Flavorist*, **22**, 19–40.

15 Hammer, K.A., Carson, C.F. and Riley, T.V. (1999) *Journal of Applied Microbiology*, **86**, 985–990.

16 Nguefack, J., Leth, V., Zollo, P.H.A. and Mathur, S.B. (2004) *International Journal of Food Microbiology*, **94**, 329–334.

17 Landolt, P.J., Hofstetter, R.W. and Biddick, L.L. (1999) *Environmental Entomology*, **28**, 954–960.

18 Turlings, T.C.J. and Wäckers, F.L. (2004) Recruitment of predators and parasitoids by herbivore-damaged plants, in *Advances in Insect Chemical Ecology* (eds R.T. Cardé and J. Millar), Cambridge University Press, Cambridge, pp. 21–75.

19 Edris, A.E. (2007) *Phytotherapy Research*, **21**, 308–323.

20 Welsh, C. (1997) *Complementary Health Practice Review*, **3**, 11–15.

21 Donoyama, N. and Ichiman, Y. (2006) *International Journal of Aromatherapy*, **16**, 175–179.

22 Harris, B. (2006) *International Journal of Aromatherapy*, **16**, 117–131.

23 Trapp, S.C. and Croteau, R.B. (2001) *Genetics*, **158**, 811–832.

24 McGarvey, D.J. and Croteau, R. (1995) *Plant Cell*, **7**, 1015–1026.

25 Cseke, L., Dudareva, N. and Pichersky, E. (1998) *Molecular Biology and Evolution*, **15**, 1491–1498.

26 Pichersky, E. and Gershenzon, J. (2002) *Current Opinion in Plant Biology*, **5**, 237–243.

27 Gershenzon, J. (2006) *Phytochemistry*, **67**, 1562–1563.

28 Wise, M.L. and Croteau, R. (1999) Monoterpene biosynthesis, in *Comprehensive Natural Products Chemistry: Isoprenoids* (ed. D.E. Cane), Elsevier Science, Oxford, pp. 97–153.

29 Bohlmann, J., Meyer-Gauen, G. and Croteau, R. (1998) *Proceedings of the National Academy of Sciences of the United States of America*, **95**, 4126–4133.

30 Davis, E.M. and Croteau, R. (2000) Cyclization enzymes in the biosynthesis of monoterpenes, sesquiterpenes, and diterpenes, in *Topics in Current Chemistry: in Biosynthesis: Aromatic Polyketides,*

Isoprenoids, Alkaloids (eds F.J. Leeper and J.C. Vederas), Springer, Heidelberg, pp. 53–95.

31 Steele, C.L., Crock, J., Bohlmann, J. and Croteau, R. (1998) *Journal of Biological Chemistry*, **273**, 2078–2089.

32 Aubourg, S., Lecharny, A. and Bohlmann, J. (2002) *Molecular Genetics and Genomics*, **267**, 730–745.

33 Lange, B.M., Wildung, M.R., Stauber, E.J., Sanchez, C., Pouchnik, D. and Croteau, R. (2000) *Proceedings of the National Academy of Sciences of the United States of America*, **97**, 2934–2939.

34 Charles, D.J. and Simon, J.E. (1990) *Journal of the American Society for Horticultural Science*, **115**, 458–462.

35 Boutekedjiret, C., Bentahar, F., Belabbes, R. and Bessiere, J.M. (2003) *Flavour and Fragrance Journal*, **18**, 481–484.

36 Coelho, J.A., Grosso, C., Pereira, A.P., Burillo, J., Urieta, J.S., Figueiredo, A.C., Barroso, J.G., Mendes, R.L. and Palavra, A.M.F. (2007) *Flavour and Fragrance Journal*, **22**, 438–442.

37 Hudaib, M. and Aburjai, T. (2007) *Flavour and Fragrance Journal*, **22**, 322–327.

38 Basta, A., Pavlovic, M., Couladis, M. and Tzakou, O. (2007) *Flavour and Fragrance Journal*, **22**, 197–200.

39 Benkaci-Ali, F., Baaliouamer, A.Y., Meklati, B.Y. and Chemat, F. (2007) *Flavour and Fragrance Journal*, **22**, 148–153.

40 Eltz, T., Zimmermann, Y., Haftmann, J., Twele, R., Francke, W., Quezada-Euan, J.J. and Lunau, K. (2007) *Proceedings Biological Sciences*, **274**, 2843–2848.

41 Moyler, D.A. (1993) *Flavour and Fragrance Journal*, **8**, 235–247.

42 Lucchesi, M.E., Chemat, F. and Smadja, J. (2004) *Journal of Chromatography A*, **1043**, 323–327.

43 Wang, Z., Ding, L., Li, T., Zhou, X., Wang, L., Zhang, H., Liu, L., Li, Y., Liu, Z., Wang, H., Zeng, H. and He, H. (2006) *Journal of Chromatography A*, **1102**, 11–17.

44 Lucchesi, M.E., Chemat, F. and Smadja, J. (2004) *Flavour and Fragrance Journal*, **19**, 134–138.

45 Theis, A.L., Waldack, A.J., Hansen, S.M. and Jeannot, M.A. (2001) *Analytical Chemistry*, **73**, 5651–5654.

46 Rohloff, J. (1999) *Journal of Agricultural and Food Chemistry*, **47**, 3782–3786.

47 Tranchida, P.Q., Presti, M.L., Costa, R., Dugo, P., Dugo, G. and Mondello, L. (2006) *Journal of Chromatography A*, **1103**, 162–165.

48 Cavalli, J.-F., Fernandez, X., Lizzani-Cuvelier, L. and Loiseau, A.-M. (2003) *Journal of Agricultural and Food Chemistry*, **51**, 7709–7716.

49 Fäldt, J., Eriksson, M., Valterova, I. and Borg-Karlson, A.-K. (2000) *Zeitschrift für Naturforschung*, **55c**, 180–188.

50 Fakhari, A.R., Salehi, P., Heydari, R., Ebrahimi, S.N. and Haddad, P.R. (2005) *Journal of Chromatography A*, **1098**, 14–18.

51 Besharati-Seidani, A., Jabbari, A. and Yamini, Y. (2005) *Analytica Chimica Acta*, **530**, 155–161.

52 Field, J.A., Nickerson, G., James, D.D. and Heider, C. (1996) *Journal of Agricultural and Food Chemistry*, **44**, 1768–1772.

53 Tholl, D., Boland, W., Hansel, A., Loreto, F., Rose, U.S.R. and Schnitzler, J.-P. (2006) *Plant Journal*, **45**, 540–560.

54 Pichersky, E., Noel, J.P. and Dudareva, N. (2006) *Science*, **311**, 808–811.

55 Baldwin, I.T., Halitschke, R., Paschold, A., von Dahl, C.C. and Preston, C.A. (2006) *Science*, **311**, 812–815.

56 Orav, A., Stulova, I., Kailas, T. and Müürisepp, M. (2004) *Journal of Agricultural and Food Chemistry*, **52**, 2582–2586.

57 Choi, H.S. and Sawamura, M. (2002) *Bioscience, Biotechnology, and Biochemistry*, **66**, 439–443.

58 Burt, S. (2004) *International Journal of Food Microbiology*, **94**, 223–253.

59 Sabulal, B., Dan, M., Anil John, J.A., Kurup, R., Chandrika, S.R. and George, V. (2007) *Flavour and Fragrance Journal*, **22**, 521–524.

60 Marriot, P.J., Shellie, R. and Cornwell, C. (2001) *Journal of Chromatography A*, **936**, 1–22.

61 Decouzon, M., Géribaldi, S., Rouillard, M. and Sturla, J.-M. (2006) *Flavour and Fragrance Journal*, **5**, 147–152.

62 Baser, K.H.C., Demirci, B., Iscan, G., Hashimoto, T., Demirci, F., Noma, Y. and Asakawa, Y. (2006) *Chemical and Pharmaceutical Bulletin*, **54**, 222–225.

63 Tswett, M.S. (1906) *Berichte der Deutschen Botanischen Gesellschaft*, **24**, 384–393.

64 Burt, S.A., Vlielander, R.P., Haagsman, H. and Veldhuizen, E.J.A. (2005) *Journal of Food Protection*, **68**, 919–926.

65 Izmailov, N.A. and Schraiber, M.S. (1938) *Farmatsiya*, **3**, 1–7.

66 Stahl, E. (1983) *Angewandte Chemie (International Edition in English)*, **22**, 507–516.

67 Wagner, H. and Bladt, S. (1996) *Plant Drug Analysis: A Thin Layer Chromatography Atlas*, 2nd edn, Springer, Heidelberg.

68 Reich, E. and Schibli, A. (2006) *High-Performance Thin-Layer Chromatography for the Analysis of Medicinal Plants*, Thieme, New York.

69 James, A.T. and Martin, A.J.P. (1954) *British Medical Bulletin*, **10**, 170–176.

70 Kojima, M., Tsunoi, S. and Tanaka, M. (2003) *Journal of Chromatography A*, **984**, 237–243.

71 Linskens, H.F. and Jackson, J.F.(eds) (1986) *Modern Methods of Plant Analysis: Gas Chromatography/Mass Spectrometry*, Vol. 3, Springer, Berlin.

72 Kovats, E. (1958) *Helvetica Chimica Acta*, **41**, 1915–1932.

73 Van den Dool, H. and Kratz, P.D. (1963) *Journal of Chromatography*, **11**, 463–471.

74 Merfort, I. (2002) *Journal of Chromatography A*, **967**, 115–130.

75 Zhao, Z., Lu, J., Leung, K., Chan, C.L. and Jiang, Z.H. (2005) *Chemical and Pharmaceutical Bulletin*, **53**, 856–860.

76 Baldovini, N., Tomi, F. and Casanova, J. (2001) *Phytochemical Analysis*, **12**, 58–63.

77 Paolini, J., Costa, J. and Bernardini, A.F. (2007) *Phytochemical Analysis*, **18**, 235–244.

78 Oumzil, H., Ghoulami, S., Rhajaoui, M., Ilidrissi, A., Fkih-Tetouani, S., Faid, M. and Benjouad, A. (2002) *Phytotherapy Research*, **16**, 727–731.

79 Duru, M.E., Öztürk, M., Ugur, A. and Ceylan, Ö. (2004) *Journal of Ethnopharmacology*, **94**, 43–48.

80 Baranska, M., Schulz, H., Walter, A., Rösch, P., Quilitzsch, R., Lösing, G. and Popp, J. (2006) *Vibrational Spectroscopy*, **42**, 341–345.

81 Schulz, H., Özkan, G., Baranska, M., Krüger, H. and Özcan, M. (2005) *Vibrational Spectroscopy*, **39**, 249–256.

82 Porter, N.G. and Wilkins, A.L. (1999) *Phytochemistry*, **50**, 407–415.

83 Sugawara, Y., Hara, C., Aoki, T., Sugimoto, N. and Masujima, T. (2000) *Chemical Senses*, **25**, 77–84.

84 Li, C., Heinemann, P. and Irudayaraj, J. (2007) *Transactions of the ASABE*, **50**, 1417–1425.

85 Ngassoum, M.B., Mapongmetsem, P.-M., Tatsadieu, L., Jirovetz, L., Buchbauer, G. and Shahabi, M. (2005) *Journal of Essential Oil Research*, **17**, 492–495.

86 Kondoh, T., Yamada, S., Shioda, S. and Torii, K. (2005) *Chemical Senses*, **30** (Suppl 1), i172–i173.

87 Sugawara, Y., Hino, Y., Kawasaki, M., Hara, C., Tamura, K., Sugimoto, N., Yamanishi, Y., Miyauchi, M., Masujima, T. and Aoki, T. (1999) *Chemical Senses*, **24**, 415–421.

88 Dobell, C. (1960) *Antony van Leeuwenhoek and His 'Little Animals'*, Dover, New York, pp. 141–142.

89 Gill, A.O. and Holley, R.A. (2004) *Applied and Environmental Microbiology*, **70**, 5750–5755.

90 Buckle, J. (1997) *Clinical Aromatherapy in Nursing*, Arnold, London.

91 Dabbah, R., Edwards, V.M. and Moats, W.A. (1970) *Applied Microbiology*, **19**, 27–31.

92 Kar, N.A. and Jain, S.R. (1971) *Plant Foods for Human Nutrition*, **20**, 231–237.

93 Larrondo, J.V., Agut, M. and Calvo-Torras, M.A. (1995) *Microbios*, **82**, 171–172.

94 Lis-Balchin, M. and Deans, S.G. (1997) *Journal of Applied Microbiology*, **82**, 759–762.

95 Boonchild, C. and Flegel, T. (1982) *Canadian Journal of Microbiology*, **28**, 1235–1241.

96 Deans, S.G. and Ritchie, G. (1987) *International Journal of Food Microbiology*, **5**, 165–180.

97 Ghannoum, M.A. (1988) *Journal of General Microbiology*, **134**, 2917–2924.

98 Kivanc, M. and Akgäul, A. (1988) *Flavour and Fragrance Journal*, **3**, 95–98.

99 Knobloch, K., Pauli, P., Iberl, B., Weigand, H. and Weiss, N. (1989) *Journal of Essential Oil Research*, **1**, 119–128.

100 Lambert, R.J.W., Skandamis, P.N., Coote, P.J. and Nychas, G.-J.E. (2001) *Journal of Applied Microbiology*, **91**, 453–462.

101 Özkan, G., Sagdic, O. and Özcan, M. (2003) *Food Science and Technology International*, **9**, 85–88.

102 Adam, K., Sivropoulou, A., Kokkini, S., Lanaras, T. and Arsenakis, M. (1998) *Journal of Agricultural and Food Chemistry*, **46**, 1739–1745.

103 Shin, S. and Kang, C.-A. (2003) *Letters in Applied Microbiology*, **36**, 111–115.

104 Kohlert, C., van Rensen, I., Marz, R., Schindler, G., Graefe, E.U. and Velt, M. (2000) *Planta Medica*, **66**, 495–505.

105 Lee, K.-W., Everts, H. and Beynen, A.C. (2004) *International Journal of Poultry Science*, **3**, 738–752.

106 Giordani, R., Trebaux, J., Masi, M. and Regli, P. (2001) *Journal of Ethno-pharmacology*, **78**, 1–5.

107 Holmberg, S., Solomon, S. and Blake, P. (1987) *Reviews of Infectious Diseases*, **9**, 1065–1078.

108 Neu, H. (1992) *Science*, **257**, 1064–1073.

109 Tomasz, A. (1994) *New England Journal of Medicine*, **330**, 1247–1251.

110 Smith, R. (1999) *Bulletin of the World Health Organization*, **77**, 862.

111 Shin, S. and Lim, S. (2004) *Journal of Applied Microbiology*, **97**, 1289–1296.

112 Smith, R.D. and Coast, J. (2002) *Bulletin of the World Health Organization*, **80**, 126–133.

113 Appelbaum, P. (1992) *Clinical Infectious Diseases*, **15**, 77–83.

114 D'Antuono, L.F., Galleti, G.C. and Bocchini, P. (2000) *Annals of Botany*, **86**, 471–478.

115 Muller-Riebau, F.J., Berger, B.M., Yegen, O. and Cakir, C. (1997) *Journal of Agricultural and Food Chemistry*, **45**, 4821–4825.

116 Randrianalijaona, J.-A., Ramanoelina, P.A.R., Rasoarahona, J.R.E. and Gaydou, E.M. (2005) *Analytica Chimica Acta*, **545**, 46–52.

117 Srivastava, N.K. and Luthra, R. (1994) *Journal of Experimental Botany*, **45**, 1127–1132.

118 Svoboda, K.P., Gough, J., Hampson, J. and Galambosi, B. (1995) *Flavour and Fragrance Journal*, **10**, 139–145.

119 Deans, S.G. and Svoboda, K.P. (1988) *Journal of Horticultural Science*, **63**, 503–508.

120 Marotti, M., Piccaglia, R., Giovanelli, E., Deans, S.G. and Eaglesham, E. (1994) *Flavour and Fragrance Journal*, **9**, 125–129.

121 Galambosi, B., Svoboda, K.P., Hampson, J.B. and Asakawa, Y. (1999) *Zeitschrift für Arznei- und Gewürzpflanzen*, **4**, 19–23.

122 Trombetta, D., Castelli, F., Sarpietro, M.G., Venuti, V., Cristani, M., Daniele, C., Saija, A., Mazzanti, G. and Bisignano, G. (2005) *Antimicrobial Agents and Chemotherapy*, **49**, 2474–2478.

123 Cox, S.D., Mann, C.M., Markham, J.L., Bell, H.C., Gustafson, J.E., Warmington, J.R. and Wyllie, S.G. (2000) *Journal of Applied Microbiology*, **88**, 170–175.

124 Carson, C.F., Hammer, K.A. and Riley, T.V. (2006) *Clinical Microbiology Reviews*, **19**, 50–62.

125 de Billerbeck, V.G., Roques, C.G., Bessière, J.-M., Fonvieille, J-L. and Dargent, R. (2001) *Canadian Journal of Microbiology*, **47**, 9–17.

126 Pinto, E., Pina-Vaz, C., Salgueiro, L., Gonçalves, M.J., Costa-de-Oliveira, S.,

Cavaleiro, C., Palmeira, A., Rodrigues, A. and Martinez-de-Oliveira, J. (2006) *Journal of Medical Microbiology*, **55**, 1367–1373.

127 Lambert, R.J.W., Skandamis, P., Coote, P.J. and Nychas, G.-J.E. (2001) *Journal of Applied Microbiology*, **91**, 453–462.

128 Ultee, A., Kets, E.P.W. and Smid, E.J. (1999) *Applied and Environmental Microbiology*, **65**, 4606–4610.

129 Ultee, A., Bennik, M.H.J. and Moezelaar, R. (2002) *Applied and Environmental Microbiology*, **68**, 1561–1568.

130 Chami, F., Chami, N., Bennis, S., Trouillas, J. and Remmal, A. (2004) *Journal of Antimicrobial Chemotherapy*, **54**, 909–914.

131 Veldhuizen, E.J.A., von Bokhoven, J.L.M.T., Zweijtzer, C., Burt, S.A. and Haagsman, H.P. (2006) *Journal of Agricultural and Food Chemistry*, **54**, 1874–1879.

132 Knowles, J.R., Roller, S., Murray, D.B. and Naidu, A.S. (2005) *Applied and Environmental Microbiology*, **71**, 797–803.

133 Ali, S.M., Khan, A.A., Ahmed, I., Musaddiq, M., Ahmed, K.S., Polasa, H., Rao, L.V., Habibullah, C.M., Sechi, L.A. and Ahmed, N. (2005) *Annals of Clinical Microbiology and Antimicrobials*, **4**, 20.

134 Ultee, A., Gorris, L.G. and Smit, E.J. (1998) *Journal of Applied Microbiology*, **85**, 211–218.

135 Ultee, A., Kets, E.P.W., Alberda, M., Hoekstra, F.A. and Smid, E.J. (2000) *Archives of Microbiology*, **174**, 233–238.

136 Ultee, A., Slump, R.A., Steging, G. and Smid, E.J. (2000) *Journal of Food Protection*, **63**, 620–624.

137 Ultee, A. and Smid, E.J. (2001) *International Journal of Food Microbiology*, **64**, 373–378.

138 Lambert, R.J.W., Skandamis, P.N., Coote, P.J. and Nychas, G.-J.E. (2001) *Journal of Applied Microbiology*, **91**, 453–462.

139 Sikkema, J., De Bont, J.A.M. and Poolman, B. (1995) *Microbiological Reviews*, **59**, 201–222.

140 Mann, C.M., Cox, S.D. and Markham, J.L. (2000) *Letters in Applied Microbiology*, **30**, 294–297.

141 Moleyar, V. and Narasimham, P. (1992) *International Journal of Food Microbiology*, **16**, 337–342.

142 Janssen, A.M., Tsai-Sioe, W.H.R., Scheffer, J.J.C. and Svendsen, A.B. (1988) *Flavour and Fragrance Journal*, **3**, 137–140.

143 Pol, I.E. and Smid, E.J. (1999) *Letters in Applied Microbiology*, **29**, 166–170.

144 Periago, P.M. and Moezelaar, R. (2001) *International Journal of Food Microbiology*, **68**, 141–148.

145 Periago, P.M., Palop, A. and Fernandez, P.S. (2001) *Food Science and Technology International*, **7**, 487–492.

146 Helander, I.M., Alakomi, H.-L., Latva-Kala, K., Mattila-Sandholm, T., Pol, I., Smid, E.J., Gorris, L.G.M. and von Wright, A. (1998) *Journal of Agricultural and Food Chemistry*, **46**, 3590–3595.

147 Rhayour, K., Bouchikhi, T., Tantaoui-Elaraki, A., Sendide, K. and Remmal, A. (2003) *Journal of Essential Oil Research*, **15**, 286–292.

148 Bennis, S., Chami, F., Chami, N., Bouchikhi, T. and Remmal, A. (2004) *Letters in Applied Microbiology*, **38**, 454–458.

149 Krist, S., Halwachs, L., Sallaberger, G. and Buchbauer, G. (2007) *Flavour and Fragrance Journal*, **22**, 44–48.

150 Bagamboula, C.F., Uyttendaele, M. and Debevere, J. (2004) *Food Microbiology*, **21**, 33–42.

151 Olasupo, N.A., Fitzgerald, D.J., Gasson, M.J. and Narbad, A. (2003) *Letters in Applied Microbiology*, **36**, 448–451.

152 Gaysinsky, S., Davidson, P.M., Bruce, B.D. and Weiss, J. (2005) *Journal of Food Protection*, **68**, 1359–1366.

153 Ooi, L.S., Li, Y., Kam, S.L., Wang, H., Wong, E.Y. and Ooi, V.E. (2006) *American Journal of Chinese Medicine*, **34**, 511–522.

154 Inouye, S., Takizawa, T. and Yamaguchi, H.J. (2001) *Journal of Antimicrobial Chemotherapy*, **47**, 565–573.

155 Sato, K., Krist, S. and Buchbauer, G. (2006) *Biological and Pharmaceutical Bulletin*, **29**, 2292–2294.

156 Sato, K., Krist, S. and Buchbauer, G. (2007) *Flavour and Fragrance Journal*, **22**, 435–437.

157 van Zyl, R.L., Seatlholo, S.T., van Vuuren, S.F. and Viljoen, A.M. (2006) *Journal of Essential Oil Research*, **18**, 129–133.

158 Jirovetz, L., Buchbauer, G., Schmidt, E., Denkova, Z., Stoyanova, A.S., Nikolova, R. and Geissler, M. (2007) *Journal of Essential Oil Research*, **19**, 288–291.

159 Duru, M.E., Öztürk, M., Ugur, A. and Ceylan, Ö. (2004) *Journal of Ethnopharmacology*, **94**, 43–48.

160 Sonboli, A., Babakhani, B. and Mehrabian, A.R. (2006) *Zeitschrift fur Naturforschung C*, **61**, 160–164.

161 Bowles, B.L., Sackitey, S.K. and Williams, A.C. (1995) *Journal of Food Safety*, **15**, 337–347.

162 Rao, A.V. and Bekheet, I.A. (1976) *Applied and Environmental Microbiology*, **32**, 479–482.

163 Alderman, D.J. and Smith, P. (2001) *Aquaculture*, **196**, 211–243.

164 National Committee for Clinical Laboratory Standards (2003) *Performance Standards for Antimicrobial Disk Susceptibility Tests, NCCLS document M2-A8*, 8th edn, NCCLS, Wayne, PA.

165 Gabhainn, S.N., Bergh, Ø., Dixon, B., Donachie, L., Carson, J., Coyne, R., Curtin, J., Dalsgaard, I., Manfrin, A., Maxwell, G. and Smith, P. (2004) *Aquaculture*, **240**, 1–18.

166 National Committee for Clinical Laboratory Standards (2000) *Methods for Dilution Antimicrobial Susceptibility Tests for Bacteria that Grow Aerobically, NCCLS document M7-A5*, 5th edn, NCCLS, Wayne, PA.

167 Mann, C.M. and Markham, J.L. (1998) *Journal of Applied Microbiology*, **84**, 538–544.

168 Hood, J.R., Wilkinson, J.M. and Cavanagh, H.M.A. (2003) *Journal of Essential Oil Research*, **15**, 428–433.

169 Moon, T., Cavanagh, H.M.A. and Wilkinson, J.M. (2006) *Journal of Essential Oil Research*, **18**, 578–580.

170 Carson, C.F., Hammer, K.A. and Riley, T.V. (1995) *Microbios*, **82**, 181–185.

171 Kazmi, S.J.A. and Mitchell, A.G. (1978) *Journal of Pharmaceutical Sciences*, **67**, 1266–1271.

172 Van Doorne, H. (1990) Interactions Between Preservatives and Pharmaceutical Components, in *Guide to Microbiological Control in Pharmaceuticals* (eds S. Denyer and R. Baird), Ellis Horwood, New York.

173 Altman, P. (1991) *Manufacture*, **12**, 23–24.

174 Remmal, A., Bouchikhi, T., Tantaoui-Elaraki, A. and Ettayebi, M. (1993) *Journal de Pharmacie de Belgique*, **48**, 352–356.

175 Weseler, A., Geiss, H.K., Saller, R. and Reichling, J. (2005) *Die Pharmazie*, **60**, 498–502.

176 Chorianopoulos, N.G., Lambert, R.J., Skandamis, P.N., Evergetis, E.T., Haroutounian, S.A. and Nychas, G.J. (2006) *Journal of Applied Microbiology*, **100**, 778–786.

177 Grassmann, J., Hippeli, S., Dornisch, K., Rohnert, U., Beuscher, N. and Elstner, E.F. (2000) *Arzneimittel-Forschung*, **50**, 135–139.

178 Silva, J., Abebe, W., Sousa, S.M., Duarte, V.G., Machado, M.I.L. and Matos, F.J.A. (2003) *Journal of Ethnopharmacology*, **89**, 277–283.

179 Liu, Z.L. and Ho, S.H. (1999) *Journal of Stored Products Research*, **35**, 317–328.

180 Huang, Y., Ho, S.-H., Lee, H-C. and Yap, Y.-L. (2002) *Journal of Stored Products Research*, **38**, 403–412.

181 Isman, M.B. (2000) *Crop Protection*, **19**, 603–608.

182 Ansari, M.A., Vasudevan, P., Tandon, M. and Razdan, R.K. (2000) *Bioresource Technology*, **71**, 267–271.

183 Schnitzler, P., Schon, K. and Reichling, J. (2001) *Die Pharmazie*, **56**, 343–347.

184 Schnitzler, P., Koch, C. and Reichling, J. (2007) *Antimicrobial Agents and Chemotherapy*, **51**, 1859–1862.

185 Minami, M., Kita, M., Nakaya, T., Yamamoto, T., Kuriyama, H. and Imanishi, J. (2003) *Microbiology and Immunology*, **47**, 681–684.

186 Duschatzky, C.B., Possetto, M.L., Talarico, L.B., Garcia, C.C., Michis, F., Almeida, N.V., de Lampasona, M.P., Schuff, C. and Damonte, E.B. (2005) *Antiviral Chemistry and Chemotherapy*, **16**, 247–251.

187 Halcon, L.L. (2002) *Minnesota Medicine*, **85**, 42–46.

188 Robins, J.L.W. (1999) *Journal of Holistic Nursing*, **17**, 5–17.

189 Lis-Balchin, M. (1997) *Journal of the Royal Society for the Promotion of Health*, **117**, 324–329.

190 Muyima, N.Y.O., Zulu, G., Bhengu, T. and Popplewell, D. (2002) *Flavour and Fragrance Journal*, **17**, 258–266.

191 Mazza, G., Ciaravolo, S., Chiricosta, G. and Celli, S. (1992) *Flavour and Fragrance Journal*, **7**, 111–116.

192 Temple, W.A., Smith, N.A. and Beasley, M. (1991) *Journal of Toxicology: Clinical Toxicology*, **29**, 257–262.

193 Bekele, J. and Hassanali, A. (2001) *Phytochemistry*, **57**, 385–391.

194 Luo, M., Jiang, L.-K. and Zou, G.-L. (2005) *Journal of Food Protection*, **68**, 581–588.

195 Orafidiya, L.O., Agbani, E.O., Iwalewa, E.O., Adelusola, K.A. and Oyedapo, O.O. (2004) *Phytomedicine*, **11**, 71–76.

196 Lis-Balchin, M. (1999) *Journal of the Royal Society for the Promotion of Health*, **119**, 240–243.

197 Villar, D., Knight, M.J., Hansen, S.R. and Buck, W.B. (1994) *Veterinary and Human Toxicology*, **36**, 139–142.

198 Hammer, K.A., Carson, C.F., Riley, T.V. and Nielsen, J.B. (2006) *Food and Chemical Toxicology*, **44**, 616–625.

199 Smith, R.L., Cohen, S.M., Doull, J., Feron, V.J., Goodman, J.I., Marnett, L.J., Portoghese, P.S., Waddell, W.J., Wagner, B.M., Hall, R.L., Higley, N.A., Lucas-Gavin, C. and Adams, T.B. (2005) *Food and Chemical Toxicology*, **43**, 345–363.

200 Doganca, S., Gurkan, E., Hurlak, F., Tuzan, O.T. and Tuzlaci, E. (1997) *Fitoterapia*, **68**, 80.

201 Meyer, B.N., Ferrigni, N.R., Putnam, J.E., Jacobsen, L.B., Nichols, D.E. and Mclaughlan, J.L. (1982) *Planta Medica*, **45**, 31–34.

202 Cheng, S.-S., Chang, H.-T., Chang, S.-T., Tsai, K.-H. and Chen, W.-J. (2003) *Bioresource Technology*, **89**, 99–102.

203 Sabulal, B., Dan, M., John, J.A., Kurup, R., Pradeep, N.S., Valsamma, R.K. and George, V. (2006) *Phytochemistry*, **67**, 2469–2473.

8

Application of Plant Extracts and Products in Veterinary Infections

Jacobus N. Eloff and Lindy J. McGaw

Abstract

The impact of veterinary infectious diseases is experienced worldwide, both in terms of losses in livestock production, and with regard to the harmful effects on companion animals such as dogs and cats. Animals are susceptible to a vast array of infectious organisms, and diseases may be caused by protozoa, viruses, bacteria and fungi amongst others. As in the human therapeutic arena, antibiotic resistance is a major problem in the treatment of animal infections and comparable strategies are required to meet the challenges faced. In the area of growth promotion of production animals, alternatives to commercial synthetic antibiotics are being sought following bans on the use of antibiotic feed additives in many regions. Structurally related antibiotics are often used in human and animal therapy. This has caused concern that resistance developed in animal pathogens may be transferred to human pathogens. The search for new antimicrobial substances in both the human and veterinary medical fields deserves coordination and cooperative research efforts. The healing power of plants has been exploited for generations in treating animals as well as humans. Herbalism is growing in popularity worldwide, and many veterinarians, particularly in European countries, appear to be using herbal medicine on their animal patients. Natural products originating from plants have an immensely diverse array of structures which may serve as possible lead compounds for further development into therapeutic treatments. An alternative to the routine pharmaceutical approach is the development of active extracts which have had inactive constituents selectively removed to increase activity. An economically significant application of active plant extracts may be as organic in-feed additives to replace antibiotic growth promoters. Plants enjoy an exceptional position as a viable source of new anti-infectives in the treatment and prevention of animal infections.

New Strategies Combating Bacterial Infection. Edited by Iqbal Ahmad and Farrukh Aqil
Copyright © 2009 WILEY-VCH Verlag GmbH & Co. KGaA, Weinheim
ISBN: 978-3-527-32206-0

8.1
Introduction

Animals are an integral part of our lives, whether as a source of food, clothing materials, transport or companionship. Infectious diseases are of great significance economically and also for the quality of life of our animals. An infection is generally taken to signify the invasion and multiplication of microorganisms in tissues of the body, with accompanying effects specific to the class of invading microorganism, such as local cell and tissue injury, inflammation, and other symptoms. Major animal diseases encountered include infections by bacteria, fungi, viruses and protozoa. Gastrointestinal helminth parasites are the cause of severe livestock production losses, and are also commonly found in dogs and cats. In the treatment of animal infections, alarming trends are developing with regard to the resistance of pathogenic causal organisms of diseases against currently available drugs, reflecting what is seen in human infectious disease control.

Antibiotics are naturally occurring, semisynthetic and synthetic substances with antimicrobial activity that can be administered orally, parenterally or topically. They are used in human and veterinary medicine to treat and prevent disease, and as prophylactic growth promoters in food production animals. Antibiotic resistance may have originated at the same time as antibiotic compounds, protecting antibiotic-producing organisms from their own products and permitting originally susceptible organisms to compete with other species. The more an antibiotic is used, especially at lower concentrations, the more probable is the likelihood of resistant bacteria developing among pathogens and among commensal bacteria of animals in an exposed population.

The search for new antibiotic compounds to fight infections is achieving new dimensions and alternative sources are being explored. This is not only in response to antibiotic resistance, but is developing as consumers become increasingly aware of using organic alternatives that are potentially less harmful to the environment and possibly less harsh on the animal being treated. Plants have been used for many thousands of years to treat a multitude of ailments in humans and animals. This is supported by the interesting and significant biological effects discovered in numerous plant extracts and natural products isolated from them. The amazing structural diversity of active plant metabolites lends additional support to the potential for discovery of useful medications against infectious diseases.

This chapter will begin with a brief description of the most significant infectious diseases in veterinary healthcare. A discussion of the rationale motivating the search for anti-infective compounds in plants will follow, leading into a brief review of biological activity studies of plants used in ethnoveterinary medicine. The important issue of antibiotic resistance and the contribution of antibiotic feed additives to this phenomenon will be addressed. The impact of plants as replacements for these growth promoters will be assessed with reference to the relevant literature. The scope of the chapter is concentrated for reasons of space constraints on the biological activity of plants used in traditional veterinary medicine and the potential of plants as alternatives for antibiotic growth promoters. Finally, the conclusion will deal with future prospects and recommendations for further research on the use of plants in the treatment of animal infections.

8.2
Veterinary Infectious Diseases and their Significance

Many infectious diseases afflict livestock, as well as companion animals, in various regions of the world. In terms of production animals, diseases have differing impacts in different production systems. These impacts vary with regard to the production losses and control costs, whether the system is a commercial one or more in line with individual rural livestock-keepers. Livestock assume vastly different roles in the developed and developing worlds [1]. In the developed world, a wide variety of available food coupled with a rising incidence of cardiovascular disease, diabetes and other disorders have led to a negative outlook on consumption of animal protein [1]. In these countries the per capita consumption of meat and other livestock products is expected to decline in the future [2]. In contrast, in developing countries where malnutrition is rife, meat and milk are crucial sources of protein and other nutrients, and the demand for such livestock products is predicted to grow by around 3% per year [2]. Besides this, livestock are able to generate income to be used for purchasing food, healthcare and education, and can provide the necessary fertilizer for crop production [1]. In Africa, growth in the livestock population and productivity is not keeping pace with the human population expansion, and reasons for this include declining livestock production and inadequate disease control services [3].

Infectious diseases of livestock animals have multiple effects, from the animal itself, to the farm, ecological zone and so on, to national and international levels [1]. Disease control strategies need to account for these multiple impacts to be effective in dealing with the relative importance of different diseases and how they should be controlled. Morris [4] described the direct effects of animal diseases on livestock productivity as ranging from reduced feed intake and altered digestion and metabolism to increased morbidity and mortality, decreased reproduction, weight gain and milk production. The control of animal diseases may involve sizeable costs, particularly if adequate investment into livestock management to avoid initial disease risk is not made.

It is not only the causal organisms of disease that are targets of prophylaxis and therapy, but also the vectors that carry and transmit these diseases. Much interesting research is being conducted involving the anti-tick, or acaricidal, activity of plant-derived preparations. In addition, diseases caused by helminth parasites, although not strictly classified as infections, are also of major economic significance and some success has been attained in studying anthelmintic activity of plants. As a result of these developments, a brief discussion of the importance of ticks and helminth parasites will also be included, as will the potential of plants as acaricides and anthelmintics.

8.2.1
Protozoal Diseases

Many groups of protozoa are responsible for animal and human diseases. Protozoa of the genus *Trypanosoma* parasitize all classes of vertebrates – mammals, birds, reptiles, amphibians and fish – causing trypanosomoses [5]. The general characteristics of trypanosomosis include the intermittent presence of parasites in the blood and sporadic fever. Affected animals often develop anemia, accompanied by loss of body condition, reduced productivity and often high mortality [5].

A common parasite of many animal species and humans worldwide is the protozoan *Toxoplasma gondii*. Although most of the livestock infections caused by the parasite are subclinical, toxoplasmosis causes reproductive disorders in sheep and goats [6]. *Neospora caninum* is a coccidian parasite closely related to *T. gondii*, and affects livestock and companion animals.

Another protozoal disease occurring in all domestic livestock species and chickens is coccidiosis, caused by infection with *Eimeria* or *Isospora* spp. The disease is characterized clinically by enteritis, following invasion of the mucosal cells of the gastrointestinal tract [7]. Subclinical infections are common and the disease is mostly encountered when animals are overcrowded, stressed or kept in unhygienic conditions. However, in well-managed animals there is a balance between the host and parasite. Economic losses attributable to coccidiosis include mortality and poor performance, and high cost of treatment or prevention [7].

Babesiosis is caused by infection with protozoan parasites belonging to the genus *Babesia*. These parasites are transmitted by ticks and are generally host-specific. Another genus of tick-borne protozoan is *Theileria*, species of which cause diseases known collectively as theilerioses. An example of a manifestation of this disease is East Coast Fever, a usually fatal disease of cattle caused by *Theileria parva*.

8.2.2
Viral Diseases

Viral diseases of livestock and humans are prevalent and frequently devastating. Paramyxoviruses affect only vertebrates and are transmitted directly from infected to susceptible hosts, causing major diseases such as canine distemper, rinderpest and Newcastle disease as well as measles and mumps in humans. Rinderpest is a highly contagious disease of cattle, and the decimation of cattle and wildlife populations caused by this virus has had a substantial impact on human history and ecology [8]. Another paramyxovirus of economic importance is bovine respiratory syncytial virus, which causes mild to severe respiratory tract disease.

Retroviruses, influenzaviruses, coronaviruses, parvoviruses and herpesviruses are other types of viral infections affecting animals. An example of a picornavirus is the highly contagious foot-and-mouth disease of cloven-hoofed animals and camelids.

8.2.3
Rickettsial and Chlamydial Diseases

Diseases caused by bacteria in the orders Rickettsiales and Chlamydiales are causes of economically important livestock infections. For example, heartwater (cowdriosis) is a tick-borne rickettsial disease of cattle, sheep and goats caused by *Ehrlichia ruminantium*, previously known as *Cowdria ruminantium*. This disease is associated with high fever, nervous signs, edema of the lungs and brain, and finally death. It is a major cause of stock losses in sub-Saharan Africa [9]. Anaplasmosis afflicts cattle, sheep, goats and some wild ruminants, and is caused by the rickettsial organism *Anaplasma*. Infections in mammals may result in a lack of visible symptoms, through to severe disease and mortality.

Chlamydiosis, which affects a wide range of hosts, is characterized by various syndromes depending on the species and strain of *Chlamydia* or *Chlamydophila*, the host species and the affected organ [10]. The severity of the disease ranges from life-threatening to asymptomatic, but usually moderate to mild disease is seen, leading to persistent infection [10].

8.2.4
Bacterial and Fungal Diseases

Bacteria and fungi have been implicated in a number of important diseases in animals. Among the Gram-negative species, important pathogenic genera of veterinary importance are *Campylobacter*, *Arcobacter*, *Helicobacter* and *Lawsonia*. These bacteria are usually found in the gastrointestinal and reproductive tracts. *Campylobacter jejuni* infection in cattle causes enteric disease, occasionally mastitis and abortion, and may also cause abortion in sheep and goats [11]. *C. jejuni* is also found in humans as an enteric pathogen and is an important zoonosis which arises mainly from poor hygiene [11].

Other important Gram-negative bacteria are *Bartonella*, *Brucella*, *Neisseria*, *Pseudomonas*, *Salmonella* and *Vibrio*. Bovine brucellosis caused by *Brucella abortus* is a very contagious disease characterized by mid- to late-term abortion in cows and this disease is an important zoonosis [12]. *Salmonella* serovars are responsible for salmonellosis in livestock, causing enteritis, septicemia or abortion and *Salmonella* is also a zoonosis.

Gram-positive bacteria are implicated in various diseases of veterinary importance, for instance species of *Staphylococcus*, *Streptococcus*, *Enterococcus*, *Bacillus* and *Clostridium*. *Staphylococcus* species (e.g. *Staphylococcus aureus* and *Staphylococcus epidermidis*) are part of the normal microbial flora on the skin and mucous membranes of healthy humans and animals. However, once factors such as trauma, a compromised immune system or disturbance of the normal microflora are introduced, these bacteria may cause disease, such as suppurative infections in livestock, especially abscesses, mastitis and pyoderma. In livestock, species of *Streptococcus* cause a wide variety of primary or secondary suppurative conditions, for example mastitis, septicaemia, genital tract infections, strangles and endocarditis.

Clostridium perfringens affects most domestic animal species as well as humans, causing a variety of enteric diseases [13]. The organism secretes toxins, resulting in food poisoning in humans, and in livestock it causes enterotoxaemia, hemorrhagic and necrotic enteritis, and postabortion septicemia amongst other ailments. *C. perfringens* is particularly significant in poultry where it may cause necrotic enteritis [14]. The disease is usually controlled by adding antimicrobials in feed and has recently re-emerged in countries which have banned antimicrobial growth promoters [15].

Tuberculosis (TB), a chronic contagious disease of many vertebrate animals, is caused by infection with acid-fast bacterial species of the genus *Mycobacterium*. Bovine TB, caused by *Mycobacterium bovis*, occurs globally and appears to be an increasing problem, even in developed countries where major difficulties in controlling and eradicating the disease are encountered [16]. Bovine TB is an important zoonotic disease and its prevalence in humans generally correlates closely with the level of infection in the cattle population of the particular country [17]. Paratuberculosis, or Johne's disease, caused by *Mycobacterium avium* ssp. *paratuberculosis*, occurs worldwide, and affects cattle, sheep, goats and wild ruminants [18].

With regard to fungal pathogens, a few of the most important genera are represented by *Aspergillus*, *Candida* and *Cryptococcus*. *Aspergillus* spp. are opportunistic pathogens, causing pulmonary aspergillosis in animals and humans with compromised immune systems, particularly in avian species which are most susceptible [19]. *Aspergillus fumigatus* is the most common fungus associated with fungal infections in livestock [20]. Cryptococcosis in humans and animals is a life-threatening disease that develops following inhalation and dissemination of *Cryptococcus neoformans* from the lungs to the central nervous system [21]. *Candida albicans* is a member of the commensal flora of the gastrointestinal and genitourinary tract of dogs, cats and humans [22], and an opportunistic pathogen in these species, potentially causing disease when the host defense is compromised.

8.2.5
Helminth Parasites

Helminths are a major cause of reduced production in livestock in many countries, particularly the tropics [23]. Several genera of parasitic worms are responsible for livestock diseases, including *Haemonchus*, *Trichostrongylus*, *Strongyloides*, *Nematodirus*, *Ostertagia*, *Toxocara* and *Ascaris*. Common symptoms of parasite infestation are diarrhea and wasting. Clinical and subclinical forms of parasitism result in direct and indirect losses, with direct losses including acute illness and death. Of much importance are the indirect consequences which involve a loss in livestock products such as milk, meat, skins and wool.

8.2.6
Ticks as Vectors of Infectious Diseases

Ticks are vectors of many economically important diseases of livestock and as such are key targets for infection control. Cattle diseases of major financial consequence transmitted by ticks include heartwater, anaplasmosis, babesiosis and theileriosis.

Sheep and goats are affected by heartwater, theileriosis, anaplasmosis and spiro-chaetosis, while horses, mules and donkeys may be infected with equine piroplas-mosis and spirochaetosis. Pigs are susceptible to porcine babesiosis and African swine fever. Apart from transmitting infectious diseases to livestock, certain tick species may be associated with toxicoses such as sweating sickness, paralysis and brown ear tick toxicosis [24].

At present, tick control on domestic animals relies on the integrated use of available technologies such as chemicals, management of acaricide resistance, host resistance and tick-borne disease vaccines [25]. This concept entails using acaricides rationally in conjunction with other strategies to minimize production losses caused by ticks and the diseases they carry in a cost-effective way [24]. Fundamentally, the control of ticks and hence of tick-borne diseases merits much attention, and the role of plants here is potentially significant.

8.3
Plants as Sources of Antimicrobial Compounds

The healing ability of plants has been exploited for thousands of years, in the treatment of both humans and animals. Of the estimated 250 000 plant species in the world, only in the region of 5–15% have been explored for potentially useful bioactive compounds [26]. In the United States, around 25–50% of dispensed pharmaceuticals originate from higher plants, but few of these are intended for use as antimicrobials, because bacterial and fungal sources have been a prolific source of these com-pounds [27]. With the occurrence of antibiotic resistance, antimicrobial plant extracts are attracting much interest owing to the potential for discovery of novel lead structures for development, as well as for the efficacy of whole-plant extracts in combating infections.

Plants produce specialized secondary metabolites which have extraordinarily diverse structures and many functions (e.g. to attract pollinators, as defensive substances against parasites and predators or as compounds to resist pests and diseases) [26, 27]. In comparison with synthetic compound libraries, natural products from plants offer sterically more complex structures [28, 29]. Metabolites in plants have different functions relating to their structures, and the major groups of chemicals in plants and their mechanisms of action in veterinary medicine have been lucidly summarized by Yarnell [30]. There are different approaches to the discovery of pharmacologically active metabolites from plants, including the random selection method, the chemotaxonomic method, following traditional medicine leads (the ethnobotanical approach) or the ecological approach [31].

8.3.1
Traditional Ethnoveterinary Medicine

In many parts of the world, there exists a strong culture of using plants to treat a diversity of ailments afflicting humans, and also the animals important to our livelihood and well-being. The use of herbal preparations in the treatment of animals is principally acknowledged to occur in areas of developing countries that lack ready access to state veterinary services, or where the medicines offered are too expensive for rural livestock-keepers. Ethnoveterinary medicine may be in a position to offer affordable and easily available remedies for common troubles, such as mild diarrhea, skin diseases, intestinal worms, wounds and reproductive ailments [32], although orthodox medicines remain drugs of choice for infectious disease outbreaks and epidemics. Ethnoveterinary medicine may also act as sources of new treatments as active plant extracts or as possible leads for novel antimicrobial compounds with commercial potential.

In animals, the most commonly treated disorders comprise diarrhea, and general gastrointestinal problems, internal parasites, eye inflammation, retained placenta, heartwater, coughing, redwater and tick infestation [33]. It is critical when evaluating herbal remedies used in animal health care to bear in mind that ethnoveterinary medicine as a whole includes diagnostic procedures, animal husbandry practices, surgical methods and traditional veterinary theory in addition to the use of ethno-veterinary plants to prevent and control disease [33, 34].

8.3.2
Evaluation of Ethnoveterinary Plants for Efficacy Against Infectious Diseases

The growing recognition of the role of ethnoveterinary medicine, particularly in developing countries, is supported by an increasing number of publications in the scientific literature reporting on ethnobotanical and biological activity studies of plants used in traditional veterinary medicine. Rigorous evaluation of the efficacy as well as safety of these remedies is vital and may supply promising leads for active extracts or pure compounds that can be developed for commercial use. The majority of studies available are *in vitro* studies, owing to economic and ethical considerations, and concentrate on the antibacterial or antifungal activity of plant extracts. Plants used in traditional veterinary medicine may be prepared in a variety of ways and may be used singly or in combination with other plants. The methods of application and dosages are also considerations to be taken into account when evaluating activity of the remedy, so lack of activity in a laboratory study does not necessarily provide conclusive evidence for inefficacy of a treatment. Ethnoveterinary remedies may not function to completely rid the body of infectious microorganisms, but may support the immune system, allowing the host to build up immunity against the remaining organisms.

Efficacy *in vivo* cannot be presumed after achieving good results with *in vitro* tests, one of the reasons being that the dose or concentration which is active *in vitro* may not be able to be extrapolated to that required to reach adequate plasma volumes in a whole organism [35]. Bioavailability is important if the remedy is orally administered, and factors such as absorption and metabolism may be responsible for discrepancies between *in vitro* and *in vivo* tests [35]. Absorption and metabolism can be affected by other compounds in the extract which may boost or restrain absorption, and other compounds may upregulate metabolic enzymes in the liver. Houghton *et al.* [35] noted that traditional preparation methods of plant-based medicines could remove or concentrate such compounds, and if the correct method is not followed in making extracts for pharmacological testing, then the extract may show activities different from the extract prepared according to traditional methods. Similarly, if the actual target organism is not used in a screen, then caution should be exercised in relating results obtained against a model organism to results against a pathogenic organism. In an extremely useful review, Cos *et al.* [36] highlighted a series of criteria that should be taken into account when embarking on investigations of anti-infective natural products. In the remainder of this section, representations of studies on plants used in animal healthcare will be discussed.

Ethnoveterinary plants used to treat infectious diseases in cattle were screened for antibacterial activity using the microplate serial dilution technique [37]. The organisms used in testing were those recommended for antibacterial testing by the National Committee for Clinical and Laboratory Standards [38], namely *Escherichia coli*, *Enterococcus faecalis*, *Pseudomonas aeruginosa* and *S. aureus*. Hexane, methanol and water extracts were prepared and tested, and were in general found to be most active against *E. faecalis* and *S. aureus*. As Vlietinck *et al.* [39] reported, Gram-positive bacteria are generally more susceptible to antimicrobial substances than are Gram-negative species because of differences in the bacterial cell wall composition. In the study of McGaw *et al.* [37], 33% of plant extracts screened had minimum inhibitory concentration (MIC) values less than 1 mg/ml and these levels of antibacterial activity may contribute towards rationalizing the use of the plants in ethnoveterinary medicine.

Ziziphus mucronata (Rhamnaceae) revealed high levels of antibacterial activity in the above-mentioned preliminary screen [37], and 2,3-dihydroxyl-up-20-en-28-oic acid and zizyberanalic acid were isolated from the leaves of this plant [40]. The first-mentioned compound was highly active against *S. aureus*, supporting claims of the efficacy of a *Z. mucronata* leaf paste in treating bacterial infections in animals as well as humans.

Bizimenyera *et al.* [41] used the same technique to identify substantial antibacterial activity against *S. aureus* and *P. aeruginosa* in *Peltophorum africanum* (Fabaceae). Pastoralists and farmers use the root and bark extracts to treat stomach ailments such as diarrhea and dysentery in cattle [41], and the antibacterial activity discovered in extracts of the plant tends to support its traditional usage against infections that may be of bacterial origin. In this study, organic solvents (ethanol, acetone, dichloromethane and hexane) were used to prepare extracts.

Similarly, *Gunnera perpensa* (Gunneraceae) was investigated to discover whether antibacterial efficacy could be answerable for its reputed activity against endometritis and retained placenta in cattle [42]. It was concluded that the relatively weak antibacterial activity of *G. perpensa* extracts against the four bacterial species mentioned above was probably only a contributory factor, together with its known uterotonic activity [43], to the general observed therapeutic effect of the plant.

The efficacy of medicinal plants used to treat wounds and retained placenta in livestock in South Africa was investigated by Luseba *et al.* [44]. Several dichloromethane extracts displayed antibacterial and anti-inflammatory activity. *Cissus quadrangularis* stem and *Jatropha zeyheri* root extracts were selectively inhibitory against cyclooxygenase-2 in the anti-inflammatory investigation. None of the extracts was mutagenic in the Ames test against *Salmonella typhimurium* strain TA98. Luseba *et al.* [44] argued that although water is traditionally a commonly available solvent to prepare medicinal extracts, the activity of organic extracts need not be disregarded. The reasons for this are that in livestock wound treatment, the whole-plant material is often processed and applied locally, and in the case of ailments such as retained placenta where the treatment mixtures are given orally (probably unfiltered), active ingredients may be released [44].

The ethnoveterinary approach was followed by Stein *et al.* [45] in identifying plant extracts with antifungal activity against pathogenic yeasts and filamentous fungi. Subsequent work on species of *Pterocaulon* with antifungal activity revealed that coumarins isolated from the extracts were inactive, leading to the conclusion that potential synergistic effects or minor active compounds were responsible for activity [46].

In an interesting study investigating antifungal activity of plant extracts against clinical isolates of the most common and important disease-causing fungi in animals, Masoko *et al.* [47] discovered significant antifungal activity in several *Terminalia* species (Combretaceae). The fungal pathogens used in the study were representative of the various morphological forms of fungi, including yeasts (*C. albicans* and *C. neoformans*), moulds (*A. fumigatus*) and thermally dimorphic fungi (*Sporothrix schenckii*). Of the hexane, dichloromethane, acetone and methanol extracts of *Terminalia* leaves, the acetone extracts displayed the highest antifungal activity.

Many reports exist on the efficacy of plants used in human traditional medicine against a number of different viruses [48–50]. Virucidal assays are commonly employed, including investigations of inhibition of viral cytopathic effect or plaque inhibition. Kudi and Myint tested plants from northern Nigeria with a history of use in both human and veterinary traditional medicine for their antiviral activity. Extracts were tested against poliovirus, astrovirus, herpes simplex viruses and parvovirus, and most of the extracts displayed some level of activity against more than one virus [51].

There is much optimism over the therapeutic possibilities of plant-based antiviral agents. Diverse compounds isolated from plants have antiviral activity (e.g. flavonoids, terpenoids, polyphenolics, saponins, proteins, alkaloids, polysaccharides, essential oils, etc.) [52, 53]. Several of these phytochemicals have complementary and overlapping mechanisms of action, including antiviral effects by either inhibiting the formation of viral DNA or RNA or inhibiting the activity of viral reproduction [53]. Some plant compounds have a unique antiviral mechanism and are promising candidates for further clinical research [52]. The conventional response to viral epidemics is mass vaccination, but where viruses mutate rapidly, vaccination development cannot keep pace and antiviral drugs or medications are essential [54]. Interestingly, contrary to antibacterial and antifungal plant chemicals, several antiviral plant compounds have shown competitive *in vitro* and *in vivo* antiviral activities with those of synthetic antiviral drugs [52].

Using a cell culture-based antibabesial test, where *Babesia caballi* cultures were exposed to plant extracts at varying concentrations, Naidoo *et al.* [55] demonstrated that the acetone extract of *Elephantorrhiza elephantina* rhizome was active. This plant is a popular component of ethnoveterinary remedies. *Urginea sanguinea*, *Rhoicissus tridentata* and *Aloe marlothii* acetone extracts were not active in this assay [55]. An *Ehrlichia ruminantium* culture system was used to evaluate the antirickettsial activity of two ethnoveterinary plants, *E. elephantina* and *A. marlothii* [56]. *E. ruminantium* cultures were incubated with the acetone extracts of the leaves, and compared to oxytetracycline and untreated controls. *E. elephantina* and *A. marlothii* both demonstrated anti-ehrlichial activity with EC_{50} values of 111.4 and 64.5 µg/ml and EC_{90} values of 228.9 and 129.9 µg/ml, respectively. The corresponding EC_{50} and EC_{90} for oxytetracycline were 0.29 and 0.08 µg/ml. The plant extracts possibly produced their inhibitory activity by a similar mechanism, unrelated to that of the tetracyclines [56].

The *in vitro* trypanocidal activity of some Nigerian medicinal plants which have either been reported in the literature as having trypanocidal activity or been used for the treatment of human and animal trypanosomiasis was evaluated [57]. Some extracts had very good activity with relatively low toxicity, providing a rationale for further investigation *in vivo*.

Livestock infestations with helminth parasites are common, especially in rural areas, and anthelmintic remedies form a major component of ethnoveterinary medicine. Although some would consider that helminthiases, or parasitic worm infestations, do not strictly fall within the definition of infections, there have been many reports on screening medicinal plants for anthelmintic activity, justifying a mention in this chapter. Synthetic anthelmintics have some serious disadvantages, such as nonavailability in some developing countries, cost, risk of misuse potentially leading to drug resistance, environmental pollution and food residues [23]. Traditionally used anthelmintic plants offer an alternative to manufactured anthelmintics that is sustainable as well as environmentally acceptable and these plants may have an important role in the future control of helminth parasites [23]. According to Hammond *et al.* [23], plant anthelmintics could play a key role in veterinary medicine in the tropics, supplementing or even replacing synthetic drugs for economic reasons, and also when commercial drugs are unavailable or where resistance has developed.

In vivo trials to test for anthelmintic activity are expensive and time consuming, and parasitic nematodes are difficult to maintain in artificial culture systems in the laboratory. The free-living helminth species, *Caenorhabditis elegans*, has been used as a model organism in basic screening tests as it is easier, cheaper and faster than parasitic nematode systems [58, 59]. Most commercially available broad-spectrum anthelmintics demonstrate activity against *C. elegans* [58], but caution should be exercised in extrapolating activity against a free-living nematode to activity against a parasitic species [60]. Many plant extracts have been shown to possess activity against free-living *C. elegans* [37, 61, 62].

In vitro assays testing plant extracts against parasitic nematode eggs and larvae have yielded some interesting results. For example, inhibition of egg hatching and larval development against the two most important livestock nematode parasites *Haemonchus contortus* and *Trichostrongylus colubriformis* by various plant extracts has been reported. Test samples incubated with the fresh nematode eggs may inhibit hatching in the egg hatch assay [63]. The larval development assay [64] detects the ability of test substances to slow or prevent development of eggs into infective larvae. The two assays together can provide a reasonable indication of anthelmintic activity of plant extracts or natural products. One such *in vitro* study using *H. contortus* as test organism [65] discovered that 50 of 115 ethanol extracts prepared from 79 plant species showed larvicidal activity and two were ovicidal. Plants were selected on the basis of ethnomedical and ethnoveterinary studies.

A popular and widely used ethnoveterinary plant, *Peltophorum africanum*, is used in treating helminth diseases, and the acetone extracts of the leaf, bark and root have been screened for activity against *H. contortus* and *T. colubriformis* in the egg hatch and larval development assays [66, 67]. The extracts were active in the assays at 0.2 mg/ml, supporting the use of this plant in traditional medicine. *In vivo* evidence of effectiveness and lack of toxicity is necessary before further development of this active plant extract. Reports have been published on anthelmintic effects of plant extracts, many with ethnoveterinary uses, in animal models [68–70].

As mentioned earlier, tick-borne diseases cause major problems for livestock farmers. The repellent and toxic effects of plant extracts against ticks have been studied by various researchers, revealing some promising results. In one such study, Nchu [71] analyzed the repellent effects and direct toxicity of extracts of *Allium* species against adults of *Hyalomma marginatum rufipes*. A high repellency index (65–79.48%) was shown by the acetone extract of *Allium porrum* and the dichloromethane extract of *Allium sativum* was toxic to 100% of ticks within 1 h of exposure. Essential oils from the aerial parts of *Lippia javanica* and *Tagetes minuta* showed a concentration-dependent repellent effect on the ticks [71]. A growth inhibition bioassay indicated that *T. minuta* essential oil delayed molting to adult stage of 60% of engorged nymphs of *H. m. rufipes*.

A related study reported that ethyl acetate extracts of *Senna italica* ssp. *arachoides* demonstrated increasing acaricidal activity against *H. m. rufipes* with increasing concentration [72]. An *in vivo* experiment suggested that aqueous extracts of *S. italica* ssp. *arachoides* ingested by guinea pigs and rabbits may interfere with the feeding performance of adult *H. m. rufipes* ticks [72]. Mkolo [73] showed that some plants used traditionally as arthropocides in South Africa, namely *Eucalyptus globoidea* and *Lavendula angustifolia*, are effective tick repellents. Zorloni [74] surveyed plants used in Ethiopia for ethnoveterinary tick control purposes, and investigated 28 of these plants for repellent and toxic effects against adult *Rhipicephalus pulchellus* ticks. Organic solvent extracts of many species showed promising repellency activities, and *Calpurnia aurea* yielded the best results in the tick toxicity assay [74].

At present, vaccination is probably the most durable and cost-effective method of preventing bacterial and viral infections in comparison to the rest of the prophylactic and therapeutic methods combined [75]. Microorganisms and arthropod vectors have survived for millennia in a constantly changing chemical environment, and although pharmaceuticals may be successful in the short term, new vaccine development is a better long-term strategy for the control of animal and human diseases [75]. However, mammalian immunological defences against infectious organisms have also evolved through the centuries and exploiting mechanisms to enhance or stimulate the immune system is a realistic way to achieve enduring control of infectious organisms [75]. Plant extracts and immune-stimulating compounds derived from plants have a significant role to play.

Our experience over the past few years in the Phytomedicine Programme (University of Pretoria) has shown that, although organic extracts are frequently active, most aqueous plant extracts that we have tested to date have little activity. Therefore, we believe that antioxidant effects, with positive effects on the immune system, may be responsible for claimed antimicrobial efficacy in traditional medicine. After isolating antibacterial and antifungal compounds from members of the Combretaceae and plants used in ethnoveterinary medicine, we found that in more than 90% of cases the antibacterial compounds were relatively nonpolar compounds. By screening acetone extracts of leaves of a large number of southern African plants selected using a taxonomic approach, we have discovered several plants with excellent antibacterial and antifungal activities. There is strong evidence for synergistic antimicrobial effects in plant extracts and we are investigating the development of anti-infective extracts as well as the identification of pure isolated compounds.

Of the first 100 tree species screened in this project, approximately 10% of the species examined had excellent antibiotic activity, with MIC values below 0.04 mg/ml against *S. aureus* and another 10% with MIC values below 0.08 mg/ml. Against *E. coli*, the percentages were 2 and 5% with the same MIC values. Many of these plants are not known to have antibacterial activity and these results demonstrate the viability of the project. To date almost 400 species have been screened and those with promising activity are being studied further to identify which compounds are responsible for efficacy, among other aims.

8.4
Antibiotic Resistance and the Impact of Antibiotic Feed Additives

In recent decades, a major source of health concern in humans and animals has been the emergence of antibiotic resistance in bacteria. This occurrence has been associated with selective pressure applied during the extensive and indiscriminate use of antibiotics in medicine, veterinary practice and as growth promoters in the animal production industry. Antibacterial drugs used as growth promoters are added to the feed of entire herds and flocks at subtherapeutic levels over an extended period of time [76]. The intensive production of food animals has led to the widespread use of antibiotics for growth-promoting purposes in addition to disease treatment and prophylaxis [77].

Antibiotic-resistant bacteria were first identified in the 1940s, but the consequences of these findings were not appreciated while new antibiotics were still being discovered [78]. Today, the rate of discovery of new antibiotic agents is drastically reduced and this has meant the problem of antibiotic resistance is rapidly becoming a global health crisis.

It is clear that misuse of antibiotics in human beings has contributed to the increasing rates of resistance, but recently the use of antibiotics in production animals and the resultant effect on resistance levels in people has been more closely analyzed. Antimicrobials are used therapeutically and prophylactically as growth promoters in animals. In Europe and North America, antimicrobial agents have been employed as in-feed growth promoters for about 50 years [79]. Some argue that the impact of use of antibiotics in animals – whether therapeutic or as growth promoters – pales by comparison with human use and that efforts should be concentrated on the misuse of antibiotics in people [75]. It is difficult to quantify the portion of resistance in humans that comes from food animals and there is rising concern over the movement of resistant bacteria through the food chain. Some authors believe it is the total use of antibiotics that matters because the bacteria do not distinguish between whether the drug was used therapeutically, prophylactically or for growth promotion [75].

As early as 1969, the Swann Committee recommended that antibiotics used to treat people and animals (and other substances that can cause resistance against them) should not be used as growth promoters in animal feeds [80]. The performance-enhancing antimicrobial drugs were incorporated into feed at small doses to improve feed conversion and reduce the formation of toxins [77]. As a result of increasing concerns over the adverse effect of some antibiotics on human or animal health, and the transfer of resistance between different bacteria and between human and animals, several antibiotics have been banned as feed additives in the European Union since 1999, including bacitracin, spiramycin, tylosin, virginiamycin [81] and olaquindox [82]. These antibiotics were banned for use as growth promoters owing to their structural similarity to antimicrobials used to treat human diseases [83]. The use of the Precautionary Principle in implementing these bans has been criticized by those who dispute the presence of evidence of actual risk to human health [84]. In spite of this, the result of the ban has been a noticeable decrease in antibiotic resistance in animals and humans in some countries in Europe [83, 85–87].

Terminating use of antibiotic growth promoters was projected to be possible in both industrialized and developing countries with minimum effect on food production, provided that adequate attention was given to implementing alternative disease-prevention strategies in food animal production [78]. In contrast to some expectations, very little change in food production was seen where the use of growth promoters was stopped. Feed conversion was mildly reduced, but the extra feed still costs less than farmers were spending on antibiotic growth promoters. Singer [78] contended that with modern farming practices, it seems that growth promoters are not as necessary as they may have been 40–50 years ago.

As an example of the correlation between the use of antibiotic growth promoters and resistance, Bager *et al.* [88] reported on a study undertaken to determine the association between use of avoparcin as a feed additive and incidence of vancomycin-resistant *Enterococcus faecium* on poultry farms and in pig herds. Strains of *E. faecium* strains with lowered susceptibility to avoparcin are also resistant to vancomycin, suggesting that resistance is mediated by the same gene [89, 90]. The investigation was conducted as a retrospective cohort study, where flocks and herds either exposed or not exposed to avoparcin previously were compared to detect differences in occurrence of vancomycin-resistant *E. faecium* and whether causes apart from avoparcin use could explain the difference. The results were strongly supportive of the proposal that the use of avoparcin as a growth promoter is associated with vancomycin-resistant enterococci in domestic animals.

A Finnish study showed that growth-promoting antimicrobials can be withdrawn successfully in piglet production and that it is possible to produce pigs without them. However, the health status of the animals, husbandry practices and management have to be good on the farms [91]. This is critical in ensuring that withdrawal of antibiotic growth promoters does not lead to drastically higher therapeutic use of antibiotics, which would in turn exacerbate the problem of antibiotic resistance.

8.5
Plants as Replacements for Antibiotic Feed Additives

The use of low concentrations of antibiotics in food animals selects for bacteria resistant to antibiotics. It is believed that these resistant bacteria might spread via the food chain to humans and cause human infection or transfer genetic material to human bacterial pathogens. This concern has led to the banning of antibiotic feed additives in the European Union, and further calls for this ban to be widened. In developed countries where modern, hygienic farming practices are rigorously adhered to, the need for antibiotic growth promoters has perhaps fallen away, but there is a definite cost factor involved. Rising consumer concern has also contributed to the search for alternatives to antibiotic growth promotion. However, in developing countries where animals may be kept in suboptimal conditions, and where animal production needs to be supported to sustain growing populations, any mechanism of promoting growth of livestock that is not harmful to the human population is welcomed. Alternatives to antibiotic growth promoters are sought to sustain levels of production in many regions of the world.

Some candidate replacements for antibiotics are competitive exclusion products, probiotics, prebiotics, organic acids, enzymes, plant extracts, hen egg antibodies, bacteriophages and vaccination [92]. The use of plant extracts in supplementation of animal feed is receiving closer attention, owing to the known antimicrobial, antiviral and antioxidant properties in many herbs, as well as the ability of some to stimulate the endocrine and immune system. Beneficial effects on the gut environment and microflora have been demonstrated by some botanical ingredients [93]. A plethora of *in vitro* studies and reviews have been published on the antibacterial activity of essential oils and other plant-derived extracts and natural products [27, 94–97]. Plant products may possess a dual role in promoting growth and protecting animals from pathogenic microorganisms.

Although the mechanisms of the growth-promoting effects of antibiotics are not fully understood, it has been suggested that they stimulate beneficial bacteria and inhibit harmful microorganisms in the gastrointestinal tract [98, 99]. Antibiotic growth promoters, already banned in Europe, are under review in Australia and the United States [100]. Recent publications record the effects on rumen microbes of various plant compounds [101] and the effect of these compounds in terms of altering some rumen microbial processes [102–106]. Hart *et al.* [107] reviewed the use in altering rumen fermentation of natural plant products including essential oils, saponins and related compounds, concluding that plant products indeed have the potential to be used as rumen manipulating agents.

For practical use, it is necessary to test these plant-based preparations in animal models. In a poultry study, Losa and Kohler [108] discovered a reduced concentration of *C. perfringens* per gram of intestinal contents after supplementing feed with a commercial preparation (Crina poultry) of essential oil components. Specific blends of essential oil components containing thymol or carvacrol and eugenol, curcumin and piperin were shown to control the proliferation of *C. perfringens* in the broiler intestine [109]. It was suggested that antibacterial effects of the essential oil, stimulation of digestive enzymes, stabilization of intestinal microflora, and inactivation of *C. perfringens* toxins could act to reduce *C. perfringens* colonization in the broiler gut.

The Chinese traditional herbal medicine Bazhen, a composition of eight herbal medicines, was investigated as a replacement for antibiotics as growth promoters in piglets [110]. Experimental results indicated that average daily weight gain of the Bazhen group was higher than that of the control group, but lower than that of the antibiotics group. However, feed intake was reduced while feed to weight gain ratio was improved in the Bazhen group compared to the control and antibiotic groups. Bazhen supplementation stimulated various immune response parameters compared to the control group, namely immunoglobulin G, γ-globulin levels, neutrophil activity and white blood cell counts following challenge with lipopolysaccharide. The immune response results of white and red blood cells, interleukin-6 and tumor necrosis factor-α levels in the Bazhen group were greater than those obtained with antibiotic supplementation. It was concluded that Bazhen could be a potential substitute for the antibiotics chlorotetracycline and oxytetracycline in weaned piglet feed [110].

Saponins are naturally occurring surface-active glycosides produced mainly by plants. These compounds have diverse effects in animal and human systems, and are the active principles in several medicinal and health products [111]. When cultured fish – common carp and Nile tilapia – were continuously fed in their diet low levels (150–300 mg/kg) of a triterpenoid saponin-rich *Quillaja* saponin (commercially available from Sigma), a growth-promoting effect was observed [111]. The practical implications of this study embrace the potential for increasing the efficiency of feed utilization of cultured fish.

The application of saponins in human and animal nutrition was analyzed by Cheeke [112]. Saponins are natural detergents because they have water-soluble carbohydrate as well as fat-soluble steroid or triterpenoid components. Two major commercial sources of saponins are *Yucca schidigera* and *Quillaja saponaria*, extracts of which are used as dietary additives for livestock and companion animals, primarily for ammonia and odor control [112]. Saponins used as feed additives have a host of beneficial effects, such as antiprotozoal activity against pathogenic protozoa (e.g. giardia), antibacterial activity against Gram-positive bacteria, antifungal and antiviral activity as well as antioxidant and immunostimulatory properties [112, 113].

Coccidiosis caused by species of the *Eimeria* protozoan parasite is one of the most important diseases in the poultry industry, resulting in losses of millions of dollars annually. Natural plant-based alternatives to using commercial coccidiostats in feed may potentially provide products functioning by mechanisms other than those of chemotherapeutics, with the additional advantage of a natural origin. Coccidial infection is associated with lipid peroxidation of the intestinal mucosa, and a recent study surmised that antioxidant compounds could hold promise for the control of *Eimeria* infections in poultry [114]. Four plant extracts with known antioxidant activity were screened for their anticoccidial activity in chickens. *Combretum woodii* at 160 mg/kg proved to be extremely toxic to the birds, while treatment with *Tulbaghia violacea* (35 mg/kg), *Vitis vinifera* (75 mg/kg) and *Artemisia afra* (150 mg/kg) resulted in Feed Conversion Ratios similar to toltrazuril, the positive control, and higher than the untreated control. *T. violacea* significantly decreased the oocyst production in the birds. It was concluded that antioxidant-rich plant extracts have potential benefits in treating and possibly also preventing coccidial infections. The promising results obtained with *T. violacea* provide impetus for further studies on the value of the plant as a therapeutic or prophylactic anticoccidial agent.

In the Phytomedicine Programme at the University of Pretoria, we are investigating combining the use of an antioxidant product known to enhance the immune system in humans with a mixture of natural antibacterial compounds occurring in extracts of plants. The antioxidant preparation can be produced at low cost from a plant waste product following a procedure that we have developed and licensed. The plants used for extracting antibacterial compounds have been used in human medicine for centuries and are considered safe. These extracts will be refined to contain different antibacterial compounds as well as antioxidant activity that may improve immune function. Mixtures of these extracts have been tested on poultry challenged with *C. perfringens* and some of the mixtures gave statistically significant better results than the widely used antibiotic feed additive [115]. These results have been submitted for a preliminary patent, and research is continuing on developing a product to be used in the poultry and other production animal industries.

The examples described above show that plants have great potential with regard to antimicrobial efficacy in animal health. Mechanisms of action of plant-based products need to be elucidated and *in vivo* studies conducted with promising candidates. The organic nature of plant antimicrobials is an attractive feature in terms of livestock production, compared with the problems experienced with antibiotic residues in animal products such as meat and milk. Naturally, potentially harmful plant residues need to be investigated prior to marketing a plant antimicrobial. Generation of antibiotic resistance may also be avoided with the use of active plant extracts, as a variety of antimicrobial compounds with differing mechanisms of action may exist in the same preparation.

8.6
Conclusions

Scientists in many disciplines, including ethnobotany, microbiology, pharmacology, toxicology and veterinary science, are studying the potential benefits of plants in the search for new antimicrobial agents. Extracts of thousands of plants have been screened, and from these plants many active pure phytochemicals have been identified and reported, mostly using *in vitro* systems.

The potential of plants as sources of anti-infective agents is promising and with this in mind, certain aspects of this field of research require close attention. As Cowan [27] recommends, it would be of great advantage to standardize methods of extraction and testing in the laboratory, to facilitate interpretation and comparison of results obtained by different research groups. Where plant extracts are reported to have interesting efficacy against microbes, a thorough characterization of the extract would assist future work on the plant species of interest. Increased attention should be devoted to bioavailability studies and evaluation of other pharmacokinetic as well as pharmacodynamic effects of medicinal preparations. Although the cost is a great deal higher than for *in vitro* tests, investigations using animals are necessary to confirm efficacy of promising treatments. Improved validation of preliminary *in vitro* models is required to enable prediction of good activity in subsequent animal tests.

Many scientists have placed their hope on isolating a single compound with excellent activity that can be developed into a single compound product against infections in humans or animals. Despite thousands of plants being investigated and hundreds of publications, there appears to be no single compound plant-based product that has entered the market yet. After isolating large numbers of compounds from many different plant species we believe that there is much more promise of delivering a useful product based on extracts rather than isolated compounds. We have abundant evidence that in antimicrobial activity of plants several compounds play a role. Even in cases where several antibacterial compounds isolated from the same plant extract were combined the expected antimicrobial activity was not obtained [116]. It appears that the activities of antibacterial compounds are potentized by other compounds that may affect bioavailability, stability or other indirect parameters. This presents a very interesting field of research.

Mechanism of action-based studies are also of great value. Apart from direct microbicidal assays, other mechanisms of infection prevention and treatment should be included in initial activity screenings, such as disruption of adhesion in anti-infection activity [27]. It is also important, particularly in the field of antimicrobial studies, to investigate the effect of plant preparations or pure compounds on beneficial organisms occurring as part of the natural flora of the body, as disruptions to the local microbial ecology may be highly detrimental to overall health. Naturally, the toxicity of treatments is a critical area of research that is essential to the further development of a potential antimicrobial medicine. It is advantageous to include toxicity screening at an early stage of activity investigations so as not to waste valuable research time and money on an extract or isolated compound that owes its supposed activity to nonspecific toxicity.

Synergistic effects of plant-derived extracts and compounds are another productive area of research. The use of medicinally active plants could be further boosted by observations of interesting increases in activity when extracts or natural products are combined together with each other or with existing antibiotics [96]. The ability of botanical natural products to stimulate the immune system is a fascinating area of research and is complementary to the direct lethal effects on microorganisms shown by many plants.

Natural products research in recent decades has supplied a wealth of information on the extraordinary diversity of chemicals in plants, as well as the biological activity of plant-derived natural products against animal and human pathogenic microorganisms. With the currently experienced problems of antibiotic resistance, the discovery of lead compounds for pharmaceuticals from plants is a realistic goal. Alternatively there is potential for producing standardized and formulated extracts to be used for treatment of microbial diseases threatening livestock production, especially in developing countries. Overall, the application of plant products in treating animal infections is an exciting and promising area of research.

References

1 Perry, B.D., McDermott, J.J. and Randolph, T.F. (2004) The control of infectious diseases of livestock: making appropriate decisions in different epidemiological and socioeconomic conditions, in *Infectious Diseases of Livestock* (eds J.A.W. Coetzer and R.C. Tustin), Oxford University Press, Cape Town, pp. 178–224.

2 Delgado, C., Rosegrant, M., Steinfeld, H., Ehui, S. and Courbois, C. (1999) Livestock to 2020: the next food revolution, Food, Agriculture and the Environment Discussion Paper 28. FPRI/ILRI/FAO, Washington, DC.

3 Thambi, E.N., Maina, O.W. and Bessin, R. (2002) *Animal and Animal Products Trade in Africa. New Development Perspectives in International Trade for Africa*, OAU-IBAR, Nairobi.

4 Morris, R.S. (1999) *OIE Scientific and Technical Review*, **18**, 305–314.

5 Connor, R.J. and VanDen Bossche, P. (2004) African animal trypanosomoses, in *Infectious Diseases of Livestock* (eds J.A.W. Coetzer and R.C. Tustin), Oxford

University Press, Cape Town, pp. 251–296.

6 Dubey, J.P. and Stewart, C.G. (2004) Toxoplasmosis, in *Infectious Diseases of Livestock* (eds J.A.W. Coetzer and R.C. Tustin), Oxford University Press, Cape Town, pp. 337–350.

7 Stewart, C.G. and Penzhorn, B.L. (2004) Coccidiosis, in *Infectious Diseases of Livestock* (eds J.A.W. Coetzer and R.C. Tustin), Oxford University Press, Cape Town, pp. 319–331.

8 Rossiter, P.B. (2004) Rinderpest in *Infectious Diseases of Livestock* (eds J.A.W. Coetzer and R.C. Tustin), Oxford University Press, Cape Town, pp. 629–659.

9 Allsopp, B.A., Bezuidenhout, J.D. and Prozesky, L. (2004) Heartwater, in *Infectious Diseases of Livestock* (eds J.A.W. Coetzer and R.C. Tustin), Oxford University Press, Cape Town, pp. 507–535.

10 Andersen, A.A. (2004) Chlamydiosis, in *Infectious Diseases of Livestock* (eds J.A.W. Coetzer and R.C. Tustin), Oxford

University Press, Cape Town,
pp. 550–564.

11 Vander Walt, M.L. (2004) *Campylobacter jejuni* infection, in *Infectious Diseases of Livestock* (eds J.A.W. Coetzer and R.C. Tustin), Oxford University Press, Cape Town, pp. 1479–1483.

12 Godfroid, J., Bosman, P.P., Herr, S. and Bishop, G.C. (2004) Bovine brucellosis, in *Infectious Diseases of Livestock* (eds J.A.W. Coetzer and R.C. Tustin), Oxford University Press, Cape Town, pp. 1510–1527.

13 Songer, J.G. (1996) *Clinical Microbiology Reviews*, **9**, 216–234.

14 Chalmers, G., Martin, S.W., Hunter, D.B., Prescott, J.F., Weber, L.J. and Boerlin, P. (2008) *Veterinary Microbiology*, **127**, 116–127.

15 Van Immerseel, F., De Buck, J., Pasmans, F., Huyghebaert, G., Haesebrouck, F. and Ducatelle, R. (2004) *Avian Pathology*, **33**, 537–549.

16 Cousins, D.V., Huchzermeyer, H.F.K.A., Griffin, J.F.T., Brückner, G.K., Van Rensburg, I.B.J. and Kriek, N.P.J. (2004) Tuberculosis, in *Infectious Diseases of Livestock* (eds J.A.W. Coetzer and R.C. Tustin), Oxford University Press, Cape Town, pp. 1973–1993.

17 Cosivi, O., Grange, J.M., Daborn, C.J., Raviglione, M.C., Fujikura, T., Cousins, D., Robinson, R.A., Huchzermeyer, H.F.A.K., DeKantor, I. and Meslin, F.-X. (1998) *Emerging Infectious Diseases*, **4**, 59–70.

18 Buergelt, C.D., Bastianello, S.S. and Michel, A.L. (2004) Paratuberculosis, in *Infectious Diseases of Livestock* (eds J.A.W. Coetzer and R.C. Tustin), Oxford University Press, Cape Town, pp. 1994–2008.

19 Jung, K., Kim, Y., Lee, H. and Kim, J-T. (2008) *Veterinary Journal*, in press.

20 Picard, J.A. and Vismer, H.F. (2004) Mycoses, in *Infectious Diseases of Livestock* (eds J.A.W. Coetzer and R.C. Tustin), Oxford University Press, Cape Town, pp. 2095–2136.

21 Uicker, W.C., Doyle, H.A., James, P.M., Langlois, M. and Buchanan, K.L. (2005) *Medical Mycology*, **43**, 27–38.

22 Greene, C.E. and Chandler, F.W. (1998) Candidiasis, in *Infectious Diseases of Dog and Cat*, Saunders, Philadelphia, PA, pp. 414–417.

23 Hammond, J.A., Fielding, D. and Bishop, S.C. (1997) *Veterinary Research Communications*, **21**, 213–228.

24 Norval, R.A.I. and Horak, I.G. (2004) Vectors: ticks, in *Infectious Diseases of Livestock* (eds J.A.W. Coetzer and R.C. Tustin), Oxford University Press, Cape Town, pp. 3–42.

25 Willadsen, P. (1997) *Tropical Animal Health and Production*, **29**, 91S–94.

26 Pieters, L. and Vlietinck, A.J. (2005) *Ethnopharmacol*, **100**, 57–60.

27 Cowan, M.M. (1999) *Clinical Microbiology Reviews*, **12**, 564–582.

28 Hogan, J.C. (1997) *Nature Biotechnology*, **15**, 328–330.

29 Henkel, T., Brunne, R.M., Müller, H. and Reichel, F. (1999) *Angewandte Chemie (International Edition in English)*, **38**, 643–647.

30 Yarnell, E. (2007) Plant chemistry in veterinary medicine: medicinal constituents and their mechanisms of action, in *Veterinary Herbal Medicine* (eds S.G. Wynn and B.J. Fougère), Mosby, St Louis, MO, pp. 159–182.

31 Eloff, J.N. (1998) Conservation of medicinal plants: selecting medicinal plants for research and gene banking. Monographs in Systematic Botany from the Missouri Botanical Garden 71, 209–222, in *Conservation of Plant Genes III: Conservation and Utilization of African Plants* (eds R.P. Adams and J.E. Adams), Missouri Botanical Garden Press, St Louis, MO.

32 Martin, M., Mathias, E. and McCorkle, C.M. (2001) *Ethnoveterinary Medicine: An Annotated Bibliography of Community Animal Healthcare*, ITDG Publishing, London.

33 Vander Merwe, D., Swan, G.E. and Botha, C.J. (2001) *Journal of the South African Veterinary Association*, **72**, 189–196.

34 Schillhorn van Veen, T.W. (1996) Sense or nonsense? Traditional methods of animal disease prevention and control in the African savannah, in *Ethnoveterinary Research and Development* (eds C.M. McCorkle, E. Mathias and T.W. Schillhorn van Veen), Intermediate Technology Publications, London, pp. 25–36.

35 Houghton, P.J., Howes, M.-J., Lee, C.C. and Steventon, G. (2007) *Journal of Ethnopharmacology*, **110**, 391–400.

36 Cos, P., Vlietinck, A.J., VandenBerghe, D. and Maes, L. (2006) *Journal of Ethnopharmacology*, **106**, 290–302.

37 McGaw, L.J., Van der Merwe, D. and Eloff, J.N. (2007) *Veterinary Journal*, **173**, 366–372.

38 Eloff, J.N. (1998) *Planta Medica*, **674**, 711–714.

39 Vlietinck, A.J., Van Hoof, L., Totté, J., Lasure, A., Van den Berghe, D., Rwangabo, P.C. and Mvukiyumwami, J. (1995) *Journal of Ethnopharmacology*, **46**, 31–47.

40 Moloto, M.P. (2004) Isolation and characterization of antibacterial, anthelmintic, antioxidant and cytotoxic compounds present in *Ziziphus mucronata*. MSc thesis, Medical University of South Africa.

41 Bizimenyera, E.S., Swan, G.E., Chikoto, H. and Eloff, J.N. (2005) *Journal of the South African Veterinary Association*, **76**, 54–58.

42 McGaw, L.J., Gehring, R., Katsoulis, L. and Eloff, J.N. (2005) *Veterinary Research*, **72**, 129–134.

43 Kaido, T.L., Veale, D.J.H., Havlik, I. and Rama, D.B.K. (1997) *Journal of Ethnopharmacology*, **55**, 185–191.

44 Luseba, D., Elgorashi, E.E., Ntloedibe, N.T. and VanStaden, J. (2007) *South African Journal of Botany*, **73**, 378–383.

45 Stein, A.C., Sortino, M., Avancini, C., Zacchino, S. and vonPoser, G. (2005) *Journal of Ethnopharmacology*, **99**, 211–214.

46 Stein, A.C., Álvarez, S., Avancini, C., Zacchino, S. and vonPoser, G. (2006) *Journal of Ethnopharmacology*, **107**, 95–98.

47 Masoko, P., Picard, J. and Eloff, J.N. (2005) *Journal of Ethnopharmacology*, **99**, 301–308.

48 Asres, K., Bucar, F., Kartnig, T., Witvrouw, M., Pannecouque, C. and DeClercq, E. (2001) *Phytotherapy Research*, **15**, 62–69.

49 Lamien, C.E., Meda, A., Mans, J., Romito, M., Nacoulma, O.G. and Viljoen, G.J. (2005) *Journal of Ethnopharmacology*, **96**, 249–253.

50 Gebre-Mariam, T., Neubert, R., Schmidt, P.C., Wutzler, P. and Schmidtke, M. (2006) *Journal of Ethnopharmacology*, **104**, 182–187.

51 Kudi, A.C. and Myint, S.H. (1999) *Journal of Ethnopharmacology*, **68**, 289–294.

52 Vlietinck, A.J. and VandenBerghe, D.A. (1991) *Journal of Ethnopharmacology*, **32**, 141–153.

53 Jassim, S.A.A. and Naji, M.A. (2003) *Journal of Applied Microbiology*, **95**, 412–427.

54 Lans, S., Khan, T.E., Curran, M.M. and McCorkle, C.M. (2007) Ethnoveterinary medicine: potential solutions for large-scale problems? in *Veterinary Herbal Medicine* (eds S.G. Wynn and B.J. Fougère), Mosby, St Louis, MO, pp. 17–32.

55 Naidoo, V., Zweygarth, E., Eloff, J.N., and Swan, G.E. (2005) *Veterinary Parasitology*, **130**, 9–13.

56 Naidoo, V., Zweygarth, E. and Swan, G.E. (2006) *Journal of Veterinary Research*, **73**, 175–178.

57 Adewunmi, C.O., Agbedahunsi, J.M., Adebajo, A.C., Aladesanmi, A.J., Murphy, N. and Wando, J. (2001) *Journal of Ethnopharmacology*, **77**, 19–24.

58 Simpkin, K.G. and Coles, G.C. (1981) *Journal of Chemical Technology and BioTechnology*, **31**, 66–69.

59 Rasoanaivo, P. and Ratsimamanga-Urverg, S. (1993) *Biological Evaluation of Plants with Reference to the Malagasy Flora*, Napreca, Madagascar.

60 Geary, T.G. and Thompson, D.P. (2001) *Veterinary Parasitology*, **101**, 371–386.

61 McGaw, L.J., Jäger, A.K. and vanStaden, J. (2000) *Journal of Ethnopharmacology*, **72**, 247–263.

62 McGaw, L.J. and Eloff, J.N. (2005) *South African Journal of Botany*, **71**, 302–306.

63 Coles, G.C., Bauer, C., Borgsteede, F.H.M., Geerts, S., Klei, T.R., Taylor, M.A. and Waller, P.J. (1992) *Veterinary Parasitology*, **44**, 35–44.

64 Coles, G.C., Tritschler, J.P., Giordano, D.J., Laste, N.J. and Schmidt, A.L. (1988) *Research in Veterinary Science*, **45**, 50–53.

65 Koné, W.M., Atindehou, K.K., Dossahoua, T. and Betschart, B. (2005) *Pharmaceutical Biology*, **43**, 72–78.

66 Bizimenyera, E.S., Githiori, J.B., Swan, G.E. and Eloff, J.N. (2006) *Journal of Animal and Veterinary Advances*, **5**, 608–614.

67 Bizimenyera, E.S., Githiori, J.B., Eloff, J.N. and Swan, G.E. (2006) *Veterinary Parasitology*, **142**, 336–343.

68 Kahiya, C., Mukaratirwa, S. and Thamsborg, S.M. (2003) *Veterinary Parasitology*, **115**, 265–274.

69 Iqbal, Z., Lateef, M., Akhtar, M.S., Ghayur, M.N. and Gilani, A.H. (2006) *Journal of Ethnopharmacology*, **106**, 285–287.

70 Jabbar, A., Zaman, M.A., Iqbal, Z., Yaseen, M. and Shamim, A. (2007) *Journal of Ethnopharmacology*, **114**, 86–91.

71 Nchu, F. (2004) Developing methods for the screening of ethnoveterinary plants for tick control. MSc thesis, Medical University of Southern Africa.

72 Thembo, M.K. (2006) The anti-tick effects of *Senna italica* subsp. *arachoides* extracts on adults of *Hyalomma marginatum rufipes*. MSc thesis, University of Limpopo.

73 Mkolo, N.M. (2008) Anti-tick properties of some of the traditionally used plant-based products in South Africa. MSc thesis, University of Limpopo.

74 Zorloni, A. (2007) Evaluation of plants used for the control of animal ectoparasitoses in southern Ethiopia (Oromiya and Somali regions). MSc thesis, Phytomedicine Programme, University of Pretoria.

75 Allsopp, B.A., Babiuk, L.A. and Babiuk, S.L. (2004) Vaccination: an approach to the control of infectious diseases, in *Infectious Diseases of Livestock* (eds J.A.W. Coetzer and R.C. Tustin), Oxford University Press, Cape Town, pp. 239–247.

76 Van Poucke, C., Dumoulin, F., Yakkundi, S., Situ, C., Elliott, C.T., Grutters, E.M., Verheijen, R., Schilt, R., Eriksson, S. and Van Peteghem, C. (2006) *Analytica Chimica Acta*, **557**, 204–210.

77 Situ, C. and Elliott, C.T. (2005) *Analytica Chimica Acta*, **529**, 89–96.

78 Singer, R.S. (2003) *Lancet Infectious Diseases*, **3**, 47–48.

79 Garofalo, C., Vignaroli, C., Zandri, G., Aquilanti, L., Bordoni, D., Osimani, A., Clementi, F. and Biavasco, F. (2007) *International Journal of Food Microbiology*, **113**, 75–83.

80 Swann, M. (1969) *Report of the Joint Committee on the Use of Antibiotics in Animal Husbandry and Veterinary Medicine, Cmnd 4190*, HMSO, London.

81 The Council of the European Union (1998) Council Regulation (EC) No 2821/98. *Official Journal of the European Communities*, L351/4.

82 The Council of the European Union (1998) Commission Regulation (EC) No 2788/98. *Official Journal of the European Communities*, L347/31.

83 Phillips, I., Casewell, M., Cox, T., DeGroot, B., Friis, C., Jones, R., Nightingale, C., Preston, R. and Waddell, J. (2004) *Journal of Antimicrobial Chemotherapy*, **53**, 28–52.

84 Phillips, I. (2007) *International Journal of Antimicrobial Agents*, **30**, 101–107.

85 Pantosti, A., DelGrosso, M., Tagliabue, S., Macri, A. and Caprioli, A. (1999) *Lancet*, **354**, 741–742.

86 Klare, I., Badstubner, D., Konstabel, C., Bohme, G., Claus, H. and Witte, W. (1999) *Microbial Drug Resistance (Larchmont, NY)*, **5**, 45–52.

87 Aarestrup, F.M., Seyfarth, A.M., Emborg, H.D., Pedersen, K., Hendriksen, R.S. and Bager, F. (2001) *Antimicrobial Agents and Chemotherapy*, **45**, 2054–2059.

88 Bager, F., Madsen, M., Christensen, J. and Aarestrup, F.M. (1997) *Preventive Veterinary Medicine*, **31**, 95–112.

89 Aarestrup, F.M. (1995) *Microbial Drug Resistance (Larchmont, NY)*, **1**, 255–257.

90 Klare, I., Heier, H., Claus, H., Reissbrodt, R. and Witte, W. (1995) *FEMS Microbiology Letters*, **125**, 165–172.

91 Laine, T., Yliaho, M., Myllys, V., Pohjanvirta, T., Fossi, M. and Anttila, M. (2004) *Preventive Veterinary Medicine*, **66**, 163–174.

92 Dahiya, J.P., Wilkie, D.C., VanKessel, A.G. and Drew, M.D. (2006) *Animal Feed Science and Technology*, **129**, 60–88.

93 Besra, S.E., Gomes, A., Chaudhury, L., Vedasiromoni, J.R. and Ganguly, D.K. (2002) *Phytotherapy Research*, **16**, 529–533.

94 Kalemba, D. and Kunicka, A. (2003) *Current Medicinal Chemistry*, **10**, 813–819.

95 Burt, S. (2004) *International Journal of Food Microbiology*, **94**, 223–253.

96 Rios, J.L. and Recio, M.C. (2005) *Journal of Ethnopharmacology*, **100**, 80–84.

97 Rosato, A., Vitali, C., DeLaurentis, N., Armenise, D. and Milillo, M.A. (2007) *Phytomedicine*, **14**, 727–732.

98 Piddock, L.J. (1996) *Journal of Antimicrobial Chemotherapy*, **38**, 1–3.

99 Wegener, H.C. (2003) *Current Opinion in Microbiology*, **6**, 439–445.

100 Durmic, Z., McSweeney, C.S., Kemp, G.W., Hutton, P., Wallace, R.J. and Vercoe, P.E. (2008) *Animal Feed Science and Technology*, **145**, 271–284.

101 McIntosh, F.M., Williams, P., Losa, R., Wallace, R.J., Beever, D.A. and Newbold, C.J. (2003) *Applied and Environmental Microbiology*, **69**, 5011–5014.

102 Cardozo, P.W., Calsamiglia, S., Ferret, A. and Kamel, C. (2004) *Journal of Animal Science*, **82**, 3230–3236.

103 Mohammed, N., Ajisaka, N., Lila, Z.A., Hara, K., Mikuni, K., Kanda, S. and Itabashi, H. (2004) *Journal of Animal Science*, **82**, 1839–1846.

104 Wallace, R.J. (2004) *Proceedings of the Nutrition Society*, **63**, 621–629.

105 Carulla, J.E., Kreuzer, M., Machmuller, A. and Hess, H.D. (2005) *Australian Journal of Agricultural Research*, **56**, 961–970.

106 Busquet, M., Calsamiglia, S., Ferret, A. and Kamel, C. (2006) *Journal of Dairy Science*, **89**, 761–771.

107 Hart, K.J., Yáñez-Ruiz, D.R., Duval, S.M., McEwan, N.R. and Newbold, C.J. (2008) *Animal Feed Science and Technology*, in press.

108 Losa, R. and Kohler, B. (2001) Prevention of colonization of *Clostridium perfringens* in broilers intestine by essential oils, in *13th European Symposium on Poultry Nutrition*, WPSA, Blankenberge.

109 Mitsch, P., Zitterl-Eglseer, K., Kohler, B., Gabler, C., Losa, R. and Zimpernik, I. (2004) *Poultry Science*, **83**, 669–675.

110 Lien, T.F., Horng, Y.M. and Wu, C.P. (2007) *Livestock Production Science*, **107**, 97–102.

111 Francis, G., Makkar, H.P.S. and Becker, K. (2005) *Animal Feed Science and Technology*, **121**, 147–157.

112 Cheeke, P.R. (1999) Actual and potential applications of *Yucca schidigera* and *Quillaja saponaria* saponins in human and animal nutrition, in *Proceedings of the American Society of Animal Science*, Indianapolis. http://www.asas.org/jas/symposia/proceedings/0909.pdf. Accessed December 2007.

113 Francis, G., Kerem, Z., Makkar, H.P.S. and Becker, K. (2002) *British Journal of Nutrition*, **88**, 587–605.

114 Naidoo, V., McGaw, L.J., Bisschop, S.P.R., Duncan, N. and Eloff, J.N. (2008) *Veterinary Parasitolology*, **153**, 2814–2819.

115 Chikoto, H. (2006) Development of a product derived from plant extracts to replace antibiotic feed additives used in poultry production. PhD thesis, Phytomedicine Programme, University of Pretoria.

116 Angeh, J.E. (2005) Isolation and characterization of anti-bacterial compounds present in members of *Combretum* section Hypocrateropsis. PhD thesis, Phytomedicine Programme, University of Pretoria.

9
Honey: Antimicrobial Actions and Role in Disease Management

Peter Molan

Abstract

The ancient treatment of dressing infected wounds with honey is rapidly becoming re-established in professional medicine, especially where wounds are infected with antibiotic-resistant bacteria. This is because of the demonstrated sensitivity of such bacteria to the antibacterial activity of honey, which is not influenced by whether or not strains are resistant to antibiotics. Honey has been found to have a very broad spectrum of activity, but its potency of antibacterial activity can vary greatly. In most honeys the antibacterial activity is due to enzymatically produced hydrogen peroxide and thus the potency of its antibacterial activity can be decreased by catalase present in an open wound. Manuka honey has an antibacterial component derived from the plant source. Manuka honey with a quality-assured level of antibacterial activity is being used by companies marketing honey products for wound care that are registered with the medical regulatory authorities in various countries. Such honey can be diluted 10-fold or more and still completely inhibit the usual wound-infecting species. There is a large amount of clinical evidence for the effectiveness of honey in clearing infection in wounds, and some clinical evidence of its effectiveness in treating other infections. Although the antibacterial potency of honey is insufficient to allow its use systemically, there are various clinical applications besides wound care in which it is used topically or where it does not get excessively diluted, such as for treatment of gastritis, enteritis, gingivitis, ophthalmological infections and bronchial infections. In most of these applications the anti-inflammatory activity of honey is of additional benefit in decreasing the inflammation resulting from infection. Additional clinical research is needed to provide better evidence of the effectiveness of honey in these therapeutic applications of honey.

New Strategies Combating Bacterial Infection. Edited by Iqbal Ahmad and Farrukh Aqil
Copyright © 2009 WILEY-VCH Verlag GmbH & Co. KGaA, Weinheim
ISBN: 978-3-527-32206-0

9.1
Introduction

An Editorial in the *Journal of the Royal Society of Medicine* in 1989 [1], entitled 'Honey – a remedy rediscovered', expressed the view that 'The therapeutic potential of uncontaminated pure honey is grossly under-utilized' and that 'The time has now come for conventional medicine to lift the blinds off this "traditional remedy" and give it its due recognition'. This Editorial noted the many papers being published reporting good results when honey was used as a dressing on infected wounds and when used in an electrolyte solution in a clinical trial on treatment of diarrhea. In many of the published reports on treatment of infected wounds honey was used where antibiotics were failing to clear the infection. The rapidly increasing number of papers published on the use of honey on wounds in more recent years is probably a reflection of the escalation of the problem of bacteria developing resistance to antibiotics. It is probably also a reflection of honey becoming available as various registered sterile wound-care products, especially ones designed for ease of use [2]. This chapter covers the nature and spectrum of the antimicrobial activity of honey, the evidence for its clinical effectiveness in clearing infection, and the other beneficial therapeutic activities that are seen when honey is used as a topical antimicrobial agent.

9.1.1
History

Honey is the oldest medicine known and in many ancient races of people was prescribed by physicians for a wide variety of ailments [3]. The ancient Egyptians, Assyrians, Chinese, Greeks and Romans all used honey, in combination with other herbs and on its own, to treat wounds and diseases of the gut [4]. Its use for the treatment of diarrhea was recommended by the Muslim prophet Mohammed [5]. In Ancient Greece, Aristotle [6] wrote of honey being a salve for wounds and sore eyes and Dioscorides around 50 AD wrote of honey being 'good for sunburn and spots on the face' and 'for all rotten and hollow ulcers' [7]. He also wrote that 'honey heals inflammation of the throat and tonsils, and cures coughs'.

The use of honey as a therapeutic agent has continued into present-day folk medicine. In India, lotus honey is used to treat eye diseases [8]. Other examples of current-day usage of honey in folk medicine are: as a traditional therapy for infected leg ulcers in Ghana [9], as a traditional therapy for earache in Nigeria [10], and as a traditional therapy in Mali for the topical treatment of measles and in the eyes in measles to prevent corneal scarring [11]. Honey also has a traditional folklore usage for the treatment of gastric ulcers [12] and its ancient usage to treat sore throats has continued into the traditional medicine of modern times [13].

However, many medical professionals are of the opinion that honey has no place in modern medicine. An Editorial in *Archives of Internal Medicine* assigned honey to the category of 'worthless but harmless substances' [14] and Editorials in other medical journals have clearly shown a lack of awareness of the research that has demonstrated the rational explanations for the therapeutic effects of honey [15, 16]. Many physicians are not even aware that honey has an antibacterial activity beyond the osmotic effect of its sugar content [16–23], yet there have been numerous publications over the past 70 years reporting that there are other components of honey that have a much more potent antimicrobial effect.

The first indication that the antimicrobial activity of honey was not just an osmotic effect was in a report by Sackett [24] who observed that the antibacterial potency was increased by limited dilution of honey – an observation that was hard to explain. More intensive study two decades later by Dold *et al.* [25] led to the discovery of an antibacterial factor which they termed 'inhibine' – a term widely used in the literature for the next 26 years until the antibacterial factor was identified as hydrogen peroxide by White *et al.* [26]. The term 'inhibine number' was also used, this being the number of dilution steps a honey could be subjected to and still have antibacterial activity. Subsequent studies have found that where a range of honeys has been tested against a single species of microorganism the minimum inhibitory concentration (MIC) of honey varied widely: 25–0.25 % [27], greater than 50–1.5 [28], 20–0.6 [29] and 50–1.5% (v/v) [30].

This discovery by microbiologists studying honey that different honeys varied markedly in their antimicrobial potency is in effect probably a rediscovery of ancient wisdom. The ancient physicians who prescribed honey for various ailments would have had no knowledge of the principles involved in its medicinal action, just an empirical knowledge gained from its effective usage. However, they were aware that some honeys were better others for medical usage: Dioscorides around 50 AD stated that a pale yellow honey from Attica was the best, being 'good for all rotten and hollow ulcers' [7]; and Aristotle [6], discussing differences in honeys, referred to pale honey being 'good as a salve for sore eyes and wounds'. Present-day folk medicine also recognizes differences in honeys: the strawberry-tree honey of Sardinia is valued for its therapeutic properties [31]; in India, lotus honey is said to be a panacea for eye diseases [8]; honey from the Jirdin valley of Yemen is highly valued in Dubai for its therapeutic properties [32]; manuka honey has a long-standing reputation in New Zealand folklore for its antiseptic properties (K. Simpson, personal communication). This knowledge that honey is not a 'generic medicine' but needs appropriate selection for therapeutic use is not widespread, so until recently most clinical treatment and microbiological studies have been done with honey with an unknown level of antimicrobial activity.

9.2
Nature of the Antimicrobial Activity of Honey

9.2.1
High Osmolarity

The osmolarity of honey alone is sufficient to prevent microbial growth. Granulated honey is a saturated solution of sugars and clear honey is a supersaturated solution.

Although honey with a high water content can spoil because some osmophilic yeasts can live in it, no fermentation occurs if the water content is below 17.1% [33]. The water content of honey is usually 15–21% by weight [34]. Of the solids in honey, 84% is comprised of the monosaccharides fructose and glucose [35]. The strong interaction of these sugar molecules with water molecules leaves very few of the water molecules available for microorganisms. The water molecules that are 'free' water are measured as the water activity (a_w). The mean values of a_w for honey have been reported as 0.562 and 0.589 [36], 0.572 and 0.607 [37], and 0.62 [38]. Many species of bacteria have their growth completely inhibited by the a_w being in the range 0.94–0.99 [39, 40]. Calculated on the basis of the concentration being proportional to $-\log a_w$, these inhibitory values of a_w correspond to solutions of a typical honey (with a_w of 0.6) of concentrations from 12 down to 2% (v/v) [40]. Fungi are generally much more tolerant than bacteria of low a_w [39]. *Staphylococcus aureus* has an exceptionally high tolerance of low a_w: for complete inhibition of growth of *S. aureus* the a_w has to be lowered below 0.86 [39, 41, 42], which would be a typical honey at 29% (v/v). There have been many reports of granulated sugar being used as a wound dressing [43], but it has been reported that infection is not cleared or new infection becomes established in cases where urine or heavy exudate from wounds dilutes the sugar [44]. With honey, the presence of other antimicrobial factors allows it to be inhibitory even when diluted down to an osmolarity that will freely allow growth of microorganisms. With a honey that has a median level of antibacterial activity it is possible to have it diluted to as low a concentration as 2% (v/v) and still have it completely inhibit the growth of *S. aureus* [45]. In a study of methicillin-resistant *S. aureus* (MRSA) [46] with honeys that had near median levels of antibacterial potency it was found that whereas the MIC of the honeys for any of the strains was below 4% (v/v), the MIC for a syrup of a mixture of sugars at the concentrations that occur in honey was above 30% (v/v). Similar studies with coagulase-negative staphylococci [47], *Burkholderia cepacia* [48], enterococci [46] and *Pseudomonas* spp. [49] found MIC values for the honeys of 3–5, 2.1–5, 3.83–9.66 and 4.33–9.0%, respectively, whereas the MIC for the syrup was 27.5–31.7, 17.5–22, 27.7–29.8 and 17–22%, respectively.

9.2.2
Acidity

The antibacterial activity of honey is partly due also to acidity. Honey is characteristically of a pH between 3.2 and 4.5 [34]. This acidity is due primarily to honey containing 0.23–0.98% (1.8–7.5 mmol/kg) gluconolactone/gluconic acid [35], which is formed by the action of the enzyme glucose oxidase which bees add to the nectar they collect to make honey. However, no correlation has been found between antibacterial activity and the pH of the honey when this has been studied [37, 50–54]. This may be because of different degrees of buffering in different honeys: the pH does not necessarily indicate the titratable acidity, but it is the titratable acidity that determines the final pH when honey is diluted by a neutralizing solution. With such a low concentration of acid in honey there is not much lowering of the pH when honey is added to culture media or serum. In work with *S. aureus* no inhibition was seen with gluconic acid added to nutrient broth at levels up to 0.25% [55]. However, in a study with *Corynebacterium diphtheriae* the MIC of the honey used was found to be 4.5%, but was 10% when the honey was neutralized [56]. The pH of the nutrient broth containing honey at 4.5% was measured and found to be 6.2. The acidity of honey was found to be of effect in the inhibition of *Bacillus cereus* also: inhibition by 50% honey in an agar diffusion assay was lost if phosphate buffer was added to bring the pH to 6.1–6.5 [57]. The low pH of honey that has not been too much diluted by a neutralizing medium would be at least partially inhibitory to many animal pathogens. The optimum pH for growth of these pathogens is normally between 7.2 and 7.4, although the minimum pH values for growth of some common wound-infecting species are: *Escherichia coli*, 4.3; *Salmonella* species, 4.0; *Pseudomonas aeruginosa*, 4.4; *Streptococcus pyogenes*, 4.5 [58]. The concentration of bicarbonate (the principle buffering ion) in the extracellular fluid of the body is 25 mmol/l, so the dilution of a honey containing a median level of gluconolactone/gluconic acid with an equal volume of extracellular fluid would raise the pH of the honey to 6.8. This means that where honey gets diluted by body fluid the acidity of honey makes a minor contribution to antibacterial activity and it is the other antibacterial components that are primarily responsible for control of infection when honey is used therapeutically.

9.2.3
Hydrogen Peroxide

Hydrogen peroxide is the major antimicrobial factor in most honeys. Adcock [59] found that the antibacterial activity of honey could be removed by the addition of catalase (which catalyzes the destruction of hydrogen peroxide), and White *et al.* [26] demonstrated a direct relationship between the hydrogen peroxide produced and the 'inhibine number' of various honeys. The hydrogen peroxide in honey is produced by the action of the enzyme glucose oxidase, which is secreted into collected nectar from the hypopharyngeal gland of the bees. A similar type of antimicrobial system was discovered when Fleming's work on the antibacterial properties of *Penicillium notatum* was followed up by Coulthard *et al.* [55]. They traced the cause of the erratic results they were obtaining to the potent activity of a second factor, notatin, which was present in addition to penicillin. They found notatin to be a combination of the enzyme glucose oxidase with glucose, and showed the activity of notatin to be due to the hydrogen peroxide produced. Oxygen needs to be available for the reaction:

β-D-glucose + oxygen → δ-gluconolactone + hydrogen peroxide

This means that that this antimicrobial activity from honey can only be of use under aerobic conditions. The production of hydrogen peroxide during the ripening of honey serves to sterilize the honey stored in the comb, but undiluted honey has a negligible level of hydrogen peroxide [26, 60, 61].

Glucose oxidase is practically inactive in full-strength honey, it giving rise to hydrogen peroxide only when the honey is diluted [26]. One explanation for this is that the activity of the enzyme is suppressed by the low pH in ripened honey. The enzyme has an optimum pH of 6.1, with a good activity from pH 5.5 to 8, but the activity drops off sharply below pH 5.5 to near 0 at pH 4 [62]. It is not a case of substrate inhibition of the enzyme, as glucose concentrations beyond those occurring in honey do not suppress the rate of reaction, the optimum substrate concentration for the glucose oxidase in honey being exceptionally high (1.5 mol/l) [62]. This high optimum concentration is well suited to the enzyme's functioning in ripening honey (the concentration of glucose in ripened honey being around 2 mol/l), but will markedly limit the rate of production of hydrogen peroxide in well-diluted honey. The need to dilute honey to get the enzyme active is most likely because of the low water activity of honey, as it is known that enzymes need a sufficiently high water activity to be active [63]. As honey is diluted the activity of glucose oxidase increases to a peak at a concentration around 30–50% (v/v) honey as the water activity is increased, then falls again as the enzyme and substrate concentrations are decreased by further dilution [64]. Honey solutions were found to maintain at least half of the maximum rate of generation of hydrogen peroxide over a wide range of dilution that is concentrations of honey from approximately 15 to 67% (v/v) [64].

Inhibition of the enzyme by high concentrations of honey is not caused by either of the products of the reaction. In a system buffered to prevent inhibition of the enzyme by low pH, no inhibition at all was seen with 10 mmol/l gluconic acid or gluconolactone [62]. Nor does hydrogen peroxide cause inhibition at the levels that are produced in honey [65]. However, studies with honey [26] and with the isolated enzyme [65] found the rate of reaction to be falling off over a short period of time. Adding ascorbic acid to remove the hydrogen peroxide as it was produced gave a fivefold increase in the rate of reaction [65]. Bang *et al.* [64] found that when 50% solutions of honey were incubated, hydrogen peroxide accumulated to a peak level then the concentration of hydrogen peroxide dropped, it becoming zero after 24–48 h. This is probably the result of damage to the enzyme by accumulated hydrogen peroxide, as it has been reported that addition of 68 mmol/l hydrogen peroxide to glucose oxidase isolated from honey caused a significant decline in the enzyme's rate of reaction after 20 min [65]. Whilst this means that honey does not have prolonged antimicrobial activity once it has been diluted, it does have the advantage of preventing hydrogen peroxide from accumulating to levels that are harmful to body tissues. The maximum concentration of hydrogen peroxide achieved when a 50% (v/v) solution of a honey with a high level of antibacterial activity was incubated was found to be 3.65 mmol/l [64], which is 242-fold lower than the 3% (882 mmol/l) solution of hydrogen peroxide typically used as an antiseptic [66]. The use of hydrogen peroxide as an antiseptic has been discouraged because it is cytotoxic [67], but at the low levels that form in honey this is not a problem. Hydrogen peroxide has gone out of common use as an antiseptic also because it causes inflammation, but the antioxidant content of honey would help prevent inflammation being caused, as it has been found that it oxidative species formed from hydrogen peroxide, rather than hydrogen peroxide itself, that are responsible for the activation of the transcription factor NF-κB involved in the inflammatory response in leukocytes [68]. This activation can be prevented by antioxidants [69].

Although only low levels of hydrogen peroxide accumulate in diluted honey, this is still an effective antimicrobial system because of its continuous production. Hydrogen peroxide has been found to be more effective when supplied by continuous generation by glucose oxidase than when added as a bolus [70]. *E. coli* exposed to a constantly replenished stream of hydrogen peroxide had their growth inhibited by as little as 0.02–0.05 mmol/l hydrogen peroxide, a concentration that was not damaging to fibroblast cells from human skin [71]. Rates of production of hydrogen peroxide in diluted honey that have been reported are: 2.2–5.6 mmol/l/h for 30% (v/v) solutions of eight honeys (three of them blends of 20–30 samples of individual honeys) [64], 0–2.12 mmol/l/h for 14% (v/v) solutions of 90 samples of honey [61], 0–4.8 mmol/l/h for 20% (v/v) solutions of 37 samples of honey [72] and 0.10–0.58 mmol/l/h for 36% (v/v) solutions of 25 samples of honey [59].

Quite low levels of hydrogen peroxide are required for antibacterial activity. It has been reported that *S. aureus* failed to grow in 24 h in nutrient broth containing hydrogen peroxide at 0.29 mmol/l, but grew at 0.15 mmol/l [55]. In other work with *S. aureus* the 20% inhibition of growth over an incubation period of 16 h that was observed corresponded with an accumulation of 0.12 mmol/l hydrogen peroxide from the glucose oxidase–glucose system used to generate it [72]. Others found growth of only one colony of *S. aureus* on a nutrient agar plate containing 0.29 mmol/l hydrogen peroxide and none at the next level tested, 0.5 mmol/l [26].

The level of hydrogen peroxide achieved in diluted honey varies from sample to sample. It can be related to the floral source, as components from some floral sources can affect the enzyme activity that gives rise to hydrogen peroxide and others affect the destruction of hydrogen peroxide. The level of hydrogen peroxide achieved is the result of there being a dynamic equilibrium between the rate of its production and the rate of its destruction [61]. Hydrogen peroxide has been found to rapidly disappear when added to dilute honey [61]. Catalase, an enzyme that destroys hydrogen peroxide, has been shown to be present in honey [73], it coming from the pollen and nectar of certain plants, more from the nectar [74]. Honeys from some floral sources have been found to have very high levels of catalase activity and these accumulate low levels of hydrogen peroxide, whereas those with low levels of catalase activity accumulate high levels of hydrogen peroxide [28, 74]. However, it has been found that hydrogen peroxide disappears when added to honey even if honey is boiled beforehand to inactivate catalase, indicating that loss though chemical reaction is involved as well as through enzymic destruction [26]. Variation between honeys occurs also in the rate of production of hydrogen peroxide. Extraction of honey from the combs and processing to remove wax and other particles requires the honey to be heated. Very large differences have been found between honeys from different floral sources in the thermal stability of the glucose oxidase in them [75] and in the sensitivity of glucose oxidase to denaturation by light [76]. Thus, the rate of production of hydrogen peroxide will depend on the exposure of honey to heat and light, particularly daylight and the light from fluorescent tubes [61], in its processing and storage, as well as it depending on the floral source of the honey.

9.2.4
Additional Antibacterial Factors

In some honeys there are antimicrobial factors additional to osmolarity, acidity and production of hydrogen peroxide. Reports of antibacterial activity in honey that is stable to heating well in excess of the variation in stability of glucose oxidase indicates that hydrogen peroxide is not the only antibacterial factor in diluted honey. In a study of some Jamaican honeys, the activity of the two most active honeys was not reduced by steam sterilizing, whereas in the others it was reduced or destroyed [77]. Conifer honeydew honey, with exceptionally high activity, was reported to contain a heat-stable as well as a heat-sensitive antibacterial factor [50]. More direct evidence for the existence of antibacterial factors additional to hydrogen peroxide is seen in reports of activity persisting in honeys treated with catalase to remove the hydrogen peroxide activity [57, 59, 72, 78–82]. In one of these studies where substantial antibacterial activity remained it was shown by direct assay of the level of hydrogen peroxide present that the catalase had been completely effective [59]. Lysozyme has been identified in honey, usually occurring at a level of 5–10 µg/ml, occasionally at 35–100 µg/ml if the honey is freshly extracted from the comb, but at much lower levels in older samples [83]. The flavonoid pinocembrin has been identified as an antibacterial component of honey, but at a level only 1–2% of what would be required to account for the observed activity not due to hydrogen peroxide [72]. Some phenolic acid components of manuka (*Leptospermum scoparium*) honey with antibacterial activity have been identified: 3,5-dimethoxy-4-hydroxybenzoic acid (syringic acid), methyl 3,5-dimethoxy-4-hydroxybenzoate (methyl syringate) and 3,4,5-trimethoxybenzoic acid [84], but these were later found to account for no more than 4% of the antibacterial activity of diluted honey not due to hydrogen peroxide [85]. In viper's bugloss (*Echium vulgare*) honey this type of activity was accounted for entirely by its content of 1,4-dihydroxybenzene [85], but the activity was very low compared with that of manuka honey [78].

9.2.5
Manuka Honey

Manuka honey, produced in large quantities in New Zealand, is very unusual in having a high level of antibacterial activity after addition of catalase to destroy hydrogen peroxide, sufficient catalase being added to remove hydrogen peroxide at a level 100 times higher than that with activity equivalent to the most active honey in the study [80]. The possibility was investigated that the activity remaining in manuka honey after the addition of catalase was the result of a component of this honey inhibiting the enzyme, but it was shown that inhibition did not occur [78]. This type of

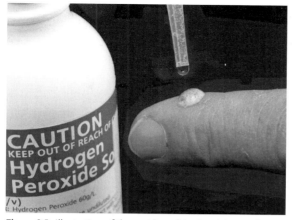

Figure 9.1 Illustration of the rapidity of the breakdown of hydrogen peroxide (to oxygen and water) when it is exposed to the catalase activity in a small drop of blood on a pricked finger.

antibacterial activity is significant for clinical applications because all cells of the body contain the enzyme catalase, so at least part of the antibacterial activity of other types of honey will be destroyed if the honey comes in contact with cells. As hydrogen peroxide freely diffuses across cell membranes this breakdown can be quite rapid, as is illustrated in Figure 9.1, where a drop of 3% hydrogen peroxide solution has been placed on a pricked finger. There will not be complete breakdown of hydrogen peroxide because the enzyme will act more slowly as the concentration of its substrate decreases, so eventually there will be an equilibrium level reached where the rate of production equals the rate of destruction. Thus, although both types of antibacterial activity in assays in agar or broth may appear to be of similar potency, where the honey is exposed to catalase activity in or on the body the activity of other honeys will be less than that of manuka honey. Also, the unusual antibacterial activity in manuka honey is fully effective in undiluted honey, whereas other types of honey need dilution before glucose oxidase becomes active and production of hydrogen peroxide begins. Figure 9.2 shows an illustration of this difference, where wound dressings were prepared from a manuka honey and a clover honey, each with the same level of antibacterial activity when compared as 25% solutions in an agar well diffusion assay. Placing the pieces of dressing against the cut edge of the agar gel seeded with *S. aureus* simulates the situation where honey dressings are placed on an infected open wound. It can be seen that antibacterial activity has diffused deeply into the agar from the manuka honey, but very little antibacterial activity has been produced in the clover honey as there has been little dilution to activate the glucose oxidase enzyme to produce hydrogen peroxide. It is for these reasons that companies marketing honey products for wound care that are registered with the medical regulatory authorities in Australia, Canada, the European Union, Hong Kong, New Zealand and the United States have chosen to use manuka honey or the equivalent honey produced from other *Leptospermum* species in Australia.

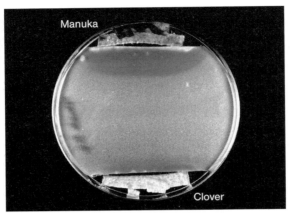

Figure 9.2 Model of honey dressings on an infected wound. One dressing pad was impregnated with manuka honey, the other with clover honey.

Each honey had the same level of antibacterial activity in an agar well diffusion assay. They were placed against the cut edge at each and of the agar which had been seeded with *S. aureus*, then the plate was incubated at 37 °C for 18 h.

9.3
Spectrum and Potency of the Antimicrobial Activity of Honey

There have been many reports published on the sensitivity of a wide range of species of bacteria and fungi to honey. However, in much of this work only a single concentration of honey has been used. Sometimes this concentration has been high enough for the inhibition of microbial growth that has been observed to have probably been due just to the osmotic effect of the honey. Also, with much of the published research, even where MIC values for honey are reported the honey has been arbitrarily chosen, so its antimicrobial potency relative to that of other honey is not known. As mentioned above, the MIC has been found to vary up to 100-fold between different honeys, which means that much of the published data is not a useful indication of the results that could be expected with other honey if the use of honey for infection control is being considered. A review of all the research on the antimicrobial activity of honey published up to 1992 is available [86, 87] for anyone wanting to see the scope of this.

In the present chapter only the findings reported which give information useful for making clinical decisions will be covered. Thus, data are presented which are either the range of MIC values found where numerous different honeys were tested or are the MIC values where the honey used in the research was selected to have a near-median level of antibacterial activity. The antibacterial potency of these selected honeys has been rated against phenol as a standard antibacterial substance, using an agar well diffusion assay with a standard strain of *S. aureus* [78]. Many companies

Table 9.1 MICs of honeys, for various species of bacteria, reported in studies where numerous different honeys were used.

No. of samples	Species	Mean MIC (% v/v)	SD	Reference
60	Staphylococcus aureus	21.6	28.1	[29]
22	Salmonella typhi H901	4.2	6.1	[27]
	Escherichia coli	4.1	5.5	
	Shigella flexneri Type I	1.4	2.1	
	Proteus morganii	6.0	5.7	
	Staphylococcus aureus Oxford 209	7.9	9.0	
	Bacillus anthracis	10.2	10.7	
27	Staphylococcus aureus	5.6	5.0	[30]
	Streptococcus pyogenes Group A	6.1	5.2	
	Streptococcus α-haemolyticus	10.6	6.2	
	Corynebacterium diphtheriae	17.7	10.2	
	Escherichia coli	40.7	20.4	
	Proteus vulgaris	57.0	10.7	
	Pseudomonas pyocyanea	28.3	19.6	
	Klebsiella pneumoniae	30.4	19.9	
	Shigella flexneri	21.4	17.5	
	Bacillus anthracis	16.6	19.6	
	Bacillus mesentericus	27.4	24.1	
	Monilia albicans	60.0	0.0	
18	Staphylococcus aureus ATCC 6538	12.7	1.5	[88]
42	Bacillus subtilis ATCC 6633	27.8	10.8	[89]
	Escherichia coli ATCC 14948	22.8	11.2	

marketing honey for use as an antibacterial agent are now rating the activity of their honeys in the same way, which allows prediction of their likely clinical effectiveness by reference to the published research findings. The findings from research with numerous samples of honey are summarized in Table 9.1 and those from research using standardized honey are summarized in Table 9.2. The data in Table 9.1 will be less representative than that in Table 9.2 of honey in general, as the studies that are in Table 9.2 have selected honeys that have antimicrobial potency that is near the median level found for honey in a survey of 345 samples of honey, from 26 different floral sources [78]. In studies with smaller numbers of samples the activity of the honeys used may have been unusually low or unusually high.

The failure to take into account the large variance in antibacterial potency of different honeys may explain some of the large differences in results reported between hospitals using honey in similar ways. Some have reported rapid clearance of infection in a range of different types of wound, with wounds all becoming sterile in 3–6 [98, 99], 7 [100–102] or 7–10 days [103]. Others have reported bacteria still present in wounds after 2 [104, 105], 3 [106–108] and 5 weeks [109].

Where antibiotic-resistant strains of bacteria have been studied, their sensitivity to honey has been found to be essentially the same as that of the antibiotic-sensitive strains of the same species [46, 47, 94]. This and the very broad spectrum of

Table 9.2 MICs of honeys for various species of bacteria and fungi, reported in studies where honeys with standardized antibacterial activity were used.

No. of strains	Species of microorganism	Mean MIC (% v/v)	SD	Reference
Manuka honey: nonperoxide activity equivalent to 13.2% phenol				[90]
1	*Escherichia coli*	3.7		
1	*Proteus mirabilis*	7.3		
1	*Pseudomonas aeruginosa*	10.8		
1	*Salmonella typhimurium*	6		
1	*Serratia marcescens*	6.3		
1	*Staphylococcus aureus*	1.8		
1	*Streptococcus pyogenes*	3.6		
Rewarewa honey: hydrogen peroxide activity equivalent to 21.5% phenol				[90]
1	*Escherichia coli*	7.1		
1	*Proteus mirabilis*	3.3		
1	*Pseudomonas aeruginosa*	6.8		
1	*Salmonella typhimurium*	4.1		
1	*Serratia marcescens*	4.7		
1	*Staphylococcus aureus*	4.9		
1	*Streptococcus pyogenes*	2.6		
Manuka honey: nonperoxide activity equivalent to 13.2% phenol				[91]
7	*Helicobacter pylori*	5	0	
Manuka honey: nonperoxide activity equivalent to 13.2% phenol				[92]
1	*Epidermophyton floccosum*	10		
1	*Microsporum canis*	25		
1	*Microsporum gypseum*	50		
1	*Trichophyton mentagrophytes* var. *interdigitale*	25		
1	*Trichophyton mentagrophytes* var. *mentagrophytes*	20		
1	*Trichophyton rubrum*	10		
1	*Trichophyton tonsurans*	25		
Pasture honey: hydrogen peroxide activity equivalent to 14.8% phenol				[92]
1	*Epidermophyton floccosum*	10		
1	*Microsporum canis*	15		
1	*Microsporum gypseum*	20		
1	*Trichophyton mentagrophytes* var. *interdigitale*	15		
1	*Trichophyton mentagrophytes* var. *mentagrophytes*	15		
1	*Trichophyton rubrum*	5		
1	*Trichophyton tonsurans*	20		
Manuka honey: nonperoxide activity equivalent to 13.2% phenol				[93]
1	*Actinomyces pyogenes*	5		
1	*Klebsiella pneumoniae*	10		
1	*Nocardia asteroides*	5		
1	*Staphylococcus aureus*	5		
1	*Streptococcus agalactiae*	5		
1	*Streptococcus dysgalactiae*	5		
1	*Streptococcus uberis*	5		

Table 9.2 (*Continued*)

No. of strains	Species of microorganism	Mean MIC (% v/v)	SD	Reference
Rewarewa honey: hydrogen peroxide activity equivalent to 21.5% phenol				[93]
1	*Actinomyces pyogenes*	5		
1	*Klebsiella pneumoniae*	10		
1	*Nocardia asteroides*	10		
1	*Staphylococcus aureus*	5		
1	*Streptococcus agalactiae*	10		
1	*Streptococcus dysgalactiae*	10		
1	*Streptococcus uberis*	10		
Manuka honey: nonperoxide activity equivalent to 13.2% phenol				[94]
1	*Enterococcus faecalis*	7		
1	*Escherichia coli*	5		
1	*Klebsiella oxytoca*	5		
1	*Pseudomonas aeruginosa*	6		
1	*Staphylococcus aureus*	3		
1	MRSA	3		
1	*Staphylococcus aureus* NCTC6571	3		
	Escherichia coli NCTC10418	4		
Pasture honey: hydrogen peroxide activity equivalent to 14.8% phenol				[94]
1	*Enterococcus faecalis*	9		
1	*Escherichia coli*	9		
1	*Klebsiella oxytoca*	8		
1	*Pseudomonas aeruginosa*	9		
1	*Staphylococcus aureus*	5		
1	MRSA	4		
1	*Staphylococcus aureus* NCTC6571	3		
	Escherichia coli NCTC10418	7		
Manuka honey: nonperoxide activity equivalent to 13.2% phenol				[95]
20	*Pseudomonas* spp. from wounds	6.9	1.3	
Pasture honey: hydrogen peroxide activity equivalent to 14.8% phenol				[95]
20	*Pseudomonas* spp. from wounds	7.1	1.0	
Manuka honey: nonperoxide activity equivalent to 13.2% phenol				[45]
58	*Staphylococcus aureus* from wounds	2.88	0.15	
1	*Staphylococcus aureus* NCTC6571	2.89		
Pasture honey: hydrogen peroxide activity equivalent to 14.8% phenol				[45]
58	*Staphylococcus aureus* from wounds	3.79	0.25	
1	*Staphylococcus aureus* NCTC6571	3.41		
Manuka honey: nonperoxide activity equivalent to 13.2% phenol				[48]
20	*Burkholderia cepacia* (multiresistant)	2.9	0.94	
Pasture honey: hydrogen peroxide activity equivalent to 14.8% phenol				[48]
20	*Burkholderia cepacia* (multiresistant)	3.6	0.77	

(*Continued*)

Table 9.2 (*Continued*)

No. of strains	Species of microorganism	Mean MIC (% v/v)	SD	Reference
Manuka honey: nonperoxide activity equivalent to 18% phenol				[46]
18	MRSA	2.98	0.14	
7	VSE (*Enterococcus faecalis*)	4.92	0.28	
1	VRE (*Enterococcus avium*)	3.83		
3	VRE (*Enterococcus faecalis*)	4.59	0.52	
15	VRE (*Enterococcus faecium*)	4.72	0.22	
1	VRE (*Enterococcus raffinosus*)	4.86		
Pasture honey: hydrogen peroxide activity equivalent to 13.7% phenol				[46]
18	MRSA	3.07	0.26	
7	VSE (*Enterococcus faecalis*)	9.66	0.46	
1	VRE (*Enterococcus avium*)	5.6		
3	VRE (*Enterococcus faecalis*)	9.43	0.21	
15	VRE (*Enterococcus faecium*)	8.33	0.52	
1	VRE (*Enterococcus raffinosus*)	9.0		
Manuka honey: nonperoxide activity equivalent to 18% phenol				[49]
17	*Pseudomonas* spp. from burns	9.71	0.69	
Pasture honey: hydrogen peroxide activity equivalent to 14.8% phenol				[49]
17	*Pseudomonas* spp. from burns	9.0	1.22	
Manuka honey: nonperoxide activity equivalent to 16.8% phenol				[47]
2	*Staphylococcus capitis*	3.3	0.5	
11	*Staphylococcus epidermidis*	3.5	0.5	
3	*Staphylococcus haemolyticus*	3.3	0.7	
1	*Staphylococcus simulans*	3		
1	*Staphylococcus warneri*	3.3		
Pasture honey: hydrogen peroxide activity equivalent to 17.5% phenol				[47]
2	*Staphylococcus capitis*	3.8	0.6	
11	*Staphylococcus epidermidis*	3.3	0.6	
3	*Staphylococcus haemolyticus*	4.2	0.8	
1	*Staphylococcus simulans*	4		
1	*Staphylococcus warneri*	3.5		
Manuka honey: nonperoxide activity equivalent to ≥18% phenol				[96]
18	*Candida albicans*	39.9	1.7	
10	*Candida glabrata*	42.6	2.7	
10	*Candida dubliniensis*	33.4	2.5	
Medihoney (blend): nonperoxide activity equivalent to ≥18% phenol				[96]
18	*Candida albicans*	38.2	2.9	
10	*Candida glabrata*	43.1	4.2	
10	*Candida dubliniensis*	34.6	2.5	
Jarrah honey: hydrogen peroxide activity equivalent to 30.2% phenol				[96]
18	*Candida albicans*	18.5	2.7	
10	*Candida glabrata*	29.9	2.8	
10	*Candida dubliniensis*	15.4	2.8	

Table 9.2 (*Continued*)

No. of strains	Species of microorganism	Mean MIC (% v/v)	SD	Reference
Manuka honey: nonperoxide activity equivalent to 15% phenol				[97]
1	Actinobacillus actinomycetemcomitans NCTC 9709	6.1		
1	Actinomyces gerencseriae ATCC 233860	7		
1	Actinomyces naeslundii NCTC 10301	9.1		
1	Eikenella corrodins ATCC 23834	4.7		
1	Fusobacterium nucleatum ATCC 25586	5.1		
1	Peptostreptococcus micros ATCC 33270	9		
1	Porphyromonas gingivalis ATCC 33277	6.2		
1	Veillonella parvula ATCC 17745	7.2		
1	Candida albicans ATCC 10261	21.5		
1	Candida glabrata CBS 138	40		
Pasture honey: hydrogen peroxide activity equivalent to 18.2% phenol				[97]
1	Actinobacillus actinomycetemcomitans NCTC 9709	4.8		
1	Actinomyces gerencseriae ATCC 233860	9		
1	Actinomyces naeslundii NCTC 10301	4		
1	Eikenella corrodins ATCC 23834	5.8		
1	Fusobacterium nucleatum ATCC 25586	6.7		
1	Peptostreptococcus micros ATCC 33270	9.3		
1	Porphyromonas gingivalis ATCC 33277	9		
1	Veillonella parvula ATCC 17745	7		
1	Candida albicans ATCC 10261	40		
1	Candida glabrata CBS 138	40		

The level of antibacterial activity of the honeys used is expressed as the concentration of phenol, w/v, with equivalent activity against *S. aureus* ATCC 25923 in an agar well diffusion assay. For the manuka honeys this was determined with catalase added to destroy hydrogen peroxide, so the antibacterial activity is recorded as 'nonperoxide'. MRSA = methicillin-resistant *S. aureus*; VRE = vancomycin-resistant enterococci; VSE = vancomycin-sensitive enterococci.

antimicrobial activity of honey are features that make honey very convenient for clinical use as a topical agent to control infections, as it is not necessary to first identify the infecting species, nor to find the sensitivity of the microorganisms to antibiotics, before effective treatment can be given.

9.4
Other Actions

The clearance of infection by honey may involve more than the antibacterial activity of honey, as research findings with leukocytes in cell culture indicate that honey may work also by stimulating the activity of the immune system. Peripheral blood B lymphocytes and T lymphocytes in cell culture have been found to be stimulated to proliferate by honey at concentrations as low as 0.1% [110]. This low concentration of honey was also found to activate phagocytes isolated from blood [110]. Others have reported that honey at a concentration of 1% stimulates monocytes in cell culture to release the cytokines tumor necrosis factor-α, interleukin-1β and interleukin-6 which are intermediates in the immune response [111, 112]. Honey has the potential to further augment the immune response by supplying glucose, which is essential for the 'respiratory burst' in macrophages. The hydrogen peroxide thus generated is the major component of the bacteria-destroying activity of these cells [113]. The functioning of macrophages would be further aided by the supply of sugars from honey as these would provide substrates for glycolysis, which is the major mechanism for energy production in these cells. This would allow macrophages to function in damaged tissues and exudates where the poor oxygen supply would limit aerobic respiration for the supply of energy [113].

Another way in which control of infection may be aided by honey is through the ability of honey to prevent attachment of bacteria to cells. It has been reported that exposure of *Salmonella interitidis* to an 11% solution of honey for 1 h prior to mixing the washed bacteria with intestinal epithelial cells decreased the number of bacteria attaching to the cells by 74% [114]. Honey, at concentrations as low as 0.00025%, has also been found to block the PA-IIL lectin of *P. aeruginosa*, which mediates biofilm formation and adhesion to animal cells by this species of bacteria [115]. It has been found that biofilm formation by *P. aeruginosa* and by a coagulase-negative *Staphylococcus* is almost completely prevented by honey at a concentration of only 20% of its MIC [116].

9.5
Clinical Uses of Honey as an Antimicrobial Agent

The major usage of honey for control of infection has been in wound care [2], but there are reports in the modern medical literature of its successful use in ophthalmology and gastroenterology (see below), and of its effectiveness in a trial on gingivitis [117]. With the reporting that inhalation of an aerosol of a 60% solution of honey causes no adverse effects [118], there is also the possibility of using honey for treatment of bronchial infections. The author is aware of anecdotal reports of such therapy being effective, also of honey being effective in the treatment of infection of the nasal sinuses and the ear canal, and for the treatment of tineas. These are applications which warrant further research.

Although the antimicrobial activity of honey is ample for control of infection where the honey is in direct contact with the site of infection and does not get excessively diluted by body fluids, there would be far too much dilution to achieve anywhere near the MIC systemically even if the antimicrobial factors entered the circulation from the gut. However, within the gut it is feasible that a bolus of honey passing through from oral dosage would retain a concentration in excess of the MIC for gut pathogens. The results from a trial where mortality rates from induced infection of mice with *E. coli* 0157:H7 and *Salmonella typhimurium* were substantially decreased by daily subcutaneous injection of 1 ml honey is more likely to be due to the honey stimulating the immune response than from a systemic direct antibacterial activity because the dilution into the 25 g of body mass of the mouse would have given a concentration below the MIC for these pathogens for all but the most potent of the honeys they used.

Honey given at a concentration of 5% (v/v) in place of glucose in a rehydration fluid was found to give a statistically significant reduction in the duration of bacterial diarrhea (58 versus 93 h), and give no increase in the duration of nonbacterial diarrhea in a clinical trial conducted on infants and children admitted into hospital with gastroenteritis [119]. In a clinical trial in which 45 patients with dyspepsia were given no medication other than honey substantial reductions were found to result in the number of patients passing blood (from peptic ulcers) in their feces, the number with dyspepsia and the number with gastritis, duodenitis or a duodenal ulcer seen on endoscopy [120]. However, this action of honey may not be by way of its antibacterial activity, as it was found in a clinical trial that it failed to clear *Helicobacter pylori* [121]. It appears to be more likely that it is the anti-inflammatory activity of honey (see below), rather than its antibacterial activity, that is involved in its beneficial effects on gastritis. A series of publications on biochemical studies on induced gastric ulcers in rats have pointed to the effect of honey to be via reduction of inflammation; this has been reviewed by Molan [122].

The anti-inflammatory activity of honey is probably a contributing factor in the effectiveness of honey in ophthalmological applications, besides control of infection. Improvement was reported in 85% of the cases, with no deterioration in any of the other, in a trial of honey on 102 patients with a variety of ophthalmological disorders not responding to conventional treatment, such as keratitis, conjunctivitis and blepharitis [123]. Remission in more than 60% of the cases was reported where honey was used to treat blepharitis, catarrhal conjunctivitis, and keratitis [124]. A review of the use of honey in ophthalmology in Russia [125] describes anti-inflammatory, antibacterial and antifungal actions being seen, honey being used for chemical and thermal burns to the eye, conjunctivitis and infections of the cornea. A transient stinging sensation and redness of the eye soon after putting honey in the eye have been reported, but never enough to stop the treatment [123, 126]. A similar effect is experienced by some patients when honey is used to treat inflamed wounds and this has been attributed to the acidity of honey [2].

9.6
Clinical Evidence for Effectiveness of Honey on Infected Wounds

The very large body of clinical evidence for the effectiveness of honey in healing wounds has been reviewed [127]. The evidence covered in that review, plus that from trials published since the review was published [128–134], is from 23 randomized controlled trials involving a total of 2257 participants, seven clinical trials of other forms involving 142 participants treated with honey, four case studies where there were multiple wounds allowing comparison of honey with other treatment and 16 trials of honey on a total of 533 wounds in animal models (which rule out a placebo effect). Mostly the wounds involved were infected. Where details were given in the reports about the clearance of infection by honey these are listed in Table 9.3.

Table 9.3 Reported details of clearance of infection in wounds when the wounds were dressed with honey.

Type of wound	Outcome of honey treatment	Reference
Superficial burns	91% of wounds treated with honey became sterile within 7 days with honey, compared with 7% treated with silver sulfadiazine	[135]
Fresh partial-thickness burns	eight cases infected after 8 days with honey, compared with 17 treated with OpSite	[136]
Superficial burns	honey gave, better control of infection than silver sulfadiazine did	[106]
Moderate burns, 1/6th total burn area being full thickness	after 7 days of honey treatment the original 44 cases giving positive swab cultures decreased to four, but with silver sulfadiazine there was no change in the 42 cases giving positive swab cultures at the start	[137]
Superficial burns	honey gave better control of infection	[138]
Severe postoperative wound infections following abdominal surgery	mean time to get negative swab cultures was 6 days with honey treatment compared with 14.8 days with washing with 70% ethanol then applying povidone-iodine	[139]
Fournier's gangrene (necrotizing fasciitis on the scrotum)	within 1 week with honey all swabs were negative: there was no need to change from the routine antibiotics to ones to which the bacteria were found to be sensitive	[100]
Large infected surgical wounds on infants	with honey treatment, marked clinical improvement was seen in all cases after 5 days, and all wounds were closed, clean and sterile after 21 days; whereas the wounds had failed to heal with treatment of at least 14 days using intravenous antibiotics (vancomycin plus cefotaxime, subsequently changed according to bacterial sensitivity), fusidic acid ointment, and wound cleaning with aqueous 0.05% chlorhexidine solution	[108]
Multiple chronic leg ulcers, on both legs	after 10 days of dressing the ulcers with honey signs of infection had cleared and the green exudate had ceased, whereas with the ulcers dressed with Aquacell there was copious leakage of green fluid.	[140]

Table 9.3 (*Continued*)

Type of wound	Outcome of honey treatment	Reference
Recalcitrant wounds and ulcers of varied etiology	no signs of healing in 1–24 months of conventional treatment (such as Eusol toilet and dressings of Acriflavine, Sofra-Tulle or Cicatrin, or systemic and topical antibiotics), but after honey treatment the 51 wounds with bacteria present became sterile within 1 week and the others remained sterile; burn wounds treated early healed quickly, not becoming colonized by bacteria.	[101]
Broken-down wounds from radical vulvectomy with lymphadectomy	wounds became free from bacteria in 3–6 days	[99]
Surgical wounds, mostly dehiscent or infected	wounds became sterile within 1–4 days, on patients with profound immunosuppression because of chemotherapy	[141]
Disrupted abdominal wounds from Caesarean section	wounds were made sterile within 1 week	[102]

9.7
Resistant Bacteria

Because of its high osmolarity honey is not a medium in which bacteria could survive and thus have evolved genes for resistance by selection of mutant individuals with genes conferring resistance to the antibacterial factors that are effective in diluted honey. The period in which bacteria could live and have strains multiply during the production of honey in the hive would be short; then the selectively bred surviving bacterial strains would be terminated by prolonged exposure to high osmolarity. In a study designed to select for resistant mutants, by continuous exposure of cultures of *P. aeruginosa* and *S. aureus* to increasing concentrations of an antibacterial agent, no increased resistance to honey was developed yet under the same experimental conditions marked increases in resistance to antibiotics were developed [116]. Similar resistance training experiments with manuka honey and several wound pathogens are being conducted elsewhere, but have not yet succeeded in recovering honey-resistant bacteria (R. Cooper, University of Wales Institute, personal communication).

Owing to the increasing problem of bacteria almost inevitably developing resistance to antibiotics where these are extensively used, the low probability of resistance to honey developing makes the use of honey an attractive alternative for topical control of infection. As an example, although the incidence of catheter-associated blood-stream infections in dialysis patients with honey-treated catheter exit sites was found in a trial [142] to be a bit higher than in those treated with mupirocin (0.97 versus 0.85 episodes per 1000 catheter-days, not significantly different), the low likelihood of selecting for resistant strains of bacteria using honey compared with the high likelihood with continuous use of mupirocin makes the use of honey for chemoprophylaxis in patients with central venous catheters a better option.

With most life-threatening infections with antibiotic-resistant bacteria being acquired by bacteria entering the bloodstream via catheters or open wounds, there is potential for preventing cross-infection in hospitals with antibiotic-resistant strains of bacteria, by dressing all open wounds or catheter exit sites with honey. As well as the trial mentioned above there has been another trial which also has shown honey to be effective in chemoprophylaxis in patients with central venous catheters [143]. In this, the incidence of bloodstream infections in dialysis patients with honey-treated exit sites was found to be a bit lower than in those treated with povidone-iodine (12 versus 19 episodes per 1000 catheter-days, not significantly different). With wounds, the reports of cases where honey was effective in clearing established infection with MRSA and vancomycin-resistant *Enterococcus* (VRE) [140, 141, 144–148] indicate that it is likely to be effective prophylactically. If such an approach to infection control were tried, even if it were not successful it would at least give the best healing conditions for the wounds because of the other features of honey which promote wound healing.

9.8
Benefits Apart from Control of Infection in Topical Treatment with Honey

Apart from its antibacterial activity honey has a potent anti-inflammatory activity, rapidly brings about autolytic debridement of slough and necrotic tissue from wounds, rapidly deodorizes malodorous wounds, speeds up the healing process, and gives healing with minimal scarring: references to the many reports of observations of these features are given by Molan [149]. Antiseptics in common use are all cytotoxic and so slow the healing process [150]. Silver also is cytotoxic [151] and can cause poisoning systemically when absorbed from wound dressings [151]. Honey, however, is not only not toxic, but actually stimulates the growth of cells involved in wound healing [152–154] and stimulates the production of the components of the extracellular matrix [155, 156].

As is so aptly stated about honey by the Muslim prophet Mohammed (around 570–632 AD) in verse 69 of *Surah 16* ('The bee') of the Holy Qu'ran: 'From its belly cometh forth a fluid of varying hues, *wherein there is healing for mankind*'.

9.9
Future Directions

More research is needed to obtain further data on the sensitivity to honeys with standardized activity of some of the multiresistant infecting species of bacteria which infect wounds and catheter exit sites, such as *Acinetobacter baumanii* and

Stenotrophomonas maltophilia. There is also a need for research to find the sensitivity to honeys with standardized activity of untested species which cause ophthalmic, bronchial and gut infections, to establish if clinical treatment of such infections with honey is worth trying. Good clinical trials are needed to establish with certainty how effective honey is for treating such infections. There is also a need for more good clinical trials to be conducted on honey as a treatment for chronic infected wounds, as much of the large body of work that has been done to date has been carried out on acute wounds and/or has had some defects in the design of the trials. There is also a need to measure in these trials the effectiveness of honey in clearing infection, as many of the trials conducted so far have assessed healing rather than specifically assessing clearance of infection, but the healing may have resulted from other bioactivities of the honey such as the anti-inflammatory and debriding actions and the stimulation of growth of repair tissues.

References

1 Zumla, A. and Lulat, A. (1989) *Journal of the Royal Society of Medicine*, **82**, 384–385.

2 Molan, P.C. and Betts, J.A. (2004) *Journal of Wound Care*, **13**, 353–356.

3 Ransome, H.M. (1937) *The Sacred Bee in Ancient Times and Folklore*, George Allen & Unwin, London.

4 Stomfay-Stitz, J. (1960) *Science Counsellor*, **23**, 110–125.

5 Al-Bukhari, M. (1976) *Sahih Al-Bukhari*, 3rd edn revised, Kazi Publications, Chicago, IL.

6 Aristotle (1910) *Historia Animalium. The Works of Aristotle*, vol. IV, (eds J.A. Smith and W.D. Ross), Oxford University Press, Oxford.

7 Gunther, R.T. (1934) *The Greek Herbal of Dioscorides (Reprinted 1959)*, Hafner, New York.

8 Fotidar, M.R. and Fotidar, S.N. (1945) *Indian Bee Journal*, **7**, 102.

9 Ankra-Badu, G.A. (1992) *East African Medical Journal*, **69**, 366–369.

10 Obi, C.L., Ugoji, E.O., Edun, S.A., Lawal, S.F. and Anyiwo, C.E. (1994) *African Journal of Medicine Science*, **23**, 257–260.

11 Imperato, P.J. and Traoré, D. (1969) *Tropical and Geographical Medicine*, **21**, 62–67.

12 Kandil, A., El-Banby, M., Abdel-Wahed, K., Abdel-Gawwad, M. and Fayez, M. (1987) *Journal of Drug Research (Cairo)*, **17**, 103–106.

13 Beck, B.F. and Smedley, D. (1944) *Honey and Your Health*, 2nd edn, McBride, New York.

14 Soffer, A. (1976) *Archives of Internal Medicine*, **136**, 865–866.

15 Editorial (1974) *South African Medical Journal*, **56**, 2300.

16 Condon, R.E. (1993) *Surgery*, **113**, 234–235.

17 Bose, B. (1982) *Lancet*, **i**, 963.

18 Green, A.E. (1988) *British Journal of Surgery*, **75**, 1278.

19 Keast-Butler, J. (1980) *Lancet*, **ii**, 809.

20 Mossel, D.A.A. (1980) *Lancet*, **ii**, 1091.

21 Seymour, F.I. and West, K.S. (1951) *Medical Times*, **79**, 104–107.

22 Somerfield, S.D. (1991) *Journal of the Royal Society of Medicine*, **84**, 179.

23 Tovey, F.I. (1991) *Journal of the Royal Society of Medicine*, **84**, 447.

24 Sackett, W.G. (1919) Honey as a carrier of intestinal diseases. *Bulletin of the Colorado State University Agricultural Experimental Station*, **252**.

25 Dold, H., Du, D.H. and Dziao, S.T. (1937) *Zeitschrift für Hygiene und Infektionskrankheiten*, **120**, 155–167.

26 White, J.W., Subers, M.H. and Schepartz, A.I. (1963) *Biochimica et Biophysica Acta*, **73**, 57–70.

27 d'Agostino Barbaro, A., La Rosa, C. and Zanelli, C. (1961) *Quaderni della Nutrizione*, **21**, 30–44.

28 Dustmann, J.H. (1979) *Apiacta*, **14**, 7–11.

29 Buchner, R. (1966) *Südwestdeutscher Imker*, **18**, 240–241.

30 Christov, G. and Mladenov, S. (1961) *Comptes Rendus de l'Academie Bulgare des Sciences*, **14**, 303–306.

31 Floris, I. and Prota, R. (1989) *Apicoltore Moderno*, **80**, 55–67.

32 Abbas, T. (1997) *Living in the Gulf Magazine*, 50–51.

33 Amor, D.M. (1978) *Composition, Properties and Uses of Honey – A Literature Survey*, The British Food Manufacturing Industries Research Association, Leatherhead.

34 White, J.W. (1975) Composition of honey, in *Honey: A Comprehensive Survey*, (ed. E. Crane), Heinemann, London, pp. 157–206.

35 White, J.W. (1975) Physical characteristics of honey, in *Honey: A Comprehensive Survey*, (ed. E. Crane), Heinemann, London, pp. 207–239.

36 Rüegg, M. and Blanc, B. (1981) *Lebensmittel-Wissenschaft und Technologie*, **14**, 1–6.

37 Bogdanov, S., Rieder, K. and Rüegg, M. (1987) *Apidologie*, **18**, 267–278.

38 Tysset, C., Rousseau, M. and Durand, C. (1980) *Apiacta*, **15**, 51–60.

39 Leistner, L. and Rödel, W. (1975) The significance of water activity for micro-organisms in meats, in *Water Relations of Foods* (ed. R.B. Duckworth), Academic Press, London, pp. 309–323.

40 Scott, W.J. (1957) *Advances in Food Research*, **7**, 83–127.

41 Chirife, J., Scarmato, G. and Herszage, L. (1982) *Lancet*, **i**, 560–561.

42 Christian, J.H.B. and Waltho, J.A. (1964) *Journal of General Microbiology*, **35**, 205–218.

43 Topham, J. (2000) *Journal of Tissue Viability*, **10**, 86–89.

44 Bhanganada, K., Kiettiphongthavorn, V. and Wilde, H. (1986) *Journal of the Medical Association of Thailand*, **69**, 358–366.

45 Cooper, R.A., Molan, P.C. and Harding, K.G. (1999) *Journal of the Royal Society of Medicine*, **92**, 283–285.

46 Cooper, R.A., Molan, P.C. and Harding, K.G. (2002) *Journal of Applied Microbiology*, **93**, 857–863.

47 French, V.M., Cooper, R.A. and Molan, P.C. (2005) *Journal of Antimicrobial Chemotherapy*, **56**, 228–231.

48 Cooper, R.A., Wigley, P. and Burton, N.F. (2000) *Letters in Applied Microbiology*, **31**, 20–24.

49 Cooper, R.A., Halas, E. and Molan, P.C. (2002) *Journal of Burn Care and Rehabilitation*, **23**, 366–370.

50 Daghie, V., Cîrnu, I. and Cioca, V. (1971) Contribution to the study of the bacteriostatic and bactericidal action of honey produced by *Physokermes* sp. in the area of coniferous trees, in Proceedings of the XXIIIrd International Apicultural Congress, pp. 593–594.

51 Lindner, K.E. (1962) *Zentralblatt für Bakteriologie, Parasitenkunde und Infektionskrankheiten*, **115**, 720–736.

52 Plachy, E. (1944) *Zentralblatt für Bakteriologie: International Journal of Medical Microbiology*, **100**, 401–419.

53 Rychlik, M. and Dolezal, M. (1961) *Pszczelnicze Zeszyty Naukowe*, **5**, 53–64.

54 Stomfay-Stitz, J. and Kominos, S.D. (1960) *Zeitschrift für Lebensmittel-Untersuchung und -Forschung*, **113**, 304–309.

55 Coulthard, C.E., Michaelis, R., Short, W.F., Sykes, G., Skrimshire, G.E.H., Standfast, A.F.B., Birkinshaw, J.H., Raistrick, H. (1945) *Biochemical Journal*, **39**, 24–36.

56 Pothmann, F.J. (1950) *Zeitschrift für Hygiene und Infektionskrankheiten*, **130**, 468–484.

57 Roth, L.A., Kwan, S. and Sporns, P. (1986) *Journal of Food Protection*, **49**, 436–441.

58 Thimann, K.V. (1963) *The Life of Bacteria,* 2nd edn, Macmillan, New York.

59 Adcock, D. (1962) *Journal of Apicultural Research,* **1**, 38–40.

60 Radwan, S.S., El-Essawy, A.A. and Sarhan, M.M. (1984) *Zentralblatt fur Mikrobiologie,* **139**, 249–255.

61 White, J.W. and Subers, M.H. (1963) *Journal of Apicultural Research,* **2**, 93–100.

62 Schepartz, A.I. and Subers, M.H. (1964) *Biochimica et Biophysica Acta,* **85**, 228–237.

63 Alston, M.J. and Freedman, R.B. (2002) *Biotechnology and Bioengineering,* **77**, 641–650.

64 Bang, L.M., Buntting, C. and Molan, P.C. (2003) *Journal of Alternative and Complementary Medicine,* **9**, 267–273.

65 Schepartz, A.I. (1966) *Biochimica et Biophysica Acta,* **118**, 637–640.

66 Lineaweaver, W., McMorris, S., Soucy, D. and Howard, R. (1985) *Plastic and Reconstructive Surgery,* **75**, 394–396.

67 Thomas, S. (1990) *Wound Management and Dressings,* The Pharmaceutical Press, London.

68 Schreck, R., Rieber, P. and Baeuerle, P.A. (1991) *EMBO Journal,* **10**, 2247–2258.

69 Grimble, G.F. (1994) *New Horizons,* **2**, 175–185.

70 Pruitt, K.M. and Reiter, B. (1985) Biochemistry of peroxidase system: antimicrobial effects, in *The Lactoperoxidase System: Chemistry and Biological Significance* (eds K.M. Pruitt and J.O. Tenovuo) Marcel Dekker, New York, pp. 144–178.

71 Hyslop, P.A., Hinshaw, D.B., Scraufstatter, I.U., Cochrane, C.G., Kunz, S., Vosbeck, K. (1995) *Free Radical Biology and Medicine,* **19**, 31–37.

72 Bogdanov, S. (1984) *Lebensmittel-Wissenschaft und Technologie,* **17**, 74–76.

73 Schepartz, A.I. (1966) *Journal of Apicultural Research,* **5**, 167–176.

74 Dustmann, J.H. (1971) *Zeitschrift fur Lebensmittel-Untersuchung und -Forschung,* **145**, 294–295.

75 White, J.W. and Subers, M.H. (1964) *Journal of Apicultural Research,* **3**, 45–50.

76 White, J.W. and Subers, M.H. (1964) *Journal of Food Science,* **29**, 819–828.

77 James, O.B.O., Segree, W. and Ventura, A.K. (1972) *West Indian Medical Journal,* **21**, 7–17.

78 Allen, K.L., Molan, P.C. and Reid, G.M. (1991) *Journal of Pharmacy and Pharmacology,* **43**, 817–822.

79 Hodgson, M.J. (1989) Investigation of the antibacterial action spectrum of some honeys, MSc thesis, University of Waikato.

80 Molan, P.C. and Russell, K.M. (1988) *Journal of Apicultural Research,* **27**, 62–67.

81 Russell, K.M. (1983) The antibacterial properties of honey, MSc thesis, University of Waikato.

82 Willix, D.J., Molan, P.C. and Harfoot, C.J. (1992) *Journal of Applied Bacteriology,* **73**, 388–394.

83 Mohrig, W. and Messner, B. (1968) *Acta Biologica et Medica Germanica,* **21**, 85–95.

84 Russell, K.M., Molan, P.C., Wilkins, A.L. and Holland, P.T. (1988) *Journal of Agricultural and Food Chemistry,* **38**, 10–13.

85 Molan, P.C., Allen, K.L., Tan, S.T. and Wilkins, A.L. (1989) Identification of components responsible for the antibacterial activity of Manuka and Viper's Bugloss honeys, in *Annual Conference of the New Zealand Institute of Chemistry,* Paper Or1.

86 Molan, P.C. (1992) *Bee World,* **73**, 5–28.

87 Molan, P.C. (1992) *Bee World,* **73**, 59–76.

88 Miorin, P.L., Levy, N.C.J., Custodio, A.R., Bretz, W.A. and Marcucci, M.C. (2003) *Journal of Applied Microbiology,* **95**, 913–920.

89 Brudzynski, K. (2006) *Canadian Journal of Microbiology,* **52**, 1228–1237.

90 Willix, D.J., Molan, P.C. and Harfoot, C.J. (1992) *Journal of Applied Bacteriology,* **73**, 388–394.

91 Al Somai, N., Coley, K.E., Molan, P.C. and Hancock, B.M. (1994) *Journal of the Royal Society of Medicine,* **87**, 9–12.

92 Brady, N.F., Molan, P.C. and Harfoot, C.G. (1997) *Journal of Pharmaceutical Sciences*, **2**, 1–3.

93 Allen, K.L. and Molan, P.C. (1997) *New Zealand Journal of Agricultural Research*, **40**, 537–540.

94 Cooper, R.A. (1998) *Journal of Medical Microbiology*, **47**, 1140.

95 Cooper, R.A. and Molan, P.C. (1999) *Journal of Wound Care*, **8**, 161–164.

96 Irish, J., Carter, D.A., Shokohi, T. and Blair, S.E. (2006) *Medical Mycology*, **44**, 289–291.

97 Svenson, E.L. (2004) Investigating the potential of using honey for therapy of periodontal disease. MSc Thesis, University of Waikato.

98 Braniki, F.J. (1981) *Annals of The Royal College of Surgeons of England*, **63**, 348–352.

99 Cavanagh, D., Beazley, J. and Ostapowicz, F. (1970) *Journal of Obstetrics and Gynaecology of the British Commonwealth*, **77**, 1037–1040.

100 Efem, S.E.E. (1993) *Surgery*, **113**, 200–204.

101 Efem, S.E.E. (1988) *British Journal of Surgery*, **75**, 679–681.

102 Phuapradit, W. and Saropala, N. (1992) *Australian and New Zealand Journal of Obstetrics and Gynaecology*, **32**, 381–384.

103 Armon, P.J. (1980) *Tropical Doctor*, **10**, 91.

104 Harris, S. (1994) *Primary Intention*, **2**, 18–23.

105 Ndayisaba, G., Bazira, L., Habonimana, E. and Muteganya, D. (1993) *Journal of Orthopaedic Surgery*, **7**, 202–204.

106 Subrahmanyam, M. (1998) *Burns*, **24**, 157–161.

107 Dumronglert, E. (1983) *Journal of the National Research Council of Thailand*, **15**, 39–66.

108 Vardi, A., Barzilay, Z., Linder, N., Cohen, H.A., Paret, G., Barzilai, A. (1998) *Acta Paediatrica*, **87**, 429–432.

109 Wadi, M., Al-Amin, H., Farouq, A., Kashef, H. and Khaled, S.A. (1987) *Arab Medico*, **3**, 16–18.

110 Abuharfeil, N., Al-Oran, R. and Abo-Shehada, M. (1999) *Food and Agricultural Immunology*, **11**, 169–177.

111 Tonks, A., Cooper, R.A., Price, A.J., Molan, P.C. and Jones, K.P. (2001) *Cytokine*, **14**, 240–242.

112 Tonks, A.J., Cooper, R.A., Jones, K.P., Blair, S., Parton, J., Tonks, A. (2003) *Cytokine*, **21**, 242–247.

113 Ryan, G.B. and Majno, G. (1977) *Inflammation*. Kalamazoo, Upjohn, MI.

114 Alnaqdy, A., Al-Jabri, A., Al Mahrooqi, Z., Nzeako, B. and Nsanze, H. (2005) *International Journal of Food Microbiology*, **103**, 347–351.

115 Lerrer, B., Zinger-Yosovich, K.D., Avrahami, B. and Gilboa-Garber, N. (2007) *ISME Journal*, **1**, 149–155.

116 Cokcetin, N. (2007) The effect of honey on problematic pathogens. BSc Thesis, University of Sydney.

117 English, H.K., Pack, A.R. and Molan, P.C. (2004) *Journal of the International Academy of Periodontology*, **6**, 63–67.

118 Al-Waili, N. (2003) *European Journal of Medical Research*, **8**, 295–303.

119 Haffejee, I.E. and Moosa, A. (1985) *British Medical Journal*, **290**, 1866–1867.

120 Salem, S.N. (1981) *Bulletin of Islamic Medicine*, **1**, 358–362.

121 McGovern, D.P.B., Abbas, S.Z., Vivian, G. and Dalton, H.R. (1999) *Journal of the Royal Society of Medicine*, **92**, 439.

122 Molan, P.C. (2001) *Bee World*, **82**, 22–40.

123 Emarah, M.H. (1982) *Bulletin of Islamic Medicine*, **2**, 422–425.

124 Popescu, M.P., Paloş, E. and Popescu, F. (1985) *Oftalmologia (Buchar)*, **19**, 53–60.

125 Mozherenkov, V.P. (1984) *Oftalmologicheskii Zhurnal*, 188.

126 Osaulko, G.K. (1953) *Vestnik Oftalmologii*, **32**, 35–36.

127 Molan, P.C. (2006) *International Journal of Lower Extremity Wounds*, **5**, 40–54.

128 Güneş, Ü.Y. and Eşer, I. (2007) *Journal of Wound, Ostomy and Continence Nursing*, **34**, 184–190.

129 Ingle, R., Levin, J. and Polinder, K. (2006) *South African Medical Journal*, **96**, 831–835.

130 McIntosh, C.D. and Thomson, C.E. (2006) *Journal of Wound Care*, **15**, 133–136.

131 Moolenaar, M., Poorter, R.L., van der Toorn, P.P., Lenderink, A.W., Poortmans, P., Gerardus Egberts, A.C. (2006) *Acta Oncologica*, **45**, 623–624.

132 Mphande, A.N., Killowe, C., Phalira, S., Jones, H.W. and Harrison, W.J. (2007) *Journal of Wound Care*, **16**, 317–319.

133 Nagane, N.S., Ganu, J.V., Bhagwat, V.R. and Subramanium, M. (2004) *Indian Journal of Clinical Biochemistry*, **19**, 173–176.

134 Tahmaz, L., Erdemir, F., Kibar, Y., Cosar, A. and Yalcyn, O. (2006) *International Journal of Urology*, **13**, 960–967.

135 Subrahmanyam, M. (1991) *British Journal of Surgery*, **78**, 497–498.

136 Subrahmanyam, M. (1993) *British Journal of Plastic Surgery*, **46**, 322–323.

137 Subrahmanyam, M., Sahapure, A.G., Nagane, N.S., Bhagwat, V.R. and Ganu, J.V. (2001) *Ann Burns Fire Disasters*, **XI**, 143–145.

138 Nagra, Z.M., Fayyaz, G.Q. and Asim, M. (2002) *Professional Medical Journal*, **9**, 246–251.

139 Al-Waili, N.S. and Saloom, K.Y. (1999) *European Journal of Medical Research*, **4**, 126–130.

140 Natarajan, S., Williamson, D., Grey, J., Harding, K.G. and Cooper, R.A. (2001) *Journal of Dermatological Treatment*, **12**, 33–36.

141 Simon, A., Sofka, K., Wiszniewsky, G., Blaser, G., Bode, U. and Fleischhack, G. (2006) *Support Care Cancer*, **14**, 91–97.

142 Johnson, D.W., van Eps, C., Mudge, D.W., Wiggins, K.J., Armstrong, K., Hawley, C.M., Campbell, S.B., Isbel, N.M., Nimmo, G.R., Gibbs, H. (2005) *Journal of the American Society of Nephrology*, **16**, 1456–1462.

143 Mutjaba Quadri, K.H. (1999) *Seminars in Dialysis*, **12**, 397–398.

144 Chambers, J. (2006) *Palliative Medicine*, **20**, 557.

145 Dunford, C., Cooper, R., Molan, P.C. and White, R. (2000) *Nursing Standard*, **15**, 63–68.

146 Dunford, C.E. (2001) *British Journal of Nursing*, **10**, 1058–1065.

147 Eddy, J.J. and Gideonsen, M.D. (2005) *Journal of Family Practice*, **54**, 533–535.

148 Visavadia, B.G., Honeysett, J. and Danford, M.H. (2008) *British Journal of Oral and Maxillofacial Surgery*, **45**, 55–56.

149 Molan, P.C. (2002) *Ostomy/Wound Management*, **48**, 28–40.

150 Tatnall, F.M., Leigh, I.M. and Gibson, J.R. (1991) *Journal of Hospital Infection*, **17**, 287–296.

151 Poon, V.K. and Burd, A. (2004) *Burns*, **30**, 140–147.

152 Bergman, A., Yanai, J., Weiss, J., Bell, D. and David, M.P. (1983) *American Journal of Surgery*, **145**, 374–376.

153 Gupta, S.K., Singh, H., Varshney, A.C. and Prakash, P. (1992) *Indian Journal of Animal Sciences*, **62**, 521–523.

154 Kumar, A., Sharma, V.K., Singh, H.P., Prakash, P. and Singh, S.P. (1993) *Indian Veterinary Journal*, **70**, 42–44.

155 Suguna, L., Chandrakasan, G., Ramamoorthy, U. and Joseph, K.T. (1993) *Journal of Clinical Biochemistry and Nutrition*, **14**, 91–99.

156 Suguna, L., Chandrakasan, G. and Thomas Joseph, K. (1992) *Journal of Clinical Biochemistry and Nutrition*, **13**, 7–12.

10
Honey: Biological Characteristics and Potential Role in Disease Management

Mohammed Shahid

Abstract

Honey is one of the oldest known medicines that has continued to be used in folk medicine and has been used as an adjuvant for accelerating wound healing. It is reported to prevent infection and promote healing since it has ingredients very similar to antibiotics. The major antibacterial properties of honey can be attributed to its low pH, high sugar content, enzymatic production of hydrogen peroxide, a thermolabile substance called inhibine and its hygroscopic properties. Several chemicals with antibacterial activity have also been identified [i.e. pinocembrin, terpenes, benzyl alcohol, 3,5-dimethoxy-4-hydroxybenzoic acid (syringic acid), methyl 3,5-dimethoxy-4-hydroxybenzoate (methyl syringate), 3,4,5-trimethoxybenzoic acid, 2-hydroxy-3-phenylpropionic acid, 2-hydroxybenzoic acid and 1,4-dihydroxybenzene]. However, the quantities of these substances are far too low to account for any significant amount of activity. Honey also serves as an important medicine because of its mild laxative, bactericidal, sedative and antiseptic characteristics. Recently, the emergence of multidrug-resistant (MDR) organisms has created a lot of chaos in the medical field. Hence, there is a need to find an alternative to counter these MDR organisms. This chapter will discuss the factors responsible for antimicrobial properties and potential use of honey as an antimicrobial agent. This chapter will also cover the effect of honey on MDR bacteria, discussing the few reports from available literature.

10.1
Introduction

10.1.1
Nature of Honey

Honey, a popular natural biological product worldwide, is a combination of nectar of flowers or from secretion of living parts of plants and is produced by honey bees of the

New Strategies Combating Bacterial Infection. Edited by Iqbal Ahmad and Farrukh Aqil
Copyright © 2009 WILEY-VCH Verlag GmbH & Co. KGaA, Weinheim
ISBN: 978-3-527-32206-0

genera *Apis* and *Meliponinae*. The bees collect these secretions, transform them by combining with very specific substances of their self, and deposit, dehydrate and store them in honeycombs to ripen and mature. Honey bees (*Apis mellifera*) use it for their food in winter [1–3]. More than 300 types of honey exist with varying flavors and colors (from pale yellow to dark amber) depending on the types of blossoms visited by the honey bees [2]. Honey contains approximately 181 substances [4], including proteins, sugars, vitamins, minerals, moisture, hydroxymethyfurfural, enzymes, flavonoids, phenolic acids, volatile compounds and so on. Honey has a high viscosity ($1.36\,N\,s/m^2$ at $25\,°C$) with a moisture content of 21.5%, and sugars [glucose (dextrose), fructose, maltose, sucrose], mineral matter and proteins as main constituents [5, 6]. It is one of the earliest forms of sweeteners discovered by humans, and long preceded the use of beet and cane sugar [7]. Honey has beneficial characteristics due to its high nutritional value (330 kcal/100 g) and fast absorption of its carbohydrates on consumption. Moreover, the significant properties of honey, which are highly advantageous to human welfare, are its antibacterial and anti-inflammatory potential in the treatment of skin wounds and many gastrointestinal diseases [8–11].

10.1.2
Medicinal History of Honey

The historical background of the use of honey for medicinal purposes dates back to many centuries before the discovery of bacteria as a main cause of infection. Ancient scrolls, books and tablets showed the use of honey as a medicine for many purposes. Aristotle (384–322 BC) used pale honey as a 'good salve for sore eyes and wounds' [12] and Hippocrates referred its uses in wound healing [13]. Around 25 AD, Celsius used honey for various diseases such as a cure for diarrhea, as a laxative, for coughs and sore throats, and for wound healing and eye diseases [14]. Dioscorides (around 50 AD) stated that pale yellow honey from Attica was the best honey, being 'good for all rotten and hollow ulcers' as well as for inflammation of tonsils and throats, and as a cure for coughs and also in sunburn [15]. The historical background of Egypt (2000 BC), clay tablets of Sumaria (6200 BC), the Veda (Hindu holy scripture), the Holy Qu'ran, The Talmud, the Old and New Testament of The Bible, The Torah and all the holy books of India, Persia, Egypt, China, Greece and Rome have shown the significant use of honey in various diseases [13, 14, 16, 17]. The evidence for the use of honey in the middle Ages is shown in a document from 1392 detailing wound-care practices [18]. Charaka and Sushruta (the great Indian Ayurvedic stalwarts) referred to honey in dressing aids to purify sores and promote healing. The method of burn treatment through honey is also given by Sushruta [19]. The use of honey has continued to be used in folk medicine for the treatment of coughs and sore throats; infected leg ulcers in Ghana; in India, lotus honey for eye diseases; and topical treatment of measles in the eyes to prevent corneal scarring, constipation and gastric ulcers [20]. In recent years medical practitioners have shown significant progress to treat burns [21, 22] and chronically infected meningococcal skin lesions [23] with honey, and recently it has become an important topic of interest for scientific and also clinical research in wound management.

10.2
Biological Characteristics of Honey

10.2.1
Ingredients of Honey

Honey is a highly viscous and supersaturated natural sugar solution. It is a rich nutritive food, including water, calories, carbohydrates, dietary fiber, protein, ash, vitamins and minerals. One hundred grams of honey produces 304 calories and makes it a more concentrated source of energy than other sweeteners (see Table 10.1 for a detailed description). Some of the recent research demonstrated that the good quality of honey showed good maturity (98%), pH 3.50–4.21 and potassium as the most prevalent mineral (mean: 643 ppm), accounting for 79% of the total mineral content. The calcium, sodium and magnesium contents account for 14, 3 and 3%,

Table 10.1 Nutrient values of honey.

Nutrients	Average content per 100 g
Water	17.10 g
Calories	304 kcal
Total carbohydrate	82.20 g
levulose (D-fructose)	38.50 g
dextrose (D-glucose)	31.00 g
maltose (and other disaccharides)	7.20 g
sucrose (table sugar)	1.50 g
other carbohydrates	4.00 g
Dietary fiber	0.20 g
Total protein	0.30 g
Ash	0.20 g
Vitamins	
riboflavin	0.04 mg
niacin	0.12 mg
pantothenic acid	0.07 mg
vitamin B6	0.02 mg
folate	2.00 µg
vitamin C	0.50 mg
Minerals	
calcium	6.00 mg
phosphorus	4.00 mg
sodium	4.00 mg
potassium	52.00 mg
iron	0.42 mg
zinc	0.22 mg
magnesium	2.00 mg
selenium	0.80 mg
copper	0.40 mg
manganese	0.08 mg

respectively, of the total mineral content, and other minerals (copper, manganese and iron) are present at very low levels [24]. Honey contains approximately 18 essential and nonessential amino acids that vary with the honey source [25]. Tyrosine, phenylalanine (aromatic amino acids), and glutamic acid and aspartic acid are also found in honey. Glutamic acid is a very important intermediate product, which is produced during the glucose oxidase reaction in honey [26].

Honey is reported to prevent infection and promote healing since it has ingredients very similar to antibiotics. The medical community has recognized the effects of honey as an antibiotic since ancient times. Therefore, its antibiotic characteristics have also been named as 'inhibine'. Honey contains high amounts of enzymatic and nonenzymatic antioxidants, including catalase, glucose oxidase that produces hydrogen peroxide, ascorbic acid, phenolic acids, flavonoids, alkaloids and many other unidentified substances that have antimicrobial effects and worked as inhibines. These are thermolabile substance and their concentration varies with floral types [3, 26–31].

10.2.1.1 Enzymes in Honey

Honey is a good source of natural enzymes, including glucose oxidase, invertase, catalase, diastase (amylase), peroxidase and acid phosphatase [32, 33]. Antioxidant properties remain in all these enzymes. Gluconic acid and hydrogen peroxide are produced by glucose oxidase from glucose in the presence of oxygen. The nonenzymatic hydrolysis of gluconic acid using molecular oxygen releases hydrogen peroxide [34]. Glucose oxidase is also extracted from the fungi *Penicillium notatum* and *Aspergilus niger* [35].

$$\text{Glucose} + \text{Water} + \text{Oxygen} \xrightarrow{\text{Glucose oxidase}} \text{Hydrogen Peroxide} + \text{Glucolactone}$$
$$\text{Glucolactone} + \text{Water} \rightarrow \text{Gluconic acid}$$

The important applications of glucose oxidase are in biotechnology, medicine, and the leather, photographic, detergent, pharmaceutical and food industries [36]. Conversion of honey sucrose to fructose and glucose is done by invertase, which is added to the honey by the bees and becomes inactivated at high temperatures [37]. The enzyme catalase produces oxygen and water from hydrogen peroxide, and remains in small quantities in honey. Catalase activity and hydrogen peroxide content show an inverse relationship; the relationship determines the hydrogen peroxide level in honey and, therefore, it is called as 'inhibine number' [26]. The enzyme diastase (amylase) is added from bees and pollen during the ripening of nectar, which has great significance to split starch chains to randomly produce maltose and dextrins [33].

Antioxidant Role of Honey The antioxidant potential of honey has the ability to reduce oxidative reactions that are caused by damage by oxidizing agents, such as oxygen. The use of natural and synthetic antioxidants as preservatives in food products (e.g. lipid oxidation in meat and enzymic browning in fruits and vegetables) and for health effects (such as various cancers, aging process and many chronic diseases) has long been documented [38–40]. The antioxidants, both enzymatic (e.g. catalase, glucose oxidase) and nonenzymatic substances (e.g. carotenoids, tocopherols, ascorbic acid, phenolics, flavonols, organic acids, proteins and amino acids) that naturally occur in honey intercept free radicals before they can cause any damage and contribute to the antioxidant capacity. Free radicals and reactive oxygen species are highly reactive compounds derived from oxygen, and are formed during metabolic processes. Therefore, a number of disorders come in existence such as stroke, cancer, cataract, arthritis, Alzheimer's and aging due to damaging reactions of biological components [38–42]. Frankel *et al.* [43] showed the antioxidant potential of honey from 19 different types of honey samples that were collected from 14 different floral sources. It is a well-documented fact that the antioxidant potential of honey positively correlates with water percentage and with the color of honey. Most of the antioxidants of honey are water soluble, such as ascorbic acid and many alkaloids. Therefore, honeys with a higher percentage of water and a darker color have higher antioxidants for a given amount of honey [43].

Various recent studies have also confirmed the results of previous studies of honey's antioxidant capacity [32, 38]. These recent studies characterized the antioxidants of different honeys and identified the components, including ascorbic acid and phenolic compounds, and the enzymes such as peroxidase, catalase and glucose oxidase. The antioxidant properties of the honeys were measured by the oxygen radical absorbance capacity and the values were observed from 3.1 to 16.3 mM Trolox equivalents/g honey. The observation showed that darker honeys contain higher values of antioxidant. The major components responsible for the antioxidant activity were phenols as opposed to that of ascorbic acid and enzyme activity [38, 42, 44].

Flavonoids (gycosides) are nonenzymatic antioxidants that are produced by nectar secretion of specialized glands of plants [45, 46]. Honey obtained from floral nectars contains different flavonoids such as isorhamnetin 3-rhamnoside, quercetin, quercetin 3-sophoroside, myricitin 3′-methylether, kaempferol 3-rhamnoside, hesperidin (in citrus honey) and kaempferol 3-sophoroside (in rosemary honey) [47–50]. Flavonoids reduce free radical generation, protect α-tocopherol (in low-density lipoprotein) from oxidation, promote chelation of divalent cations and regenerate oxidized ascorbic acid [51].

Flavonoids have been claimed to have antithrombotic, hypolipolenic, vasoprotective and hypocholesterolemic effects, and also antibacterial, anti-inflammatory and antiviral activities [51, 52]. Their application has been reported in the prevention and cure of allergic diseases, gastric ulcers, cardiovascular diseases and cancers [51]. Flavonoids are found in a concentration of approximately 20 mg/kg in honey [53]. The antioxidants of honey are responsible for antimicrobial activities and thus have a potential use in bacterial disease management. Further investigations are needed to elucidate other medicinal properties due to the presence of antioxidants in honey.

10.2.1.2 Antibiotics in Honey

Since antibiotics are used in apiculture, they have been the main contaminants in honey. A variety of antimicrobial compounds and antibiotics, which have been reported in pollen, nectar and other floral parts of plants, accumulate in honey through honeybees. Antibiotics such as sulfonamides (sulfathiazole, sulfamethazine, sulfamethoxazole, sulfanilamide), aminoglycosides (streptomycin), tetracyclines (oxytetracycline, chlortetracycline) and amphenicols (chloramphenicol) have been detected in honey [54–64].

10.3
Antibacterial Potential of Honey

The antibacterial properties of honeys from different floral sources have been identified for more than a century, although as a medicine it has been used since ancient times. Now, it is well documented and proved by medical practitioners and also scientists that the effectiveness of honey for many medical purposes is due to its antimicrobial activity. A number of properties of honey contribute to its antibacterial activity – osmotic effects, acidity, hydrogen peroxide [65] and phytochemical factors (Table 10.2). The following subsections deal with theses factors with special reference to nonperoxide chemicals in more detail, since details on other primary factors may be found in other significant chapters of this book.

10.3.1
Osmotic Effect of Honey

A number of factors are responsible for the antibacterial activity of honey but it has been assumed that the osmotic effect is one of the most important, which is due to its highly saturated solution of sugars (glucose and fructose mixture, approximating 84%, and water approximating 15–21% by weight) [66–74]. The osmotic effect means withdrawing water away from the microbes, which in turn inhibits their growth [75].

The molecules of supersaturated sugars solution leave very few water molecules for microbes due to the strong interactions and have low water activity. Water activity measures the quantity of 'free' water responsible for the growth of bacteria and yeast. The water activity of ripened honey has been reported in many research articles to be from 0.5 to 0.62 depending on the water content, floral source and temperature. This is too low to support microbial growth of any type. However, many species of bacteria and yeasts can survive and grow as the water activity rises from 0.94 to 0.99. Thus, the water activity of honey depends on the dilution. A higher water activity of diluted honey will not be effective to inhibit bacterial species and biomass has a maximum growth rate at a water activity of 0.99 [25, 74, 76, 77].

10.3.2
Acidity of Honey

The slightly acidic characteristic of honey make it an antibacterial agent. Some proven studies [76, 78] stated that the hygroscopic properties and the acidity due to low pH (around 3.6) of honey are antibacterial, and this acidity is due to the presence of gluconic acid. Undiluted honey is quite acidic; its acidity varies from pH 3.2 to 4.5, and is too low to allow the growth of many bacterial pathogens. An optimum pH range of 7.2–7.4 is required for the growth of some bacterial species. However, some

Table 10.2 Nonperoxide substances noticed in honey from different floral sources.

Floral source	Nonperoxide markers	References
Citrus honey	hesperitin (0.28–0.84 mg/kg) methyl anthranilate (1.44–3.60 mg/kg)	[48, 53, 89]
Heather honey	flavonoids (0.06–0.5 mg/100 g): myricetin, myricetine 3-methyl ether, myricetin 3'-methyl ether and tricetin phenolic acid: ellagic (0.1–0.6 mg/100 g), p-hydroxybenzoic, syringic, o-coumaric and p-coumaric acid abscisic acid: (2.5–16.6 mg/100 g) (ellagic, abscisic acids and myricetin 3'-methyl ether are potential markers)	[49, 86, 87, 90–92]
Sunflower honey	phenolic acid: 37% cinnamic acid, 21% benzoic acid flavonoids: pinocembrin, chrysin, galangin, quercetin and pinobanksin (quercetin is a potential marker)	[93–95]
Eucalyptus honey	flavonoids: myricetin, tricetin, quercetin, luteolin and kaempferol propolis-derived flavonoid: pinobanksin, pinocembrin and chrysin (myricetin, tricetin and luteolin are potential markers)	[96, 97]
Rosemary honey	kaempferol 3-sophoroside concentration: (0.4–1.2 mg/kg) (0.33–2.48 mg/kg)	[47, 88, 97, 98] [97] [88, 98]
Other unifloral honey		
Calluna honey	ellagic acid	[98]
Alder honey	8-methoxykaempferol	[98]
Thyme honey	rosmarinic acid	[91, 92, 99]
Lavender honey	luteolin, naringenin and gallic acid	[48, 53, 91, 92, 99–101]
Rape honey	phenylpropanoic acid	[102]
Buck wheat honey	4-hydroxybenzoic acid	[102]
Fir honey	methyl ferulate	[103]
Strawberry tree honey	homogentistic acid	[103]

common wound-infecting bacterial species, such as *Escherichia coli*, *Salmonella* species, *Pseudomonas aeruginosa* and *Streptococcus pyogenes*, can even grow at a pH around 4.5. The effectiveness of honey with respect to pH to inhibit bacterial species is totally dependent on its dilution [74].

10.3.3
Hydrogen Peroxide Effect of Honey

Many biological reactions produce hydrogen peroxide when catalyzed by oxidase enzymes. It is proven that hydrogen peroxide produced enzymatically is one of the major factors responsible for the antibacterial activity of honey. The enzyme glucose oxidase is secreted from the hypopharyngeal gland of the honeybee and produces hydrogen peroxide. The enzyme has been found to be practically inactive in full-strength honey and generates hydrogen peroxide only when the honey is diluted. Full-strength honey has a negligible level of hydrogen peroxide because this substance has a short life due to the presence of the transition metal ions and ascorbic acid in honey that catalyze hydrogen peroxide decomposition to oxygen and water. It should be noted that the antibacterial activity of honey due to the presence of hydrogen peroxide varies with origin and processing [74, 76, 77, 79].

Thus, honey can act as a continuous producer of hydrogen peroxide at a level that is antibacterial, but not tissue damaging; Manzoori *et al.* [80] successfully determined the hydrogen peroxide content of several honey samples and showed that dilution of honey leads to continuous production of hydrogen peroxide. Therefore, the hydrogen peroxide determines important characteristics to select honey samples for use as an antimicrobial agent.

10.3.4
Activity Due to the Presence of Nonperoxide Compounds

In recent years, many investigators have identified a number of phytochemicals in honey and also in various foods. They have reported that the peroxide-generating system does not account for all of the observed antibacterial activity in honey and it has been found that heating honeys inactivates the glucose oxidase, causing loss of peroxide activity against some species whilst it is retained against others [74]. The direct evidence for the existence of nonperoxide factors in honey was seen when hydrogen peroxide was removed by treating with catalase and antibacterial activity still was shown by some honeys [81]. This activity was independent of enzyme activity, being referred to as nonperoxide antibacterial activity. Lysozyme, phenolic acid and flavonoids are nonperoxide factors of honey [79]. Bogdanov [65] suggested that the major part of the nonperoxide antibacterial activity is produced by honeybees that which feed on plant extracts and have variations in the antibacterial activity of their different unifloral honeys. Wahdan [82] also suggested that phenolic acids and flavonoids might be a significant factor of the antibacterial activities of honey. The nonperoxide substances found in honey from different floral sources are shown in Table 10.2.

Manuka honey obtained from *Leptospermum scobarium* in New Zealand has the greatest antibacterial activity due entirely to the nonperoxide components, including several phenolic compounds (methyl syringate and syringic acid) [83, 84]. Methyl syringate was found to possess significant antibacterial activity against *Staphylococcus aureus* [50]. An Australian honey obtained from *Leptospermum polygalifolium* was also found to possess a high level of nonperoxide antibacterial activity [85]. The nonperoxide antibacterial activity is heat and light insensitive in comparison to hydrogen peroxide, and is found to be intact after storage of honey for long periods [50].

Several researchers have identified various chemicals with antibacterial activity, including pinocembrin, terpenes, benzyl alcohol, 3,5-dimethoxy-4-hydroxybenzoic acid (syringic acid), methyl-3,5-dimethoxy-4-hydroxybenzoate (methyl syringate), 3,4,5-trimethoxybenzoic acid, 2-hydroxy-3-phenylpropionic acid, 2-hydroxybenzoic acid, 1,4-dihydroxybenzene and many others (Table 10.2). However, the quantities of these substances were far too low to account for any significant amount of activity [74]. On the other hand, some investigators have found nonperoxide antibacterial activity to be more significant than hydrogen peroxide with respect to the antibacterial effect [79]. However, the exact cause and role of the nonperoxide antibacterial activity is still not very clear, and requires further investigations.

10.4
Potential Use of Honey as an Antibacterial Agent

The potential use of honey in medicine as an antimicrobial agent has been recognized since antiquity. A number of bacterial species are susceptible to the antimicrobial activity of honey. Recent research on wound dressing, peptic and nonhealing ulcers including diabetic foot ulcers, gastroenteritis and tinea infections in humans, as well as mastitis in dairy animals and also its effect on multidrug-resistant (MDR) bacteria has been conducted to confirm the efficacy of honey against pathogenic bacterial agents.

10.4.1
Use of Honey in Wound Management

The use of honey in wound healing has been documented from many centuries. In modern medicine, it has been rediscovered that topical use of honey as an antibacterial agent is effective for the treatment of wound healing, burns and skin ulcers [78, 104–124]. The investigations documented that inflammation, swelling and pain quickly diminished, unpleasant odors ceased, sloughing of necrotic tissue occurred without the need for debridement, dressings were removed without pain and without any damage to regrowing tissue, and the healing occurred quickly with minimal scaring and grafting being unnecessary. In many cases, where standard antibiotics and antiseptics failed to heal infected lesions, honey did it successfully [74]. It is well documented that many bacterial species such as *E. coli, Proteus species, P. aeruginosa, Salmonella typhimurium, Serratia marcescens, S. aureus* and *S. pyogenes* are responsible for wound infections, and they are susceptible to the antibacterial activity of honey [76]. Various types of wounds (abscesses, amputations, abrasions, burns, cuts, bed sores, cancrum, diabetic foot ulcers and other ulcers, cervical ulcers, burst abdominal wounds such as caesarian delivery, fistula, cracked nipples, foot ulcers in lepers, chilblains, large septic wounds, leg ulcers, malignant ulcers, tropical ulcers, skin ulcers, wounds of the abdominal wall and perineum, sickle cell ulcers and varicose ulcers) have successfully been treated with honey therapy [10]. While the exact mechanism(s) for the wound healing potential of honey is not clearly defined, Molan [10], Cooper [125] and Jones [126] suggested possible mechanisms which are summarized as follows:

- *Antibacterial activity: anticipated clinical outputs.* Sterilization of wounds, inhibition of potential wounds, pathogen- and protein-digesting enzymes that destroy tissues, deodorization of foul-smelling wounds, and protective barrier to prevent cross-contamination.

- *Anti-inflammatory activity: anticipated clinical outputs.* Resolution of edema and exudates, reduction of pain, and reduction in keloids and scarring.

- *Clinical outputs of wound healing activity.* Increased phagocytosis, increased autolytic debridement, and increased angiogenesis, promotion of granulation tissue, cell proliferation, collagen synthesis and re-epithelialization with less need for skin grafts.

It is being noticed that the clearing of infections of wounds with honey reflects more than just the antibacterial properties. The proliferation of B and T lymphocytes and also phagocytes in cell cultures is stimulated by honey at concentrations as low as 0.1% [21]. Honey also stimulates monocytes at a concentration of 1% in cell cultures to release cytokines, tumor necrosis factor-α, interleukin-1 and interleukin-6, which activate the immune response to infection [127, 128]. Bangroo *et al.* [116] concluded that honey acts mainly as a hyperosmolar medium and prevents bacterial growth because of its high viscosity, forms a physical barrier, and the catalase enzyme present in honey acts as an antioxidant. Its high nutrient content improves substrate supply in the local environment, promoting epithelialization and angiogenesis. Thus, the antimicrobial, anti-inflammatory and wound-healing stimulation effect of honey may well be an easy answer to the question of a cheap, easily available, nontoxic, nonirritant and antibacterial agent for wound and burn dressings in a developing country like India [116], and world wide.

10.4.2
Use of Honey in Peptic Ulcers

Honey is used as a traditional remedy for the treatment of peptic ulcers, gastritis and dyspepsia [20, 129]. The discovery that *Helicobacter pylori* is a causative agent in many cases of peptic ulcers and dyspepsia raised the possibility that research on the antibacterial activity of honey may explain its therapeutic action. It is confirmed through earlier research that *H. pylori* was susceptible to honey at a concentration of 20% (v/v) and to manuka honey obtained from *L. scobarium* of New Zealand, with nonperoxide antibacterial activity, at a concentration of 5% (v/v), but none showed sensitivity to a 50% (v/v) solution of honey in which the antibacterial activity was primarily due to its hydrogen peroxide content. However, the clinical trials did not demonstrate the effective potency of manuka honey on *H. pylori* [74, 129].

10.4.3
Use of Honey in Gastroenteritis

Gastroenteritis in infants is an infectious bacterial disease, and honey has been found to be more effective to treat it in comparison to glucose and electrolyte in an oral rehydration solution [130]. Honey was found to be as effective as glucose in achieving rehydration and its antibacterial activity additionally treated the infection in bacterial diarrhea.

Little information is available on the sensitivity of honey in gastroenteritis and on the antibacterial potential of honey against a panel of bacteria. Therefore, the relative antibacterial potency of honey was tested against all the bacterial species causing gastroenteritis, and comparison of manuka honey (nonperoxide activity) and honey with the usual hydrogen peroxide activity as well as with an artificial honey (i.e. a sugar solution that mimics the composition of honey) was made to assess how much of the antibacterial properties was due to the acidity and the osmotic effect of the honey's sugar [74]. Both the honeys (manuka honey and honey with hydrogen peroxide) showed bacteriostatic and bactericidal activity. Molan [74] showed that pasture honey (honey with an average level of hydrogen peroxide activity) at 4–8% (v/v) is bacteriostatic and at 5–10% (v/v) is bactericidal. Manuka honey (honey with nonperoxide activity) at 5–11% (v/v) is bacteriostatic and at 8–15% (v/v) bactericidal. The artificial honey at 20–30% (v/v) was only bacteriostatic, demonstrating that the acidity and osmotic effect were not the exact factors causing the antibacterial activity of honey [131].

10.4.4
Use of Honey in Dermatophytoses

Tineas (commonly called ring worm infections) are one of the most common diseases of humans, identified as cutaneous or superficial mycoses and caused by an important group of fungi called dermatophytes. These cutaneous infections (dermatophytoses) are often difficult to treat. Honey has been reported to possess antimicrobial properties that offer the potential to treat both fungal (by antifungal activity) and bacterial (by antibacterial) infections [74, 131]. The fungal species causing dermatophytes, such as *Epidermophyton floccosum, Microsporum* (*Microsporum canis, Microsporum gypsium*), *Trichophyton* (*Trichophyton rubrum, Trichophyton tonsurans, Trichophyton mentagrophytes* var. *mentagrophytes*) regularly infect humans and cause tineas. These dermatophytes were inhibited by honeys in agar wells, giving a clear zone around the wells in an agar well diffusion assay. A more recent study investigated two types of honey: pasture honey (hydrogen peroxide activity with average antibacterial activity) and manuka honey (average level of nonperoxide antibacterial activity), and demonstrated antifungal activity of honey against clinical isolates of dermatophytes. The results demonstrated that fungal growth was inhibited by the pasture honey due to the hydrogen peroxide and by the nonperoxide activity in manuka honey [74, 131]. Therefore, higher concentrations or lesser dilutions of honey are needed to inhibit dermatophytes responsible for tineas [74, 131].

10.4.5
Role of Honey in Diabetes

Diabetes is a malfunction in the body's ability to convert carbohydrates (sweet and starchy foods) into energy to power the body. The medical name for this is diabetes mellitus, meaning 'honey sweet diabetes'. Diabetes is characterized by an abnormally high and persistent concentration of sugar in the bloodstream. It generally accepted that 'controlling blood sugar is critically important for diabetics, and maintaining good insulin sensitivity reduces the risk for diabetes in at-risk people' and that 'experimental evidence suggests that consumption of honey compared to some other sweeteners may improve blood sugar control and insulin sensitivity; fructose found in honey may play an important role in mediating this potential health benefit'. The sugar in honey is predigested and cannot be compared to the artificial, refined sugar sold in the market. The sugar in honey is only 7%, even though it has a very sugary taste. Mad honey (obtained from *Rhododendron ponticum*, grown extensively on the mountains of the eastern Black Sea area of Turkey) is used traditionally in the management of diabetes mellitus in east Anatolia, Turkey, and in a recent study it was documented that mad honey caused significant decreases in blood glucose and lipid (cholesterol, triglyceride and very-low-density lipoprotein) levels. It is further suggested that mad honey in small doses can be used as a dietary supplement, especially by patients with type II diabetes mellitus, because B cells of the islets of Langerhans in the pancreas can secrete insulin in patients with type II diabetes mellitus [132]. Thus, honey decreased the plasma and urinary glucose concentration, while artificial honey increased them; this reduction in plasma glucose and consequently in urinary glucose could be attributed to the stimulatory effects of honey on insulin production [133–135].

10.4.6
Role of Honey in Diabetic Foot Ulcers

Honey has been used to treat infections in a wide range of wound types (venous leg ulcers of mixed etiology, burns, unhealed graft donors, diabetic foot ulcers, necrotizing fasciitis) [121, 136]. Foot ulcers remain a frequent complication of diabetes. Delayed or inadequate treatment of foot infections in diabetic patients often results in limb loss and the management of the complicated lesions can be challenging. Diabetic foot complications are the most common cause of nontraumatic lower extremity amputations in the industrialized world; the risk of lower extremity amputations is approximately 15–46 times higher in diabetics than in nondiabetics. The foot complications were the most frequent reasons for hospitalizations in patients with diabetes, accounting for up to 25% of all diabetic admissions in the United States and Great Britain [137]. It has been showed that several types of wounds and skin ulcers, which had not responded to conventional methods of treatment such as antibiotics and medicated dressings (dressing agents as iodine compounds (povidone-iodine and cadexomer-iodine, chlorhexidine, hydrogen peroxide, acetic acid and silver compounds), responded favorably to topical honey treatment [138].

The hydrogen peroxide concentration in honey is around 1 mmol/l, while it is around 1000 mmol/l in standard 3% solution, which has been found to be harmful to wounds when added as a rinse solution. On the other hand, honey was proved to prevent bacterial growth through its acidic pH. A recent study documented that the use of honey/normal saline significantly reduced amputations, wound dressing irritation, adhesion and treatment costs, and strongly recommended the use of honey/normal saline for successful treatment of diabetic foot ulcers [139].

10.4.7
Role of Honey in Ophthalmology

Honey from Attica had a special reputation as a curative substance for eye diseases. 'White honey is good as a salve for sore eyes' is demonstrated by Aristotle in 350 BC in section 627a 3 of *Historia Animalium* [12]. Lotus honey in more recent times in India was said to be a panacea for eye diseases [140]. Honey is involved in a traditional therapy in Mali for measles and is being put in the eyes to prevent the scarring of the cornea that occurs in this infection [141]. Honey has been referred to treat eyes discharging pus [142]. The use of honey to treat blepharitis (inflammation of the eyelids), catarrhal conjunctivitis and keratitis (inflammation of the cornea) has also been documented [143]. It has also been used for chemical and thermal burns to the eye, conjunctivitis, and infections of the cornea, being applied undiluted or as a 20–50% solution in water. It is well documented when treating 102 patients with a variety of ophthalmological disorders not responding to conventional treatment, such as keratitis, conjunctivitis and blepharitis [144]. The honey was applied under the lower eyelid as eye ointment and improvement was seen in 85% of cases, with no deterioration seen in any of the other 15%. There was a report of a transient stinging sensation and redness of the eye soon after putting honey in the eye, but never enough to stop the treatment in the 102 cases in the trial. A significant study has reported on Ophthacare brand eye drops (The Himalaya Drug Co., Bangalore, India), a preparation containing *Carum copticum* (seeds) 0.60% w/v, *Terminalia belerica* (fruits) 0.65% w/v, *Emblica officinalis* (fruits) 1.3% w/v, *Curcuma longa* (rhizome) 1.3% w/v, *Ocimum sanctum* (leaves) 1.3% w/v, *Rosa damascena* (petals) 1.1% w/v, *Cinnamomum camphora* 0.05% w/v and honey 3.7% w/v, evaluating its anti-inflammatory, antioxidant and antimicrobial activity using *in vivo* and *in vitro* experimental models, and the findings reveal the usefulness of Ophthacare brand eye drops in the treatment of various ophthalmic disorders such as acute and chronic conjunctivitis, eye strain, dacryocystitis, and pterygium [145].

10.4.8
Honey in the Treatment of Viral Diseases

In a recent study, 16 adult patients with a history of recurrent attacks of herpetic lesions, eight labial and eight genital, were treated by topical application of honey for one attack and acyclovir cream for another attack. For labial herpes, the mean duration of attacks and pain, occurrence of crusting, and mean healing time with honey treatment were 35, 39, 28 and 43% shorter, respectively, than with acyclovir treatment. For genital herpes, the mean duration of attacks and pain, occurrence of crusting, and mean healing time with honey treatment were 53, 50, 49 and 59% shorter, respectively, than with acyclovir. Two cases of labial herpes and one case of genital herpes remitted completely with the use of honey. No side-effects were observed with repeated applications of honey, whereas three patients developed local itching with acyclovir. This study concluded that topical honey application is safe and effective in the management of the signs and symptoms of recurrent lesions from labial and genital herpes [146]. A similar study has been reported recently, in which children and adults with recurrent herpetic lesions on the lips were treated with medical honey, as soon as a new lesion was developing. In addition, the use of topical medical honey in addition to systemic acyclovir in immunocompromised patients with zoster to prevent secondary bacterial skin infection and to accelerate healing of the herpetic lesions has been reported [147].

10.4.9
Use of Honey in Mastitis in Dairy Animals

Mastitis is a bacterial infection, and occurs in dairy animals including cows and goats. Bacterial species such as *Actinomyces pyogenes*, *Klebsiella pneumoniae*, *Nocardia asteroides*, *S. aureus*, *Streptococcus agalactiae*, *Streptococcus dysgalactiae* and *Streptococcus uberis* most commonly cause mastitis in dairy animals, and were tested for antibacterial susceptibility against honey. There are many reports of the effectiveness of honey in clearing bacterial infections in ulcers and abscesses, which suggest that honey may be suitable for the intramammary treatment of mastitis. Generally, antibiotics are used as the standard treatment for the infected udder. As honey is harmless to tissues, has antibacterial properties due to hydrogen peroxide activity and also has nonperoxide activity, it could be suitable for the treatment of mastitis when inserted into the infected udder via the teat canal. Studies have been performed to investigate two types of natural honeys and an artificial honey. The natural honeys used were rewarewa honey with an average level of hydrogen peroxide activity and no detectable nonperoxide activity, and a manuka honey with an average level of nonperoxide activity and no detectable peroxide activity. The artificial honey was used to detect the sensitivity of the bacteria to the osmotic action and acidity of honey. The results demonstrated that nonperoxide chemical substances in natural honeys are important and they could be used in the treatment of mastitis [74, 131].

10.5
Use of Honey against Multidrug Resistance

Several types of wounds such as burn wounds, chronic nonhealing ulcers, post-traumatic wounds and so on show resistance to treatment by antimicrobial therapy. Various agents, like silver sulfadiazine, silver nitrate (5%) and sulfamylon, and many methods of grafting have been used to eliminate bacterial growth from the burn surface, but none of the methods has effectively reduced the problem of infection [123].

It is a well-reported fact that *P. aeruginosa* develops more rapidly in extensive burns than in minor burns and is detected in 4% of wounds in hospital patients [120]. Honey could be used as an alternative to combat the problem of antibiotic resistance in these organisms. The antibacterial and antifungal activities of honey have been documented to enhance the healing of wounds and pressure sores [82]. Subrahmanyam *et al.* conducted studies [21, 148] on superficial, partial-thickness and deep burns, and observed that honey has a mopping up effect on free radicals, which leads to diminished scarring and contractures. Honey also plays a role in the initial phase by limiting lipid peroxidation, and at a later phase by controlling infection and promoting healthy granulation. *P. aeruginosa* can cause infection in burnt skin, sepsis and pneumonia, and occupies multiple ecological niches in nature through its minimal growth requirements, and its potency to develop a large number of extracellular protective and toxic compounds. These compounds, such as hemolysins, slime glycolipoproteins, fibrinolysins, elastase, lecithinase, DNase and phospholipase, may contribute to the pathogenicity of *P. aeruginosa* [149]. The presence of these compounds in burns and other wounds also contributes to extended hospital stays, increased costs, delayed healing rates and failed skin grafts, and promotes the risks of septicemia. Most of the infecting bacterial isolates are multidrug resistant and thus the search for new potentially effective antimicrobial agents continues [149].

Globally, the treatment of burns with honey has been documented in 5.5% of instances, while silver sulfadiazine has been documented in 1% cases for the treatment of partial-thickness and mixed burns [150]. Minimum inhibitory concentrations (MICs) of honeys against *Pseudomonas* spp. have been reported to range from 3 to 30% (v/v) depending on the nature and source of honeys [76]. Subrahmanyam *et al.* [151] investigated the antibacterial potential of honey against MDR *Pseudomonas* infection and reported the MIC against *Pseudomonas* as 25% (v/v) for all strains.

10.6
Conclusions

It is thus concluded that in the present era of increasing MDR organisms, honey could be used as an effective, cheap, easily available, nontoxic and adjuvant without any adverse side-effects in bacterial disease management. However, the antibacterial potential due to nonperoxide chemicals needs further investigation in order to understand the mechanism(s) and disease-management potential in greater detail.

References

1 Molan, P.C. (1996), *Authenticity of Honey*, in *Food Authentication* (eds P.R. Ashurst and M.J. Dennis), Blackie, London, pp. 259–303.

2 Solaomon, J., Santhi, S.V. and Jayaraj, V. (2006) *Integrated Bioscience*, **10**, 163–724.

3 Namias, N. (2003) *Surgical Infections (Larchmont)*, **4**, 219–226.

4 Al-Manary, M., Al-Meeri, A. and Al-Habori, M. (2002) *Nutrition Research*, **22**, 1041–1047.

5 Kirk, R.S. and Sawyer, R. (1991) *Sugar and Preserves in Pearson's Composition and Analysis of Foods*, 9th edn, Longman Scientific & Technical, Harlow.

6 White, J.W. Jr (1979) *Physical Characteristics of Honey* in *Honey – A Comprehensive Survey* (ed. C. Eva), Heinemann, London, pp. 207–239.

7 Coulston, A.M. (2000) *Nutrition Today*, **35**, 96–100.

8 Greenwood, D. (1993) *Lancet*, **341**, 90–91.

9 Taormina, P.J., Niemira, B.A. and Beuchat, L.R. (2001) *International Journal of Food Microbiology*, **69**, 217–225.

10 Molan, P.C. (2001) *Bee World*, **82**, 22–40.

11 Cooper, R.A., Molan, P.C. and Harding, K.G. (2002) *Journal of Applied Microbiology*, **93**, 857–863.

12 Smith, J.A. and Ross, W.D. (1910) in *Historia Animalium: The Works of Aristotle, Volume IV* (translated by D.A.W. Thompson), Oxford University Press, Oxford, pp. 631–632.

13 Jones, R. (2001) *Honey and healing through the ages* in *Honey and Healing* (eds P. Munn and R. Jones), International Bee Research Association, Cardiff, pp. 1–4.

14 Beck, D.F. and Smedley, D. (1938) *Honey and Your Health: A Nutrimental, Medicinal and Historical Commentary*, Health Resources, Silver Springs, MD.

15 Gunther, R.T. (1959) *The Greek Herbal of Dioscorides*, Hafner, New York.

16 Mcintosh, E.N. (1995) *American Food Habits in Historical Perspective*, Praeger, Westport, CT.

17 Námias, N. (2003) *Surgical Infections (Larchmont)*, **4**, 219–226.

18 Naylor, I.L. (1999) *Journal of Wound Care*, **8**, 208–212.

19 Grover, S.K. and Prasad, C.S. (1985) *Journal of NIMA*, **10**, 7–10.

20 Molan, P.C. (2001) *Why Honey is effective as a Medicine. 1. Its Use in Modern Medicine*, in *Honey and Healing* (eds P. Munn and R. Jones), International Bee Research Association, Cardiff, pp. 5–13.

21 Subrahmanyam, M. (1998) *Burns*, **24**, 157–161.

22 Subrahmanyam, M. (1999) *Burns*, **25**, 729–731.

23 Dunford, C., Cooper, R. and Molan, P.C. (2000) *Nursing Times Plus*, **96**, 7–9.

24 Conti, M.E., Stripeikis, J., Campanella, L., Cucina, D. and Tudino, M.B. (2007) *Chemistry Central Journal*, **1**, 14.

25 USDA Data Obtained From Genesis® R&D Nutrition Analysis Program Version 7.01, ESHA Research, Salem, OR.

26 White, J.W., Subers, M.H. and Schepartz, A.I. (1963) *Biochimica et Biophysica Acta*, **73**, 57–79.

27 Schepartz, A. and Subers, M.H. (1966) *Journal of Apicultural Research*, **5**, 37–43.

28 Subrahmanyam, M. (1996) *Annals of Burns and Fire Disasters*, **9**, 93–95.

29 Ladas, S.D. and Rapitis, S.A. (1999) *Nutrition*, **15**, 591–592.

30 Bogdanow, S. (1989) *Journal of Apicultural Research*, **28**, 55–57.

31 Berenbaum, M., Robinson, G. and Unnevehr, L. (1995/1996) *Antioxidant Properties of Illinois Honey: Grant Proposal for National Honey Board*, University of Illinois at Urbana-Champaign, Champaign, IL.

32 McKibben, J. and Engeseth, N.J. (2002) *Journal of Agricultural and Food Chemistry*, **50**, 592–595.

33 Crane, E. (1976) *Honey: A Comprehensive Survey*, International Bee Research Association, Heinemann, London.

34 Chaplin, M.F. and Bucke, C. (1990) *Enzyme Technology*, Cambridge University Press, Cambridge.

35 Uhlig, H. (1990) *Application of Technical Enzyme Preparations, in Industrial Enzymes and Their Applications*, John Willey & Sons Inc., New York.

36 Crueger, A. and Crueger, W. (1990) *Microbial Enzymes and Biotechnology* 2nd, (eds W.M. Fogarty and C.T. Kelly), Elsevier Applied Science, London.

37 Ensminger, A.H., Ensminger, M.E., Konlande, M.E. and Robson, J.R.K. (1983) *Food Nutrition Encyclopedia*, Pegus Press, Clovis, CA.

38 Gheldof, H. and Engeseth, N.J. (2002) *Journal of Agricultural and Food Chemistry*, **50**, 3050–3055.

39 Cross, C.E., Halliwell, B., Forish, E.T., Pryor, W.A., Ames, B.N. and Saul, R.L.D. (1987) *Annals of Internal Medicine*, **107**, 526–545.

40 Hensley, K. and Floyd, R.A. (2002) *Archives of Biochemistry and Biophysics*, **397**, 377–383.

41 Kinsella, J.E., Frankel, E., German, B. and Kanner, J. (1993) *Food Technology*, **47**, 85–89.

42 Gheldof, N., Wang, X. and Engeseth, N.J. (2002) *Journal of Agricultural and Food Chemistry*, **5**, 5870–5877.

43 Frankel, S., Robinson, G.E. and Berenbaum, M.R. (1998) *Journal of Apicultural Research*, **37**, 27–31.

44 Gheldof, N., Wang, X. and Engeseth, N.J. (2001) Characterization of the antioxidants in honeys from different floral sources, Presented at the Annual Meeting of the Institute Of Food Technologists, New Orleans, LA.

45 Gojmerac, W.L. (1880) *Bees, Bee Keeping, Honey and Pollination*, AVI Publishing Company, Westport, CT.

46 Rowland, C.Y., Blackman, A.J., D'Arcy, B.R. and Rintoul, G.B. (1995) *Journal of Agricultural and Food Chemistry*, **43**, 753–763.

47 Gil, M.I., Ferreres, F., Ortiz, A., Subra, E. and Tomás-Barberán, F.A. (1995) *Journal of Agricultural and Food Chemistry*, **43**, 2833–2838.

48 Ferreres, F., Garcia-Viguera, C., Tomás-Lorente, F. and Tomás-Barberán, F.A. (1993) *Journal of the Science of Food and Agriculture*, **61**, 121–123.

49 Ferreres, F., Andrade, P. and Tomás-Barberán, F.A. (1996) *Journal of the Science of Food and Agriculture*, **44**, 2053–2056.

50 D'Arcy, B. (2005) *Antioxidants in Australian Floral Honeys – Identification of Health-Enhancing Nutrient Components*, Rural Industries Research and Development Corporation, Barton.

51 Brovo, L. (1998) *Nutrition Reviews*, **56**, 317–333.

52 Cook, N.C. and Samman, S. (1996) *Journal of Nutritional Biochemistry*, **7**, 66–76.

53 Ferreres, F., Tomás-Barberán, F.A., Soler, C., Garciá-Viguera, C., Ortiz, A. and Tomás-Lorente, F. (1994) *Apidologie*, **25**, 21–30.

54 Bogdanov, S. (2003) *Trakia Journal of Sciences*, **1**, 19–22.

55 Kaufmann, A., Pacciarelli, B., Prijic, A., Ryser, B. and Roth, S. (1999) *Mitteilungen aus Lebensmitteluntersuchung und Hygiene*, **90**, 167–176.

56 Schwaiger, I. and Schuch, R. (2000) *Deutsche Lebensmittel Rundschau*, **96**, 93–98.

57 Diserens, J.M. and Savoy-Perroud, M.C. (2002) *Report of The NRC*, Nestle Research Center, Lausanne.

58 Klementz, D. and Pestemer, W. (1996) *Nachrichtenblatt des Deutschen Pflanzenschutzdienstes*, **48**, 280–284.

59 Edder, P., Cominoli, A. and Corvi, C. (1998) *Mitteilungen aus dem Gebiete der, Lebensmittelemtersuchung und Hygiene*, **89**, 369–382.

60 Sporns, P., Kwan, S. and Roth, L.A. (1986) *Journal of Food Protection*, **49**, 383–388.

61 Argauer, R.J. and Moats, W.A. (1991) *Apidologie*, **22**, 109–115.

62 Knaggs, M. and Powell, J. (2002) FAPAS (Food Analysis Performance Assessement Scheme), Honey Analysis, Report No 2802, Ring Trial: Determination of Tetracycline in Honey.

63 Iwaki, K., Okomura, N. and Yamazaki, M. (1992) *Journal of Chromatography*, **623**, 153–158.

64 Kaufmann, A., Roth, S., Ryser, B. and Widmer, M. (2001) *Journal of AOAC International*, **85**, 853–860.

65 Bogdanov, S. (1997) *Lebensmittel-Wissenschaft und -Technolgie*, **30**, 748–753.

66 Bose, B. (1982) *Lancet*, **i** (8278), 963.

67 Condon, R.E. (1993) *Surgery*, **113**, 234–235.

68 Green, A.E. (1988) *British Journal of Surgery*, **75**, 1278.

69 Keast-Butler, J. (1980) *Lancet*, **ii** (8198), 809.

70 Mossel, D.A. (1980) *Lancet*, **ii** (8203) 1091

71 Seymour, F.I. and West, K.S. (1951) *Medical Times*, **79**, 104–107.

72 Somerfield, S.D. (1991) *Journal of The Royal Society of Medicine*, **84**, 179.

73 Tovey, F.I. (1991) *Journal of The Royal Society of Medicine*, **84**, 447.

74 Molan, P.C. (2002) *Honey as an Antimicrobial Agent*, Waikato Honey Research Unit, University of Waikato, (http://honey.bio.waikato.ac.nz/honey_intro.shtml and http://honey.bio.waikato.ac.nz/honey_1.shmtl).

75 Chirife, J., Herszage, L., Joseph, A. and Kohn, E.S. (1983) *Antimicrobial Agents and Chemotherapy*, **23**, 766–773.

76 Molan, P.C. (1992) *Bee World*, **73**, 5–28.

77 Molan, P.C. (1992) *Bee World*, **73**, 59–76.

78 Efem, S.E.E. (2005) *British Journal of Surgery*, **75**, 679–681.

79 Taormina, P.J., Niemira, B.A. and Beuchat, L.R. (2001) *International Journal of Food Microbiology*, **69**, 217–225.

80 Manzoori, L.J., Amjadi, M. and Orooji, M. (2006) *Analytical Sciences*, **22**, 1201–1206.

81 Allen, K.L., Molan, P.C. and Reid, G.M. (1991) *Journal of Pharmacy and Pharmacology*, **43**, 817–822.

82 Wahdan, H.A.L. (1998) *Infection*, **26**, 26–31.

83 Russell, K.M., Molan, P.C., Wilkins, A.L. and Holland, P.T. (1990) *Journal of Agricultural and Food Chemistry*, **38**, 10–13.

84 Weston, R.J., Mitchell, K.R. and Allen, K.L. (1999) *Food Chemistry*, **64**, 295–301.

85 Weston, R.J. (2000) *Food Chemistry*, **71**, 235–239.

86 Ferreres, F., Andrade, P. and Tomás-Barberán, F.A. (1994) *Zeitschrift für Lebensmittel-Unterssuchng und- Forschung*, **199**, 32–37.

87 Ferreres, F., Andrade, P., Gil, M.I. and Tomás-Barberán, F.A. (1996) *Zeitschrift für Lebensmittel-Unterssuchng und-Forschung*, **202**, 40–44.

88 Ferreres, F., Juan, T., Pérez-Arquillué, C., Herrera-Marteache, A., García-Viguera, C. and Tomás-Barberán, F.A. (1998) *Journal of the Science of Food and Agriculture*, **77**, 506–510.

89 Serra-Bonvehi, J., Ventura-Coll, F. and Escola-Jorda, R. (1994) *Journal of the American Oil Chemists Society*, **71**, 529–532.

90 Anklam, E. (1998) *Food Chemistry*, **63**, 549–562.

91 Andrade, P., Ferreres, F. and Amaral, M.T. (1997) *Journal of Liquid Chromatography and Related Technologies*, **20**, 2281–2288.

92 Andrade, P., Ferreres, F., Gill, M.I. and Tomas-Barberan, F.A. (1997) *Food Chemistry*, **60**, 79–84.

93 Amiot, M.J., Aubert, S., Gonnet, M. and Tacchini, M. (1989) *Apidologie*, **20**, 115–125.

94 Sabatier, S., Amiot, M.J., Tacchini, M. and Aubert, S. (1992) *Journal of Food Science*, **57**, 773–775.

95 Tomás-Barberán, F.A., Martos, I., Ferreres, F., Radovic, B.S. and Anklam, E. (2001) *Journal of the Science of Food and Agriculture*, **81**, 485–496.

96 Martos, I., Ferreres, F. and Tomás-Barberán, F.A. (2000) *Journal of Agricultural and Food Chemistry*, **48**, 1498–1502.

97 Martos, I., Ferreres, F., Yao, L.H., D'Arcy, B.R., Caffin, N. and Tomás-Barberán, F.A. (2000) *Journal of Agricultural and Food Chemistry*, **48**, 4744–4748.

98 Soler, C., Gil, M.I., García-Viguera, C. and Tomás-Barberán, F.A. (1995) *Apidologie*, **26**, 53–60.

99 Guyot, C., Bouseta, A., Scheirman, V. and Collin, S. (1998) *Journal of Agricultural and Food Chemistry*, **46**, 625–633.

100 Ferreres, F., Blazquez, M.A., Gil, M.I. and Tomás-Barberán, F.A. (1994) *Journal of Chromatography*, **669**, 268–274.

101 Delgado, C., Tomás-Barberán, F.A., Talou, T. and Gaset, A. (1994) *Chromatographia*, **38**, 71–78.

102 Steeg, E. and Montag, A. (1988) *Deutsche Lebensmittel-Rundschau*, **84**, 103–108.

103 Cabras, P., Angioni, A., Tuberose, C., Floris, I., Reniero, F., Guillou, C. and Ghelli, S. (1999) *Journal of Agricultural and Food Chemistry*, **47**, 4064–4067.

104 Bulman, M.W. (1955) *Middlesex Hospital Journal*, **55**, 188–189.

105 Hutton, D.J. (1996) *Nursing Times*, **62**, 1533–1534.

106 Cavanagh, D., Beazley, J. and Ostapowicz, F. (1970) *Journal of Obstetrics and Gynaecology of the British Commonwealth*, **77**, 1037–1040.

107 Blomfield, R. (1973) *Journal of the American Medical Association*, **224**, 905.

108 Burlando, F. (1978) *Minerva Dermatologica*, **113**, 699–706.

109 Armon, P.J. (1980) *Tropical Doctor*, **10**, 91.

110 Bose, B. (1982) *Lancet*, **i**, 963 Lancet Infectious.

111 Dumronglert, E. (1983) *Journal of the National Research Council of Thailand*, **15**, 39–66.

112 Kandil, A., Elbanby, M., Abd-Elwahed, K., Abou Sehly, G. and Ezzat, N. (1987) *Journal of Drug Research (Cairo)*, **17**, 71–75.

113 Wadi, M. (1988) *International Journal of Crude Drug Research*, **26**, 161–168.

114 Green, A.E. (1988) *British Journal of Surgery*, **75**, 1278.

115 Mclnerney, R.J.F. (1990) *Journal of the Royal Society of Medicine*, **83**, 127.

116 Bangroo, A.K., Khatri, R. and Chauhan, S. (2005) *Journal of Indian Association of Pediatric Surgeons*, **10**, 172–175.

117 Lusby, P.E., Coomber, A. and Wilkinson, J.M. (2002) *Journal of Wound, Ostomy and Continence Nursing*, **29**, 295–300.

118 McIntosh, C.D. and Thomson, C.E. (2006) *Journal of Wound Care*, **15**, 133–136.

119 Doner, L.W. (2006) *Journal of the Science of Food and Agriculture*, **5**, 443–456.

120 Cooper, R.A., Halas, E. and Molan, P.C. (2002) *Journal of Burn Care and Rehabilitation*, **23**, 366–370.

121 Dunford, C., Cooper, R., Molan, P. and White, R. (2000) *Nursing Standard*, **15**, 63–68.

122 Simon, A., Sofka, K., Wiszniewsky, G., Blaser, G., Bode, U. and Fleischhack, G. (2006) *Support Care Cancer*, **14**, 91–97.

123 Subrahmanyam, M. (2005) *British Journal of Surgery*, **78**, 497–498.

124 Ingle, R., Levin, J. and Plinder, K. (2006) *South African Medical Journal*, **96**, 831–835.

125 Cooper, R. (2001), *How does Honey heal Wounds?*, in *Honey and Healing* (eds P. Munn and R. Jones), International Bee Research Association, Cardiff, pp. 27–34.

126 Jones, K.P. (2001) *The role of Honey in Wound Healing and Repair*, in *Honey and Healing* (eds P. Munn and R. Jones), International Bee Research Association, Cardiff.

127 Tonks, A.J., Cooper, R.A., Jones, K.P., Blair, S., Parton, J. and Tonks, A. (2003) *Cytokine*, **21**, 242–247.

128 Koshio, O., Akanuma, Y. and Kasuga, M. (1988) *Biochemical Journal*, **50**, 95–101.

129 Al Somai, N., Coley, K.E., Molan, P.C. and Hancock, B.M. (1994) *Journal of the Royal Society of Medicine*, **87**, 9–12.

130 Haffejee, I.E. and Moosa, A. (1985) *British Medical Journal*, **290**, 1866–1867.

131 Molan, P.C. (2002) *Selection of Honey as a Wound Dressing* Waikato Honey Research Unit, University of Waikato (http://honey.Bio.waikato.ac.nz/selection.shtmt).

132 Öztaşan, N., Altinkaynak, K., Akçay, F., Göçer, F. and Dane, Ş. (2005) *Turkish Journal of Veterinary and Animal Sciences*, **29**, 1093–1096.

133 Al-Waili, N. (2003) *Journal of Medicinal Food*, **6**, 231–247.

134 Al-Waili, N. (2003) *FASEB Journal*, **17**, 272.

135 Al-Waili, N. (2003) *European Journal of Medical Research*, **8**, 295–304.

136 Subrahmanyam, M. (1991) *British Journal of Surgery*, **78**, 497–498.

137 Armstrong, D.G. and Lavery, L.A. (1998) *American Family Physician*, **57**, 1325–1332.

138 Drosou, A., Falabella, A. and Kirsner, R. (2003) *Wounds*, **15**, 149–166.

139 Hammouri, K.S. (2004) *Journal of Research in Medical Sciences*, **11**, 20–22.

140 Fotidar, M.R. and Fotidar, S.N. (1945) *Indian Bee Journal*, **7**, 102.

141 Imperato, P.J. and Traoré (1969) *Tropical and Geographical Medicine*, **21**, 62–67.

142 Meier, K.E. and Freitag, G. (1955) *Zeitschrift für Hygiene und Infektionskrankheiten*, **141**, 326–332.

143 Popescu, M.P., Palos, E. and Popescu, F. (1985) *Revista de Chirurgie Oncologie Radiologie ORL, Oftalmologie Stomatologie Seria Oftalmologie*, **29**, 53–61.

144 Emarah, M.H. (1982) *Bulletin of Islamic Medicine*, **2**, 422–425.

145 Mitra, S.K., Sundaram, R., Venkataranganna, M.V., Gopumadhavan, S., Prakash, N.S., Jayaram, H.D. and Sarma, D.N. (2000) *Phytomedicine: International Journal of Phytotherapy and Phytopharmacology*, **7**, 123–127.

146 Al-Waili, N.S. (2004) *Medical Science Monitor: International Medical Journal of Experimental and Clinical Research*, **10**, MT94–MT98.

147 Simon, A., Traynor, K., Santos, K., Blaser, G., Bode, U. and Molan, P. (2007) *eCAM*, **175**, 1–9.

148 Subrahmanyam, M., Shahapure, A.G., Nagne, N.S., Bhagwat, V.R. and Ganu, J.V. (2001) *Annals of Burns Fire and Disasters*, **14**, 143–145.

149 Torregrossa, M.V., Valentino, L., Cucchiara, P., Masellis, M. and Sucameli, M. (2000) *Annals of Burns and Fire Disasters*, **13**, 143–147.

150 Hermans, M.H.E. (1998) *Burns*, **24**, 539–551.

151 Subrahmanyam, M., Hemmady, A.R. and Pawer, S.G. (2003) *Annals of Burns and Fire Disasters*, **16**, 2.

11
Probiotics: Benefits in Human Health and Bacterial Disease Management

María Carmen Collado and Yolanda Sanz

Abstract

The gastrointestinal tract microbiota constitutes a complex ecosystem that plays an important role in host health due to its involvement in nutritional, immunologic and physiological functions. Microbial imbalances have been associated with enhanced risk of specific diseases. Specific strains of *Lactobacillus*, *Bifidobacterium* and, more recently, *Propionibacterium* have been introduced as probiotics in food products due to their demonstrated health-promoting effects. The most extensive studies and clinical applications of probiotics have been related to the management of gastrointestinal infections caused by pathogenic microorganisms. The possible mechanisms underlying these antagonistic effects include competition for adhesion sites and nutritional sources, secretion of antimicrobial substances, toxin inactivation, and immune stimulation. The development of adjuvant or alternative therapies based on bacterial replacement is considered important due to the rapid emergence of antibiotic-resistant pathogenic strains and the adverse consequences of antibiotic therapies on the protective microbiota. Probiotic strains have been shown to exert a significant protective effect against pathogen infections, rotavirus diarrhea and antibiotic-associated diarrhea, and also in inflammatory bowel diseases and allergies. Probiotics could be useful to correct alterations in the composition of the intestinal microbiota associated with specific diseases. However, these processes are highly specific. It is important to characterize the properties of both the probiotic strains and the specific pathogens involved with these disorders in order to select the best probiotic strains for preventive or therapeutic purposes.

11.1
Gut Microbiota

The human gut is inhabited by an enormous, complex and dynamic population of different microorganisms. The gut microbiota influences human health through an

New Strategies Combating Bacterial Infection. Edited by Iqbal Ahmad and Farrukh Aqil
Copyright © 2009 WILEY-VCH Verlag GmbH & Co. KGaA, Weinheim
ISBN: 978-3-527-32206-0

impact on the gut defence barrier, immune function and nutrient utilization [1, 2]. Although bacteria are distributed throughout the gastrointestinal tract, the major concentration of microorganisms and the highest metabolic activity is found in the large intestine. Up to 500–600 species of microorganisms may be present in the adult human intestine and it has been estimated that bacteria account for 35–50% of the volume content of the human colon [1]. These include *Bacteroides, Lactobacillus, Clostridium, Fusobacterium, Bifidobacterium, Eubacterium, Peptococcus, Peptostreptococcus, Escherichia* and *Veillonella*. The bacterial strains with identified beneficial properties include mainly *Bifidobacterium* and *Lactobacillus*.

Microbiota deviations have been associated with enhanced risk of specific diseases including allergy, inflammatory bowel diseases (IBDs), obesity, diabetes and others [3–6]. There is a growing interest in beneficial microbes with specific functions in the human gut, which can be used in foods or supplements that increase health and prevent and treat diseases. Specific strains of *Lactobacillus, Bifidobacterium* and *Propionibacterium* have been introduced as probiotics in food products due to their observed health-promoting effects [7–9]. The protective role of probiotic bacteria against gastrointestinal pathogens and diseases, and the underlying mechanisms, have received special attention. The development of adjuvant or alternative therapies based on bacterial replacement is considered important due to the rapid emergence of antibiotic-resistant pathogenic strains and the adverse consequences of antibiotic therapies on the protective microbiota [10].

11.1.1
Functions of the Intestinal Microbiota

The intestinal microbiota develops important metabolic activities and immunologic functions that serve to maintain symbiotic relationships with the host. The activity of the intestinal microbiota acts as an active 'organ', which is involved in (i) the improvement of nutrient bioavailability and degradation of nondigestible dietary compounds, (ii) the supply of new nutrients, and (iii) the removal of harmful and antinutritional compounds. These metabolic functions have important implications in human nutrition and health, although they depend on the composition of the microbiota, and its complex interactions with the diet and the host. The dynamic environment of the human gut demands effective adaptation of the bacteria to the availability of certain nutritional components. Some studies have reported that the microbiota can efficiently adapt to the availability of different carbohydrates [11] and some species can actively indicate how much nutrients they need [12, 13]. The host provides nutrients to the microbiota in the amounts necessary to avoid their production of high quantities of nutrient. Thus, no nutrients are left for undesirable species and these are inhibited.

Gut microbiota is essential for processing dietary polysaccharides affecting energy harvest from the diet. It provides additional energy in the form of short-chain fatty acids, including acetate, propionate and butyrate. Acetate is taken up primarily by peripheral tissues [14] and can also be used by adipocytes for lipogenesis [15]. Propionate is an important precursor for gluconeogenesis in the liver and butyric acid is mostly metabolized by the intestinal epithelium as a main energy source for the intestinal epithelium, providing between 60 and 70% of all the energy and helping to maintain colonic mucosal health. Moreover, the microbiota has been reported to be involved in fat storage in the host [16]. A 60% increase in body fat content in germ-free rats upon colonization with conventional microbiota has been reported [16] and some gut microbes have also been reported to exert cholesterol-lowering effects [17].

Several members of the intestinal microbiota produce vitamins and provide them to the host, mainly vitamin K and also those of the vitamin B family [12, 18, 19]. The significance of the microbiota in salvaging energy and producing vitamins is most clearly seen in germ-free animals. Compared to conventional animals, germ-free animals require 30% more energy in their diet, supplementation of which with vitamins K and B is mandatory to maintain their body weight.

The absorption of calcium, magnesium and phosphorus is also improved by carbohydrate fermentation and production of short-chain fatty acids [20] and pH reduction. In addition, the acidification of intestinal environmental also inhibits the development and colonization of pathogens or undesirable bacteria, as well as the production of toxic elements derived from their metabolism (ammonia, phenol compounds, amines, etc.).

The microbiota is reported to contribute to human protein homeostasis. At least some requirements for amino acids are met by microbial synthesis. In contrast, fermentation of amino acids may lead to the production of a variety of toxic substances such as tumor inducers and promoters.

The intestinal immune system constitutes the primary immune organ of the human body and an essential element of the host defense against pathogenic microorganisms [21–23]. Studies have demonstrated that animals bred in a germ-free environment are highly susceptible to infections; thus, the intestinal microbiota is considered an important constituent in the mucosal defence barrier. The phenomenon is termed colonization resistance: bacteria in the gut mucosa compete for the same attachment sites as pathogenic bacteria, use the same nutrients, and produce several antimicrobial compounds inhibiting the growth of pathogens and other transient incoming bacteria, which are not members of the resident intestinal microbiota [24–26].

The intestinal microbiota also provides an important stimulus for the development of the host immune system, and regulates innate and adaptive immunity. At birth, the immune system is immature and develops upon exposure to the intestinal microbiota [27], increasing the number of Peyer's patches and immunoglobulin producing cells [28]. The innate immune system allows the host to sense a concrete microbial environment in order to promote the release of signaling molecules (cytokines and chemokines) and initiate the immune response. In general, it is considered that epithelial and monocytic cells recognize the signature molecules called pathogen-associated molecular patterns, which activate the host defense mechanisms. In contrast, the commensal bacteria share signature molecules called microbe-associated molecular patterns, which do not tiger proinflammatory responses [29].

11.1.2
Development of the Gut Microbiota

The intestinal colonization that follows birth represents the host's earliest contact with microbes [30]. During birth, and rapidly thereafter, microbes from the mother and the environment colonize the gastrointestinal tract of the infant. Initially, the microbiota of a newborn is strongly dependent on the mother's microbiota, the mode of birth and hygiene; subsequently, it is influenced by feeding practices and the environment of the infant [30].

Initial microbial studies were based on culture-dependent methodologies, limiting our understanding of intestinal microbiota species composition [30, 31]; however, recent studies using culture-independent methodologies have shown that the microbiota of infants develops rapidly during the first week and remains unstable for the first year of life, becoming more stable afterwards. The intestinal microbiota forms a natural defence barrier. Thus, early establishment of a healthy microbiota provides the first key step in long-term well-being later in life. The bacteria identified include members of the genera *Bifidobacterium*, *Enterococcus*, *Clostridium* and *Enterobacter* among others [32]. Lactic acid bacteria represent less than 1% of the total microbiota in infants, but *Bifidobacterium* makes up 60–90% of the total fecal microbiota in breast-fed infants [30, 33].

11.1.3
Factors Influencing the Composition of the Gut Microbiota

The distribution of the microbiota in the gastrointestinal tract is the result of dynamic processes. Competition for nutrients the inhibition by antimicrobial substances secreted by microbiota, predation, and competitive exclusion from binding sites contribute to the regulation of microbial populations in different gastrointestinal sections. In addition, the intestinal microbiota is influenced by other parameters related to the host such as its genotype and as the diet.

Some studies demonstrated that specific genes such as those of the major histocompatibility complex may have an impact on the development of a specific murine fecal microbiota [34, 35]. The demonstration of the genetic influence opens a new gate to analyze the pathogenesis of diseases induced by specific bacteria and genetic factors such as Crohn's disease [34].

Among environmental factors, the diet may exert a major effect on the composition and activity of the gut microbiota. As long as infants are breast-fed and/or formula-fed, the microbiota will be dominated or not by bifidobacteria in early life. Breast-feeding contributes to increased levels of bifidobacteria [36, 37]; this observation is consistent with data indicating that breast milk contains substances – including peptides and oligosaccharides – that promote the growth of bifido-bacteria [37]. Some studies reported that more lactobacilli and streptococci were present in breast-fed infants whereas formula-fed infants contain higher levels of *Staphylococcus*, *Escherichia coli* and *Clostridium difficile* [38]. A recent study shows that the prevalence and levels of *C. difficile* and *E. coli* were significantly lower in breast-fed than formula-fed infants [39], whereas bifidobacterial counts were similar. With the introduction of solid foods, the microbiota undergoes a more dramatic change and becomes diverse. This diversity results in an adult-like microbiota by the age of approximately 2 years [28, 32].

Recent molecular studies have identified *Bifidobacterium infantis*, *Bifidobacterium longum* and *Bifidobacterium breve* as the species most often found in infants [40]. The greatest differences between the microbiota of breast-fed and formula-fed infants appear to be in lactic acid bacteria colonization and the species of bifidobacteria present. *B. breve* has been reported as the most common species present in breast-fed infants, while other authors found that *B. infantis*, *B. longum* and *Bifidobacterium bifidum* predominate in breast-fed infants [30]. More recent molecular studies indicate that the *Lactobacillus acidophilus* group of organisms are the most common lactobacilli in both breast-fed and formula-fed infant feces [41], whereas *Lactobacillus gasseri* was also the most common species in breast-fed infants [41]. In addition, modifications in gut physiology and function occur with ageing, which can be linked to changes in the composition and metabolic activities of the microbiota. Several studies have reported a decrease in the numbers of beneficial microorganisms such as *Bifidobacterium* and an increase in the numbers of potentially pathogens such as the *Clostridium* group [42, 43].

Due to the environmental changes and different conditions along the gastrointes-tinal tract, different microbes can be found at different sites. The flow in the upper part of the gut, as well as secretions from the stomach, liver and pancreas, does not allow the presence of a large microbial population. However, in the lower part of the gut, the flow of the digesta becomes slower, allowing the establishment of a larger microbiota [44].

Important differences also appear between the luminal and the mucosal milieu. While the environment in the lumen may be anaerobic, at the mucosa some oxygen may leak from the tissue, creating a micro-aerobic environment. Furthermore, Paneth cells in the mucosa secrete antimicrobial substances such as lysozyme and defensins [45, 46], whereas enterocytes transport secretory immunoglobulin A from the lamina propria to the lumen [47]. This confers a different composition to the luminal and mucosal microbiota [48].

Alterations in composition of the gut microbiota are linked with certain diseases. Imbalances in the composition of the microbiota have also been reported in active diseases patients as compared to controls in infection, allergies, irritable bowel diseases and celiac [3–6, 49–51].

11.2
Probiotics, Intestinal Microbiota and Host Health

11.2.1
Probiotic Concept and Selection Criteria

A probiotic has been defined as a 'live microorganism which when administered in adequate amounts confers a health benefit to the host' [52]. Probiotics were originally used to influence both animal and human health through modulation of the intestinal microbiota. At present, specific live microbial food ingredients and their effects on human health are studied both within food matrices and as single- or mixed-culture preparations [53, 54]. Several well-characterized strains of lactobacilli and bifidobacteria are available for human use to reduce the risk of gastrointestinal infections or treat such infections [9, 23].

There was a need for guidelines to set out a systematic approach for the evaluation of probiotics. In an attempt to identify and establish the minimum requirements needed for a probiotic status the Food and Agriculture Organization of the United Nations–World Health Organization (FAO-WHO) Expert Consultation published the guidelines of Evaluation of Health and Nutritional Properties of Probiotics in Foods [52]. A scheme outlining these guidelines for the evaluation of probiotics for food use is shown in Figure 11.1.

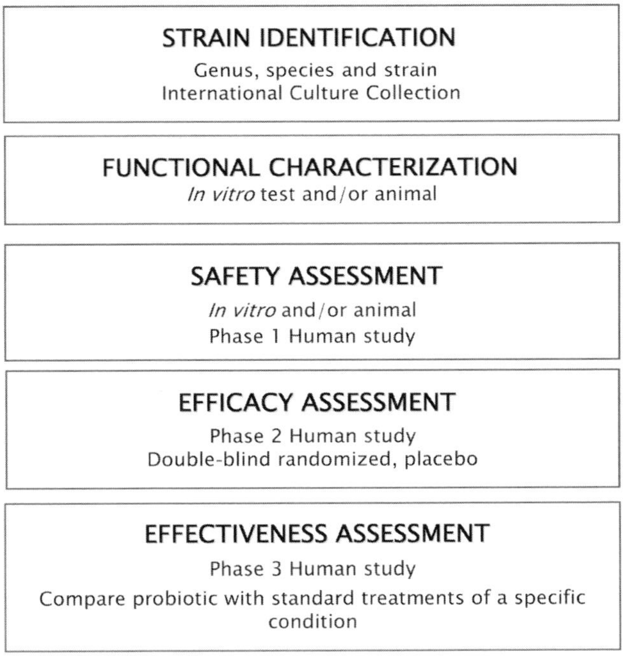

STRAIN IDENTIFICATION
Genus, species and strain
International Culture Collection

FUNCTIONAL CHARACTERIZATION
In vitro test and/or animal

SAFETY ASSESSMENT
In vitro and/or animal
Phase 1 Human study

EFFICACY ASSESSMENT
Phase 2 Human study
Double-blind randomized, placebo

EFFECTIVENESS ASSESSMENT
Phase 3 Human study
Compare probiotic with standard treatments of a specific condition

---

Figure 11.1 Guidelines for the evaluation of probiotics for food use. Adapted from [52].

11.2.1.1 Strain and Specie Identification

Strain identification is important to link a strain to a specific health effect as well as to enable accurate surveillance and epidemiological studies [52]. A possible exception is the general ability of *Streptococcus thermophilus* and *Lactobacillus delbrueckii bulgaricus* to enhance lactose digestion in lactose-intolerant individuals. In this case individual strain identity is not critical [52].

A reliable probiotic product requires correct identification of the bacterial species used and its announcement on the product label. Probiotics should be identified using currently available molecular methods, such as 16S rDNA sequencing or DNA/DNA hybridization, randomly amplified polymorphic DNA polymerase chain reaction or pulsed-field gel electrophoresis and up-to-date taxonomical nomenclature [55–57]. However, numerous studies have shown that the identity of recovered microorganisms is not always the same as that indicated on the product label [55, 57–59].

11.2.1.2 Functional Characterization

Physiological and Metabolic Characteristics   Several methods are available to study physiological characteristics of probiotic strains. Carbohydrate fermentation and enzymatic activity profiles have been widely used [55, 57, 59]. The metabolic activities of strains have been evaluated in gastrointestinal model systems. In this regard it is important to select the specific substrates or enzymatic activities relevant to the expected functional effects of the strain. Other specific tests are the ability to hydrolyze bile salts [17] and the ability to produce antimicrobial substances [25, 60].

Tolerance to Gastrointestinal Conditions   The viability of probiotic strains is considered to be important in order to ensure their optimal functionality. It has been suggested that they should reach the intestine alive and in a sufficient number (10^6–10^7 cells/ml) for their benefits to be appreciated [61, 62]. This requires their survival in the food vehicle during its shelf-life, and their resistance to the gastrointestinal conditions, the acidic environment of the stomach and bile secreted in the duodenum. To guarantee their survival during passage through the gastrointestinal tract, probiotic strains are primarily screened for their tolerance to acid pH and bile [63]. The effect of gastrointestinal conditions on probiotic survival has been assessed and different bacteria show different levels of tolerance [64–68]. Different techniques have been used for this purpose [65–68] including traditional plate counting method and also staining with fluorochromes such as the LIVE/DEAD BacLight bacterial viability kit [66–68]. This kit has been used for total counts of the bacterial population and also for differentiating live from dead bacterial cells on the basis of membrane integrity. The lack of standard procedures for the analysis of tolerance to gastrointestinal conditions makes comparisons difficult. In general, tolerance to gastrointestinal conditions is low; as a consequence, several methodologies including those based on stress adaptation mechanisms of probiotic bacteria have been reported as possible strategies to enhance their acid and bile resistance [63, 69, 70].

Adhesion Adhesion to the intestinal mucosa is regarded as a prerequisite for colonization and is an important characteristic related to the ability of strains to modulate the immune system [4, 22, 71]. Thus, adhesion has been one of the main selection criteria for new probiotic strains [4, 9, 23, 26, 54]. Many different intestinal mucosa models have been used to assess the adhesive ability of probiotics; among them, the adhesion to human intestinal mucus has been widely used [26, 54, 72–78] and good correlations have been reported with other models [21, 73, 79]. The effects of gastrointestinal conditions (pH, bile, digestive enzymes) and the effects of acid and bile resistance acquisition on the adhesion of probiotic bacteria have also been documented [77, 80]. In addition, recent reports describe how the presence of difference substances such as calcium ions or exopolysaccharides produced by probiotic bacteria can modify the bacterial adhesion to intestinal mucus [81, 82].

The adhesion levels of the probiotic and pathogenic strains showed great variability depending on the strain, species and genus [26, 54, 77, 78, 80, 83, 84]. In *in vitro* trials, the probiotic properties have mainly been tested alone or in combination with yoghurt bacteria such as *L. delbrueckii* and *L. acidophilus* [76], but rarely combined with other probiotics. However, few studies are available on the interactions of probiotics regarding adhesion properties in the intestinal mucus system [76, 85].

Probiotic bacteria can competitively inhibit the adhesion of pathogenic microorganisms and displace the previously adhered pathogens, such as *Salmonella*, *E. coli*, *Listeria monocytogenes*, *Staphylococcus aureus*, *Bacteroides vulgatus*, *C. difficile* and *Clostridium perfringens* [26, 54, 77, 78, 80, 86, 87]. It has been shown that some lactobacilli and bifidobacteria share carbohydrate-binding specificities with some enteropathogens [88], providing the rationale for the use of these microorganisms to prevent infection at an early stages by inhibiting the adhesion of enteropathogens by competitive exclusion. In general, it is considered that probiotic strains are able to inhibit the attachment of pathogenic bacteria by means of steric hindrance at enterocyte pathogen receptors. In addition, it has been observed that proteinaceous components are involved in the adhesion of probiotic strains to intestinal cells [26, 77, 86, 89]. The adhesion properties of probiotics widely vary depending on the strain, and high adherence ability in one strain not always guarantee an *in vivo* persistence and protective effect. This should always be corroborated by studies in animal models and humans [85, 87]. No relation was found between the results obtained for the adhesion inhibition and displacement of pathogens, suggesting that different mechanisms could be implied in both processes. A direct correlation has not been found between the overall adhesion level of probiotic strains and their abilities to inhibit or displace pathogens [26, 54, 78, 87, 90]. Probiotic strains and combinations that inhibit and displace pathogens may be excellent candidates for their use as new combinations in fermented milk products or as new therapies to prevent and treat specific diseases. Recent reports [54, 77, 78, 84] demonstrate that all probiotic strains and combinations tested showed abilities to inhibit, displace and compete with pathogens, but it is important to take into account the high specificity of these processes to characterize the properties of the strains in order to select the best strain combinations to prevent or treat infection by a specific pathogen.

Antimicrobial Substances   The antimicrobial metabolites produced by lactic acid bacteria can be divided into two major groups: (i) low-molecular-mass compounds (bellow 1000 Da) such as organic acids, which have a broad spectrum of action, and (ii) antimicrobial proteins, termed bacteriocins (over 1000 Da), which have a relatively narrow specificity of action against closely related organisms and other Gram-positive bacteria [91, 92].

The acids produced from fermentative metabolism of carbohydrates by probiotics have been considered to be the main antimicrobial compounds responsible for their inhibitory activity against pathogenic microorganisms [93, 94]. Bacteriocins are proteins or protein complexes that show bactericidal activity against bacterial species that are closely related to the producer species. Most of the studies related to the characterization of bacteriocins or bacteriocin-like compounds from lactic acid bacteria have been focused on species of the genera *Lactobacillus*, *Pediococcus* and *Enterococcus*, because of the diversity of their species and their potential applications as natural preservatives in foods [94]. Probiotics have also exhibited antagonistic effects against pathogens belonging to the genera *Listeria*, *Clostridium*, *Salmonella*, *Shigella*, *Escherichia*, *Helicobacter*, *Campylobacter* and *Candida* [60, 84, 95–98]. Due to the potential interest of these antimicrobial proteins in novel therapeutic developments, further studies should be carried out on their genetics, biochemistry and mechanisms of action.

11.2.1.3   Safety Considerations

Safety assessment is an essential phase in the selection and evaluation of probiotics [52]. Probiotic strains have been tested for safety, but the long history of safe consumption of some probiotics of *Lactobacillus* and *Bifidobacterium* genera could be considered the best proof of their safety [52]. Their occurrence as normal commensals of the mammalian flora and their established safe use in a diversity of foods and supplement products worldwide supports this conclusion. Although some strains have been associated with rare cases of bacteremia, usually in patients with severe underlying diseases, the safety of members of these genera is generally recognized [52, 99]. However, probiotics may theoretically be responsible for systemic infections, deleterious metabolic activities and gene transfer, and also excessive immune stimulation in susceptible individuals [52]. Regarding other bacteria, such as *Enterococcus*, *Saccharomyces boulardii* or some members of the genus *Bacillus* which have been used as probiotics, the situation is more complicated, even when they have been used for some time, and requires further case-by-case analysis.

According to the FAO-WHO guidelines for probiotic use [52], the following traits of probiotic strains must be characterized: (i) determination of antibiotic resistance patterns, (ii) assessment of deleterious metabolic activities (e.g. D-lactic acid production), (iii) assessment of side-effects during human studies, (iv) epidemiological surveillance of adverse incidents in consumers (postmarket), and (v) toxin production and determination of hemolytic activity.

11.2.1.4   Functional *In Vitro* and *In Vivo* Studies Using Animals and Humans

In vitro studies are important to assess the safety of probiotic strains and are also useful to expand the knowledge of strains. However, it has been realized that the currently available tests including identification of strains and characterization of some activities are not accurate to predict the potential use and functionality of probiotic strains *in vivo*. Thus, *in vitro* data are not enough to assure probiotic characteristics of specific strains. For this reason, probiotic strains for human use will require human studies. The principal outcome of efficacy studies on probiotics should be proven benefits in human studies, such as statistically and biologically significant improvements in conditions, reduction of symptoms, improved well-being or quality of life, reduced risk of disease, or prolonged time for a relapse or faster recovery from illness. Definite evidence for efficacy should be proven by *in vivo* tests in humans.

Probiotics have been tested for an impact on a variety of clinical conditions [52]. Standard methods for clinical evaluations comprise phase 1 (safety), phase 2 (efficacy), phase 3 (effectiveness) and phase 4 (surveillance) studies. Phase 1 studies focused on safety, phase 2 studies are generally in the form of a randomized, double-blind, placebo-controlled design, measure efficacy compared with placebo or controls. Probiotics delivered in food generally are not tested in phase 3 studies, which are concerned with comparison with a standard therapy. When a claim is made for a probiotic altering a disease state, the claim should be made based on sound scientific evidence in human subjects.

11.2.1.5 Health Claims

So far, in most countries, only general health claims were made on foods containing probiotics. The FAO-WHO [52] recommended that specific health claims on foods be allowed relating to the use of probiotics, where sufficient scientific evidence is available, as per the guidelines set forth in their report. Such specific health claims should be permitted on the product label.

As different probiotics may interact with the host in different ways, the data available for the most common probiotics and their health benefits need to be assessed, on a strain-by-strain basis, before any health-related product claims could be approved [99]. Specific protocols for probiotic efficacy assessment are needed, although the application of the efficacy assessment protocols normally used in the pharmaceutical industry has provided a standard for probiotic studies. By using this approach it can be said that certain specific probiotics have scientifically proven benefits which can be attributed to specific products [52, 99].

11.2.2
Probiotic Benefits on Human Health

Some of the beneficial effects of probiotic consumption include: improvement of intestinal tract health by means of regulation of microbiota, and stimulation and development of the immune system, synthesizing and enhancing the bioavailability of nutrients, reducing symptoms of lactose intolerance, and reducing the risk of certain diseases. The mechanisms by which probiotics exert their effects are largely unknown, but may involve modifying gut pH, antagonizing pathogens through production of antimicrobial compounds, competing for pathogen binding, receptor sites, nutrients and growth factors, stimulating immunomodulatory cells, and producing lactase. See Figure 11.2 [100].

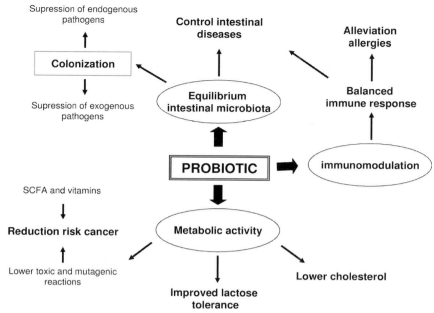

Figure 11.2 Beneficial effects of probiotics on human health. Adapted from [100].

Microbiota deviations have been associated with an enhanced risk of specific diseases as well as with the presentation of these diseases [3, 4, 6, 101]. Therefore, modulation of an unbalanced indigenous microbiota forms the rationale of probiotic therapy. During the last few decades, a number of studies have been carried out on the effects of probiotic microorganisms, using different formulae and with numerous purposes of preventing or treating diseases. Probiotic strains have been shown to exert a protective effect against acute diarrhea, rotavirus diarrhea and antibiotic-associated diarrhea [102–106], as well as *Helicobacter pylori* infection [107], and they alleviate symptoms of gastrointestinal diseases such as irritable bowel syndrome (IBS) [108–110]. In addition, probiotics have shown other health benefits related to pathogen infection and immune system stimulation [37, 105]. The use of probiotics should be further investigated for their possible benefits and side-effects if any [100, 111].

11.3
Studies on the Management and Prevention of Diseases

11.3.1
Pathogen Infection

The most extensive studies and clinical applications of probiotics have been related to the management of gastrointestinal infections caused by pathogenic microorganisms.

11.3.1.1 *H. pylori* Infection

H. pylori is a Gram-negative spiral-shaped, microaerophilic rod colonizing the human gastric mucosa. This microorganism is a specific inhabitant of the human stomach present in 70–90% of the population in developing countries and in 25–50% in developed countries [112]. Urease, an important product produced by *H. pylori*, hydrolyzes urea to ammonium, leading to increased pH in the stomach that promotes colonization of the microorganism. It has been demonstrated that the eradication of *H. pylori* from the stomach requires a combination of therapies. Antibiotics (e.g. amoxicillin, clarithromycin or nitroimidazoles) are used together with acid suppression drugs (proton pump inhibitors or H_2-receptor antagonists), in triple or quadruple combinations [113, 114]. Treatment aimed at *H. pylori* eradication has been reported to give rise to ecological disturbances with suppression of the normal microbiota and the emergence of antibiotic-resistant microbes [115]. Several *in vitro* and *in vivo* studies on the role of probiotics in the treatment of *H. pylori* infections have been performed during recent years, and they have demonstrated that the probiotic strains inhibit the growth or the attachment of *H. pylori* by means of organic acid production, antimicrobial substances of proteinaceous origin, competitive inhibition for the binding sites of mucus cells, and immunomodulation [98, 116–120]. It has been observed in mice that *Lactobacillus salivarius* inhibits the colonization of *H. pylori* and *L. salivarius* given after *H. pylori* implantation also eliminates the colonization by *H. pylori* [121]. In humans, some studies reported the potential use of probiotics for *H. pylori* treatment and eradication [122–125]. No study could demonstrate eradication of *H. pylori* infection by probiotic treatment [125], but long-term consumption of probiotics may have a favorable effect on *H. pylori* infection in humans, particularly by reducing the risk of developing disorders associated with high degrees of gastric inflammation.

11.3.1.2 Diarrhea in Children

The ability of probiotics to decrease the incidence or duration of certain diarrhea illnesses is perhaps the most substantiated health effect of probiotics. A number of specific strains, including *Lactobacillus* GG, *Lactobacillus reuteri*, *S. boulardii*, *Bifidobacterium* strains and others, have been shown to have significant benefit for acute diarrhea [46, 100, 126–130] including travellers' diarrhea and diarrhea disease in young children caused by rotaviruses [100, 130–134].

Probiotic strains have been tested for their ability to prevent or treat diarrhea induced by enteropathogenic microorganisms like enterotoxigenic *E. coli*, *Shigella* spp. and *Salmonella* spp. These pathogens are the main causes of acute diarrhea in travellers accounting for about 80% of the cases with an identified pathogen [135]. The strongest evidence of a beneficial effect of defined strains of probiotics has been established using *Lactobacillus rhamnosus* GG and *Bifidobacterium lactis* BB-12 for prevention [128, 135] and treatment [127, 131, 132, 136–138] of acute diarrhea. In addition, rotavirus infection is a major cause of severe diarrhea in infants and young children both in developed and developing countries [139]. Several studies are now indicating that rehydration combined with probiotic strains shortens the duration of diarrhea [140]. Several clinical studies indicated that a fermented milk containing the probiotic *Lactobacillus casei* DN-114001 was able to reduce the severity and duration of acute diarrhea in young children [141]. It was shown that *Lactobacillus* GG administered in an oral rehydration solution to children with acute diarrhea resulted in a shorter duration of the disease [127]. Reduction of the risk of contracting diarrhea and a reduction of rotavirus infection were demonstrated in children receiving a bifidobacteria-supplemented formula [142].

11.3.1.3 Antibiotic- and *C. difficile*-Associated Diarrhea

Some probiotic strains have been shown to exert a protective effect against *C. difficile* diarrhea and antibiotic-associated diarrhea (AAD) [143, 144]. A common complication of treatments with antimicrobial agents is the development of AAD in 5–25% of patients [145]. Antimicrobial treatment alters the ecological balance of the healthy microbiota, which can result in diarrhea and the emergence of some pathogens like *C. difficile*. The immune response of the host to *C. difficile* toxins has further been shown to be a determinant of susceptibility [135].

Lactobacillus, Bifidobacterium and Streptococcus strains have all been evaluated for prevention or treatment of diarrhea associated with antibiotic use, and found to be safe [146]. Several reviews support benefit but still call for large placebo-controlled trials to determine species and dose effectiveness for prevention, and to establish effects on length of hospital stay and cost effectiveness [104, 140]. A recent meta-analysis of probiotics in the prevention of AAD that evaluated nine controlled trials indicated that *Lactobacillus* and *S. boulardii* have the potential to prevent AAD [147]. The preventive effects of *Lactobacillus* strains LGG or DN-114001 on AAD have recently been examined in children receiving various antimicrobial therapies and they showed beneficial effects [138, 148].

11.3.2
Irritable Bowel Disease (IBD) and Irritable Bowel Syndrome (IBS)

IBD is a chronic and recurrent inflammation generally affecting the colon or the small intestine, and includes ulcerative colitis and Crohn's disease [149]. The etiology of IBD remains unclear, but genetic predisposition and alterations in microbiota are involved. There is evidence that the immune system reacts abnormally towards the endogenous microbiota [150]. Patients with IBDs have higher levels of specific bacteria such as *Bacteroides* attached to epithelial cells than healthy people [151]. The microbiota in subjects with IBS has been shown to be less stable compared to that of healthy adults [108, 152]. No single deviance has been identified in IBS microbiota, but various and different alterations in the bacterial composition have, nonetheless, been characterized by a range of techniques [108, 153]. Therapeutic manipulation of the normal gastrointestinal microbiota using probiotics, alone or in different combinations, has been regarded as a treatment option [108, 154, 155]. Researchers are assessing the use of probiotic strains and combinations in the treatment of IBDs [100, 156–158], ulcerative colitis [159], chronic pouchitis [126] and Crohn's disease [160]. The best results have been obtained so far on ulcerative colitis [159] and chronic pouchitis [161].

11.3.3
Allergic Disease

The prevalence of allergic diseases continues to increase in the developed world and constitutes a common health problem among children [162]. The nomenclature and definition of atopy and allergic diseases have been standardized by the European Academy of Allergy and Immunology [163, 164]. According to this nomenclature, 'atopy' is to be defined as a 'personal or familial tendency to produce immunoglobulin E antibodies in response to low doses of allergens, usually proteins, and to develop typical symptoms such as asthma or eczema/dermatitis'. 'Allergy' is defined as a 'hypersensitivity reaction initiated by immunologic mechanisms'; based on the immunological mechanisms involved, allergy is classified into immunoglobulin E-mediated allergy or non-immunoglobulin E-mediated allergy [163].

Microbiota aberrations have been related to the development of allergic diseases [3, 165]. A reduced ratio of bifidobacteria to clostridia has appeared in infants developing atopy [3], and allergic patients have been shown to be more often colonized with *Clostridium* and *Staphylococcus*, and have fewer *Enterococcus* and *Bifidobacterium* than do nonallergic patients [3, 166]. Gut microbiota composition may precede the development and manifestation of atopic outcomes, while early colonization with *E. coli* has been associated with a higher risk for developing eczema, and *C. difficile* has been associated with eczema, recurrent wheeze and allergic sensitization in infancy [167]. The hygiene hypothesis supports the rapid increase of atopy related to lower microbe exposure in early life and subsequent lower number of infections during infancy [168]. Recent studies reported microbiota differences between allergic and healthy infants in countries with high and low allergy prevalence [3, 169]. These changes on microbiota may be counterbalanced by probiotic bacteria, and also positive clinical effects on the prevention and treatment of atopic diseases of probiotics have been reported [3, 170–175]. The preventive potential of probiotics in atopic diseases has been demonstrated in a double-blind, randomized, placebo-controlled trial [3, 174]. *L. rhamnosus* GG was given to pregnant women for 4 weeks prior to delivery, then to newborns at high risk of allergy for 6 months, with the result that there was a significant reduction in early atopic disease [172, 176]. This study illustrates the potential for probiotic microorganisms to modulate the immune response and prevent allergic diseases. In other clinical studies with infants allergic to cow's milk, atopic dermatitis was alleviated by ingestion of probiotic strains *L. rhamnosus* GG and *B. lactis* Bb12 [170, 171]. However, no beneficial effects were found in adolescents with pollen allergy [177].

11.4
Clinical Prospects of Gut Microbiota Research

Recent advances have been made in the understanding of when to use probiotics and how they impact specific pathological states. Current evidence supports the concept that consumption of probiotics may have benefits in a multitude of disorders. The effects of probiotics in the gut are well documented, and they include the upregulation of immunoglobulins, downregulation of inflammatory cytokines and enhancement of gut barrier function, among others. In addition, new evidence supports the use of probiotics in the prevention and treatment of a number of diseases, including atopic diseases, immune disorders, obesity and diabetes.

New knowledge about intestinal microbiota, nutrition, immunity and genetics in health and disease will help to develop new probiotic strains with disease-specific functions.

Acknowledgments

This work was supported by grants AGL 2007-66126-C03-01/ALI and Consolider Fun-C-Food CSD 2007-00063 from the Spanish Ministry of Science and Innovation. The I3P-CSIC Postdoctoral Contract from the European Social Fund to M.C. Collado is fully acknowledged.

References

1 Hooper, L.V. and Gordon, J.I. (2001) *Science*, **292**, 1115–1118.

2 Guarner, F. and Malagelada, J.R. (2003) *Best Practice and Research Clinical Gastroenterology*, **17**, 793–804.

3 Kalliomäki, M., Salminen, S., Arvilommi, H., Kero, P., Koskinen, P. and Isolauri, E. (2001) *Lancet*, **357**, 1076–1079.

4 Juntunen, M., Kirjavainen, P.V., Ouwehand, A.C., Salminen, S. and Isolauri, E. (2001) *Clinical and Diagnostic Laboratory Immunology*, **8**, 293–296.

5 Ley, R.E., Backhed, F., Turnbaugh, P., Lozupone, C.A., Knight, R.D. and Gordon, J.I. (2005) *Proceedings of the National Academy of Sciences of the United States of America*, **102**, 11070–11075.

6 Turnbaugh, P.J., Ley, R.E., Mahowald, M.A., Magrini, V., Mardis, E.R. and Gordon, J.I. (2006) *Nature*, **444**, 1027–1031.

7 Lee, L.K. and Salminen, S. (1995) *Trends in Food Science and Technology*, **6**, 241–246.

8 Huis in't veld, J.H.J., Bosschaert, M.A.R. and Shortt, C. (1998) *Food Science and Technology*, **12**, 46–49.

9 Salminen, S., Ouwehand, A.C., Benno, Y. and Lee, Y.K. (1999) *Trends in Food Science and Technology*, **10**, 107–110.

10 Forestier, C., De Champs, C., Vatoux, C. and Joly, B. (2001) *Research in Microbiology*, **152**, 167–173.

11 de Vos, W.M., Bron, P.A. and Kleerebezem, M. (2004) *Current Opinion in Biotechnology*, **15**, 86–93.

12 Hooper, L.V., Midvedt, T. and Gordon, J.I. (2002) *Annual Review of Nutrition*, **22**, 273–283.

13 Xu, J., Chiang, H.C., Bjursell, M.K. and Gordon, J.I. (2004) *Trends in Microbiology*, **12**, 21–28.

14 Cummings, J.H. and McFarlane, G.T. (1997) *Journal of Parenteral and Enteral Nutrition*, **21**, 357–365.

15 Bergman, E.N. (1990) *Physiological Reviews*, **70**, 567–590.

16 Backhed, F., Ding, H., Wang, T., Hooper, L.V., Koh, G.Y., Nagy, A., Semenkovich, C.F. and Gordon, J.I. (2004) *Proceedings of the National Academy of Sciences of the United States of America*, **101**, 15718–15723.

17 Lim, H.J., Kim, S.Y. and Lee, W.K. (2004) *Journal of Veterinary Science*, **5**, 391–395.

18 Conly, J.M., Stein, L., Worobetz, L. and Rutledge-Harding, S. (1994) *American Journal of Gastroenterology*, **89**, 915–923.

19 Ballonge, J. (1998) *Bifidobacteria and probiotic action*, in *Lactic Acid Bacteria: Microbiology and Functional Aspects* (eds S. Salminen and A. Von Wright), Marcel Dekker, New York, pp. 519–587.

20 Teitelbaum, J.E. and Walker, W.A. (2002) *Annual Review of Nutrition*, **22**, 107–138.

21 Ouwehand, A.C. and Salminen, S. (2003) *Microbial Ecology in Health and Disease*, **15**, 175–184.

22 Schiffrin, E.J., Brassart, D., Servin, A.L., Rochat, F. and Donnet-Hughes, A. (1997) *American Journal of Clinical Nutrition*, **66**, 515S–520.

23 Salminen, S., Bouley, C., Boutron-Ruault, M.-C., Cummings, J.H., Franck, A., Gibson, G.R., Isolauri, E., Moreau, M.-C., Roberfroid, M. and Rowland, I. (1998) *British Journal of Nutrition*, **80**, S147–S171.

24 Tannock, G.W. (2001) *American Journal of Clinical Nutrition*, **73**, 410–414.

25 Lievin, V., Peoffer, I., Hudault, S., Rochat, F., Brassart, D., Nesser, J.R. and Servin, A.L. (2000) *Gut*, **47**, 646–652.

26 Collado, M.C., Gueimonde, M., Hernández, M., Sanz, Y. and Salminen, S. (2005) *Journal of Food Protection*, **68**, 2672–2678.

27 Grönlund, M.M., Arvilommi, H., Kero, P., Lehtonen, O.P. and Isolauri, E. (2000) *Archives of Disease in Childhood*, **83**, F186–F192.

28 Isolauri, E. (2004) *Journal of Food Science*, **69**, 135–137.

29 Didierlaurent, A., Sirard, J.C., Kraehenbuhl, J.P. and Neutra, M.R. (2001) *Cellular Microbiology*, **4**, 61–72.

30 Salminen, S., Gueimonde, M. and Isolauri, E. (2005) *Journal of Nutrition*, **135**, 1294–1298.

31 Suau, A., Bonnet, R., Sutren, M., Godon, J.-J., Gibson, G.R., Collins, M.D. and Doré, J. (1999) *Applied and Environmental Microbiology*, **65**, 4799–4807.

32 Favier, C.F., Vaughan, E.E., de Vos, W.M. and Akkermans, A.D.L. (2002) *Applied and Environmental Microbiology*, **68**, 219–226.

33 Vaughan, E., de Vries, M., Zoetendal, E., Ben-Amor, K., Akkermans, A. and de Vos, W.M. (2002) *Anthonie van Leeuwenhoek*, **82**, 341–352.

34 Vaahtovuo, J., Toivainen, P. and Eerola, E. (2001) *Antonie van Leeuwenhoek*, **80**, 35–42.

35 Toivanen, P., Vaahtovuo, J. and Eerola, E. (2001) *Infection and Immunity*, **69**, 2372–2377.

36 Rinne, M.M., Gueimonde, M., Kalliomäki, M., Hoppu, U., Salminen, S. and Isolauri, E. (2005) *FEMS Immunology and Medical Microbiology*, **43**, 59–65.

37 Rautava, S., Arvilommi, H. and Isolauri, E. (2006) *Pediatric Research*, **60**, 221–224.

38 Harmsen, H.J.M., Wildeboer-Veloo, A.C.M., Raangs, G.C., Wagendorp, A.A., Klijn, N., Bindels, J.G. and Welling, G.W. (2000) *Journal of Pediatric Gastroenterology and Nutrition*, **30**, 61–67.

39 Penders, J., Vink, C., Driessen, C., London, N., Thijs, C. and Stobberingh, E.E. (2005) *FEMS Microbiology Letters*, **243**, 141–147.

40 Matsuki, T., Watanabe, K., Tanaka, R., Fukuda, M. and Oyaizu, H. (1999) *Applied and Environmental Microbiology*, **65**, 4506–4512.

41 Satokari, R.M., Vaughan, E.E., Akkermans, A.D., Saarela, M. and de Vos, W.M. (2001) *Applied and Environmental Microbiology*, **67**, 504–513.

42 Isolauri, E., Rautava, S., Kalliomäki, M., Kirjavainen, P. and Salminen, S. (2002) *Current Opinion in Allergy and Clinical Immunology*, **2**, 263–271.

43 Woodmansey, E.J., McMurdo, M.E., MacFarlane, G.T. and MacFarlane, S. (2004) *Applied and Environmental Microbiology*, **70**, 6113–6122.

44 Tannock, G.W. (1999) *Anthonie van Leeuwenhoek*, **76**, 265–278.

45 Hornef, M.W., Wick, M.J., Rhen, M. and Normark, S. (2002) *Nature Immunology*, **3**, 1033–1040.

46 Isolauri, E., Salminen, S. and Ouwehand, A.C. (2004) *Best Practice and Research Clinical Gastroenterology*, **18**, 299–313.

47 Lloyd, M. (2003) *Pediatrics*, **111**, 1595–1600.

48 Zoetendal, E., Wright, A., Vilpponen-Salmela, T., Ben-Amor, K., Akkermans, A. and de Vos, W.M. (2002) *Applied and Environmental Microbiology*, **68**, 3401–3407.

49 Collado, M.C., Calabuig, M. and Sanz, Y. (2007) *Current Issues in Intestinal Microbiology*, **8**, 9–14.

50 Sanz, Y., Sánchez, E., Marzotto, M., Calabuig, M., Torriani, S. and Dellaglio, F. (2007) *FEMS Immunology and Medical Microbiology*, **51**, 562–568.

51 Nadal, I., Donant, E., Ribes-Koninckx, C., Calabuig, M. and Sanz, Y. (2007) *Journal of Medical Microbiology*, **56**, 1669–1674.

52 FAO/WHO Working Group (2002) *Guidelines for the Evaluation of Probiotics in Food*. Food and Agricultural Organization of the United Nations and World Health Organization, Geneva.

53 Timmerman, H.M., Koning, C.J., Mulder, L., Rombouts, F.M. and Beynen, A.C. (2004) *International Journal of Food Microbiology*, **96**, 219–233.

54 Collado, M.C., Meriluoto, J. and Salminen, S. (2007) *Food Research International*, **40**, 629–636.

55 Collado, M.C., Moreno, Y., Hernández, E. and Hernández, M. (2006) *European Journal of Food Research and Technology*, **222** (1–2), 112–117.

56 Collado, M.C. and Hernández, M. (2007) *Microbiological Research*, **162**, 86–92.

57 Gueimonde, M. and Salminen, S. (2006) *Digestive and Liver Disease*, **38**, S242–S247.

58 Temmerman, R., Pot, B., Huys, G. and Swings, J. (2003) *International Journal of Food Microbiology*, **81**, 1–10.

59 Gueimonde, M., Delgado, S., Mayo, B., Ruas-Madiedo, P., Margolles, A. and de los Reyes-Gavilan, C.G. (2004) *Food Research International*, **37**, 839–850.

60 Collado, M.C., Hernández, M. and Sanz, Y. (2005) *Journal of Food Protection*, **68**, 1034–1040.

61 Kurmann, J.A. and Rasic, J.L. (1991) *Health potential of products containing bifidobacteria*, in *Therapeutic Properties of Fermented Milks* (ed. R.K. Robinson), Elsevier Applied Food Sciences, London, pp. 117–158.

62 Bouhnik, Y. (1993) *Le Lait*, **73**, 241–247.

63 Collado, M.C. and Sanz, Y. (2006) *Journal of Microbiological Methods*, **66**, 560–563.

64 Charteris, W.P., Kelly, P.M., Morelli, L. and Collins, J.K. (1998) *Journal of Applied Microbiology*, **84**, 759–768.

65 Shing, H., Lee, J., Pestka, J.J. and Ustunol, Z. (2000) *Journal of Food Protection*, **63**, 327–331.

66 Collado, M.C., Moreno, Y., Hernández, E., Cobo, J.M. and Hernández, M. (2005) *International Journal of Food Science and Technology*, **11**, 307–314.

67 Moreno, Y., Collado, M.C., Ferrús, M.A., Cobo, J.M., Hernández, E. and Hernández, M. (2006) *International Journal of Food Science and Technology*, **41**, 275–280.

68 Masco, L., Crockaert, C., Van Hoorde, K., Swings, J. and Huys, G. (2007) *Journal of Dairy Science*, **90**, 3572–3578.

69 Noriega, L., Gueimonde, M., Sanchez, B., Margolles, A. and de los Reyes-Gavilan, C.G. (2004) *International Journal of Food Microbiology*, **94**, 79–86.

70 Collado, M.C. and Sanz, Y. (2007) *Journal of Applied Microbiology*, **103**, 1147–1157.

71 Beachey, E.H. (1981) *Journal of Infectious Diseases*, **143**, 325–345.

72 Kirjavainen, P.E., Ouwehand, A.C., Isolauri, E. and Salminen, S. (1998) *FEMS Microbiology Letters*, **167**, 185–189.

73 Aissi, E.A., Lecocq, M., Brassart, C. and Buoquelet, S. (2001) *Microbial Ecology in Health and Disease*, **13**, 32–39.

74 Tuomola, E.M., Ouwehand, A.C. and Salminen, S. (2000) *International Journal of Food Microbiology*, **60**, 75–81.

75 Rinkinen, M., Westermarck, E., Salminen, S. and Ouwehand, A.C. (2003) *Veterinary Microbiology*, **97**, 55–61.

76 Ouwehand, A.C., Isolauri, E., Kirjavainen, P.V., Tolkko, S. and Salminen, S. (2000) *Letters in Applied Microbiology*, **30**, 10–13.

77 Gueimonde, M., Noriega, L., Margolles, A., de los Reyes-Gavilan, C.G. and Salminen, S. (2005) *International Journal of Food Microbiology*, **101**, 341–346.

78 Gueimonde, M., Jalonen, L., He, F., Hiramatsu, M. and Salminen, S. (2006) *Food Research International*, **39**, 467–471.

79 Vesterlund, S., Paltta, J., Karp, M. and Ouwehand, A.C. (2005) *Journal of Microbiological Methods*, **60**, 225–233.

80 Collado, M.C., Gueimonde, M., Sanz, Y. and Salminen, S. (2006) *Journal of Food Protection*, **69**, 1675–1679.

81 Ruas-Madiedo, P., Gueimonde, M., Margolles, A., de los Reyes-Gavilan, C.G. and Salminen, S. (2006) *Journal of Food Protection*, **69**, 2011–2015.

82 Larsen, N., Nissen, P. and Willats, W.G. (2007) *International Journal of Food Microbiology*, **114**, 113–119.

83 Tuomola, E.M., Ouwehand, A.C. and Salminen, S. (1999) *FEMS Immunology and Medical Microbiology*, **26**, 137–142.

84 Lee, Y.J., Yu, W.K. and Heo, T.R. (2003) *International Journal of Antimicrobial Agents*, **21**, 340–346.

85 Collado, M.C., Meriluoto, J. and Salminen, S. (2007) *Journal of Dairy Science*, **90**, 2710–2716.

86 Bernet, M.F., Brassart, D., Nesser, J.R. and Servin, A.L. (1993) *Applied and Environ-mental Microbiology*, **59**, 4121–4128.

87 Bibiloni, R., Perez, P.F. and de Antoni, G.L. (1999) *Anaerobe*, **5**, 519–524.

88 Fujiwara, S., Hashiba, H., Hirota, T. and Forstner, J.F. (2001) *International Journal of Food Microbiology*, **67**, 97–106.

89 Fujiwara, S., Hashiba, H., Hirota, T. and Forstner, J.F. (1997) *Applied and Environmental Microbiology*, **63**, 506–512.

90 Ouwehand, A.C., Salminen, S. and Isolauri, E. (2002) *Antonie van Leeuwenhoek*, **82**, 279–289.

91 Niku-Paavola, M.L., Laitila, A., Mattila-Sandholm, T. and Haikara, A. (1999) *Journal of Applied Microbiology*, **86**, 29–35.

92 Chen, H. and Hoover, D.G. (2003) *Comprehensive Reviews in Food Science and Food Safety*, **2**, 82–100.

93 Ibrahim, S.A. and Bezkorovainy, A. (1993) *Journal of the Science of Food and Agriculture*, **62**, 351–354.

94 Bruno, F.A. and Shah, N.P. (2002) *Milchwissenschaft*, **57**, 617–621.

95 Toure, R., Kheardr, E., Lacroix, C., Moroni, O. and Fliss, I. (2003) *Journal of Applied Microbiology*, **95**, 1058–1069.

96 Servin, A.L. (2004) *FEMS Microbiology Reviews*, **28**, 405–440.

97 Gagnon, M., Kheadr, E.E., Le Blay, G. and Fliss, I. (2004) *International Journal of Food Microbiology*, **92**, 69–78.

98 Collado, M.C., González, A., González, R., Hernández, M., Ferrús, M.A. and Sanz, Y. (2005) *International Journal of Antimicrobial Agents*, **25**, 385–391.

99 Gueimonde, M., Ouwehand, A.C. and Salminen, S. (2004) *Scandinavian Journal of Nutrition*, **48**, 42–49.

100 Parvez, S., Malik, K.A., Ah Kang, S. and Kim, H.Y. (2006) *Journal of Applied Microbiology*, **100**, 1171–1185.

101 Isolauri, E., Kirjavainen, P.V. and Salminen, S. (2002) *Gut*, **50**, 54–59.

102 Chouraqui, J.P., van Egroo, L.D. and Fichot, M.C. (2004) *Journal of Pediatric Gastroenterology and Nutrition*, **38**, 288–292.

103 Gaudier, E., Michel, C., Segain, J.P., Cherbut, C. and Hoebler, C. (2005) *Journal of Nutrition*, **135**, 2753–2761.

104 McFarland, L.V. (2006) *American Journal of Gastroenterology*, **101**, 812–822.

105 Santosa, S., Farnworth, E. and Jones, P.J. (2006) *Nutrition Reviews*, **64**, 265–274.

106 Sazawal, S., Hiremath, G., Dhingra, U., Malik, P., Deb, S. and Black, R.E. (2006) *Lancet Infectious Diseases*, **6**, 374–382.

107 Gotteland, M., Brunser, O. and Cruchet, S. (2006) *Alimentary Pharmacology and Therapeutics*, **23**, 1077–1086.

108 Kajander, K., Hatakka, K., Poussa, T., Farkkila, M. and Korpela, R. (2005) *Alimentary Pharmacology and Therapeutics*, **22**, 387–394.

109 Kim, H.J., Vazquez Roque, M.I., Camilleri, M., Stephens, D., Burton, D.D., Baxter, K., Thomforde, G. and Zinsmeister, A.R. (2005) *Neurogastroenterology and Motility*, **17**, 687–696.

110 Camilleri, M. (2006) *Journal of Clinical Gastroenterology*, **40**, 264–269.

111 Bengmark, S. (2000) *Nutrition*, **16**, 611–615.

112 Go, M.F. (2002) *Alimentary Pharmacology and Therapeutics*, **16**, 1–15.

113 Malfertheiner, P., Mégraud, F., O'Morain, C., Hungin, A.P., Jones, R., Axon, A., Graham, D.Y. and Tytgat, G. and European *Helicobacter Pylori* Study Group (EHPSG) (2000) *Alimentary Pharmacology and Therapeutics*, **16**, 167–180.

114 Hamilton-Miller, J.M.T. (2003) *International Journal of Antimicrobial Agents*, **22**, 360–366.

115 Adamsson, I., Edlund, C. and Nord, C.E. (2000) *Journal of Chemotherapy*, **12**, 5–16.

116 Bernet, M.F., Brassart, D., Neeser, J.R. and Servin, A.L. (1994) *Gut*, **35**, 483–489.

117 Sgouras, D., Maragkoudakis, P., Petraki, K., Martinez-Gonzalez, B., Eriotou, E., Michopoulos, S., Kalantzopoulos, G., Tsakalidou, E. and Mentis, A. (2004) *Applied and Environmental Microbiology*, **70**, 518–526.

118 Mukai, T., Asasaka, T., Sato, E., Mori, K., Matsumoto, M. and Ohori, H. (2002) *FEMS Immunology and Medical Microbiology*, **32**, 105–110.

119 Pinchuk, I.V., Bressollier, P., Verneuil, B., Fenet, B., Sorokulova, I.B., Mégraud, F. and Urdaci, M.C. (2001) *Antimicrobial Agents and Chemotherapy*, **45**, 3156–3161.

120 Cremonini, F., Di Caro, S., Nista, E.C., Bartolozzi, F., Capelli, G., Gasbarrini, G. and Gasbarrini, A. (2002) *Alimentary Pharmacology and Therapeutics*, **16**, 1461–1467.

121 Kabir, A.M.A., Aiba, Y., Takagi, A., Kamiya, S., Miwa, T. and Koga, Y. (1997) *Gut*, **41**, 49–55.

122 Sakamoto, I., Igarashi, M., Kimura, K., Takagi, A., Miwa, T. and Koga, Y. (2001) *Journal of Antimicrobial Chemotherapy*, **47**, 709–710.

123 Myllyluoma, E., Veijola, L., Ahlroos, T., Tynkkynen, S., Kankuri, E., Vapaatalo, H., Rautelin, H. and Korpela, R. (2005) *Alimentary Pharmacology and Therapeutics*, **21**, 1263–1272.

124 de Bortoli, N., Leonardi, G., Ciancia, E., Merlo, A., Bellini, M., Costa, F., Mumolo, M.G., Ricchiuti, A., Cristiani, F., Santi, S., Rossi, M. and Marchi, S. (2007) *American Journal of Gastroenterology*, **102**, 951–956.

125 Lesbros-Pantoflickova, D., Corthésy-Theulaz, I. and Blum, A.L. (2007) *Journal of Nutrition*, **137**, 812–818.

126 Gionchetti, P., Rizzello, F., Venturi, A., Brigidi, P., Matteuzzi, D., Bazzocchi, G., Poggioli, G., Miglioli, M. and Campieri, M. (2000) *Gastroenterology*, **119**, 305–309.

127 Guandalini, S., Pensabene, L., Zikri, M.A., Dias, J.A., Casali, L.G., Hoekstra, H., Kolacek, S., Massar, K., Micetic-Turk, D., Papadopoulou, A., de Sousa, J.S., Sandhu, B., Szajewska, H. and Weizman, Z. (2000) *Journal of Pediatric Gastroenterology and Nutrition*, **30**, 54–60.

128 Saavedra, J.M., Bauman, N.A., Oung, I., Perman, J.A. and Yolken, R.H. (1994) *Lancet*, **344**, 1046–1049.

129 Kyne, L. and Kelly, C.P. (2001) *Gut*, **49**, 152–153.

130 Marteau, P.R., de Vrese, M., Cellier, C.J. and Schrezenmeir, J. (2001) *American Journal of Clinical Nutrition*, **73**, 430–436.

131 Shornikova, A.V., Casas, I.A., Isolauri, E., Mykkanen, H. and Vesikari, T. (1997) *Journal of Pediatric Gastroenterology and Nutrition*, **24**, 399–404.

132 Shornikova, A.V., Isolauri, E., Burkanova, L., Lukovnikova, S. and Vesikari, T. (1997) *Acta Paediatrica*, **86**, 460–465.

133 Vanderhoof, J.A., Whitney, D.B., Antonsson, D.L., Hanner, T.L., Lupo, J.V. and Young, R.J. (1999) *Journal of Pediatrics*, **135**, 564–568.

134 Adachi, J.A., Ostrosky-Zeichner, L., DuPont, H.L. and Ericsson, C.D. (2000) *Clinical Infectious Diseases*, **31**, 1079–1083.

135 Szajewska, H., Kotowska, M., Mrukowicz, J.Z., Armanska, M. and Mikotajczyk, W. (2001) *Journal of Pediatrics*, **138**, 361–365.

136 Isolauri, E., Juntunen, M., Rautanen, T., Sillanaukee, P. and Koivula, T. (1991) *Pediatrics*, **88**, 90–97.

137 Pedone, C.A., Bernabeu, A.O., Postaire, E.R., Bouley, C.F. and Reinert, P. (1999) *International Journal of Clinical Practice*, **53**, 179–184.

138 Isolauri, E. (2003) *Gut*, **52**, 436–437.

139 Ciarlet, M. and Estes, M.K. (2001) *Current Opinion in Microbiology*, **4**, 435–441.

140 Sullivan, Å. and Nord, C.E. (2002) *Journal of Antimicrobial Chemotherapy*, **50**, 625–627.

141 Colbère-Garapin, F., Martin-Latil, S., Blondel, B., Mousson, L., Pelletier, I., Autret, A., François, A., Niborski, V., Grompone, G., Catonnet, G. and van de Moer, A. (2007) *Microbes and Infection/ Institut Pasteur*, **9**, 1623–1631.

142 Picard, C., Fioramonti, J., Francois, A., Robinson, T., Neant, F. and Matuchansky, C. (2005) *Alimentary Pharmacology and Therapeutics*, **22**, 495–512.

143 Katz, J.A. (2006) *Journal of Clinical Gastroenterology*, **40**, 249–255.

144 Guandalini, S. (2006) *Journal of Clinical Gastroenterology*, **40**, 244–248.

145 Sullivan, Å. and Nord, C.E. (2005) *Journal of Internal Medicine*, **257**, 78–92.

146 Boyle, R.J., Robins-Browne, R.M. and Tang, M.L. (2006) *American Journal of Clinical Nutrition*, **83**, 1256–1264.

147 D'Souza, A.L., Rajkumar, C., Cooke, J. and Bulpitt, C.J. (2002) *British Medical Journal*, **324**, 1361.

148 Hickson, M., D'Souza, A.L., Muthu, N., Rogers, T.R., Want, S., Rajkumar, C. and Bulpitt, C.J. (2007) *British Medical Journal*, **335**, 80.

149 Linskens, R.K., Huijsdens, X.W., Savelkoul, P.H.M., Vandenbroucke-Grauls, C.M.J.E. and Meuwissen, S.G.M. (2001) *Scandinavian Journal of Gastroenterology*, **36**, 29–40.

150 Shanahan, F. (2002) *Lancet*, **359**, 62–69.

151 Swidinski, A., Ladhoff, A., Pernthaler, A., Swidinski, S., Loening-Baucke, V., Ortner, M., Weber, J., Hoffmann, U., Schreiber, S., Dietel, M. and Lochs, H. (2002) *Gastroenterology*, **122**, 44–54.

152 Mättö, J., Maunuksela, L., Kajander, K., Palva, A., Korpela, R., Kassinen, A. and Saarela, M. (2005) *FEMS Immunology and Medical Microbiology*, **43**, 213–222.

153 Malinen, E., Rinttilä, T., Kajander, K., Mättö, J., Kassinen, A., Krogius, L., Saarela, M., Korpela, R. and Palva, A. (2005) *American Journal of Gastroenterology*, **100**, 373–382.

154 Shanahan, F. (2001) *Gastroenterology*, **120**, 622–635.

155 Mutlu, E.A., Farhadi, A. and Keshavarzian, A. (2002) *Expert Opinion on Investigational Drugs*, **11**, 365–385.

156 Brigidi, O., Vitali, B., Swennen, E., Bazzocchi, G. and Matteuzzi, D. (2001) *Research in Microbiology*, **152**, 735–741.

157 Hamilton-Miller, J.M.T. (2001) *Microbial Ecology in Health and Disease*, **13**, 212–216.

158 Borowiec, A.M. and Fedorak, R.N. (2007) *Current Gastroenterology Reports*, **9**, 393–400.

159 Venturi, A., Gionchetti, P., Rizzello, F., Johansson, R., Zucconi, E., Brigidi, P., Matteuzzi, D. and Campieri, M. (1999) *Alimentary Pharmacology and Therapeutics*, **13**, 1103–1108.

160 Guslandi, M., Mezzi, G., Sorghi, M. and Testoni, P.A. (2000) *Digestive Diseases and Sciences*, **45**, 1462–1464.

161 Gionchetti, P., Amadini, C., Rizzello, F., Venturi, A. and Campieri, M. (2002) *Alimentary Pharmacology and Therapeutics*, **16**, 13–19.

162 Asher, M.I., Montefort, S., Björkstén, B., Lai, C.K., Strachan, D.P., Weiland, S.K.

and Williams, H. (2006) *Lancet*, **368**, 733–743.

163 Johansson, S.G., Hourihane, J.O., Bousquet, J., Bruijnzeel-Koomen, C., Dreborg, S., Haahtela, T., Kowalski, M.L., Mygind, N., Ring, J., van Cauwenberge, P., van Hage-Hamsten, M. and Wüthrich, B. (2001) *Allergy*, **56**, 813–824.

164 Johansson, S.G., Bieber, T., Dahl, R., Friedmann, P.S., Lanier, B.Q., Lockey, R.F., Motala, C., Ortega Martell, J.A., Platts-Mills, T.A., Ring, J., Thien, F., Van Cauwenberge, P. and Williams, H.C. (2004) *Journal of Allergy and Clinical Immunology*, **113**, 832–836.

165 Kalliomäki, M. and Isolauri, E. (2003) *Current Opinion in Allergy and Clinical Immunology*, **3**, 15–20.

166 Björkstén, B. (2004) *Clinical Reviews in Allergy and Immunology*, **26**, 129–138.

167 Penders, J., Thijs, C., van den Brandt, P.A., Kummeling, I., Snijders, B., Stelma, F., Adams, H., van Ree, R. and Stobberingh, E.E. (2007) *Gut*, **56**, 661–667.

168 Strachan, D.P. (1989) *British Medical Journal*, **299**, 1259–1260.

169 Ouwehand, A.C., Isolauri, E., He, F. and Salminen, S. (2001) *Journal of Allergy and Clinical Immunology*, **108**, 144–145.

170 Majamaa, H. and Isolauri, E. (1997) *Journal of Allergy and Clinical Immunology*, **99**, 179–185.

171 Isolauri, E., Arvola, T., Sutas, Y., Moilanen, E. and Salminen, S. (2000) *Clinical and Experimental Allergy*, **30**, 1604–1610.

172 Kalliomäki, M., Salminen, S., Poussa, T., Arvilommi, H. and Isolauri, E. (2003) *Lancet*, **361**, 1869–1871.

173 Rosenfeldt, V., Benfeldt, E., Nielsen, S., Michaelsen, K., Jeppesen, D., Valerius, N. and Paerregaard, A. (2003) *Journal of Allergy and Clinical Immunology*, **111**, 389–395.

174 Viljanen, M., Savilahti, E., Haahtela, T., Juntunen-Backman, K., Korpela, R., Poussa, T., Tuure, T. and Kuitunen, M. (2005) *Allergy*, **60**, 494–500.

175 Abrahamsson, T., Jakobsson, T., Böttcher, M., Fredriksson, M., Jenmalm, M., Björkstén, N. and Oldaeus, G. (2007) *Journal of Allergy and Clinical Immunology*, **119**, 1174–1180.

176 Rautava, S., Kalliomaki, M. and Isolauri, E. (2002) *Journal of Allergy and Clinical Immunology*, **109**, 119–121.

177 Helin, T., Haahtela, S. and Haahtela, T. (2002) *Allergy*, **57**, 243–246.

Index

New Strategies Combating Bacterial Infection. Edited by Iqbal Ahmad and Farrukh Aqil
Copyright © 2009 WILEY-VCH Verlag GmbH & Co. KGaA, Weinheim
ISBN: 978-3-527-32206-0